LIGHTNING

Col USAF (Ret)

John W. Mitchell
Col. USAF (Ret.)

P-38

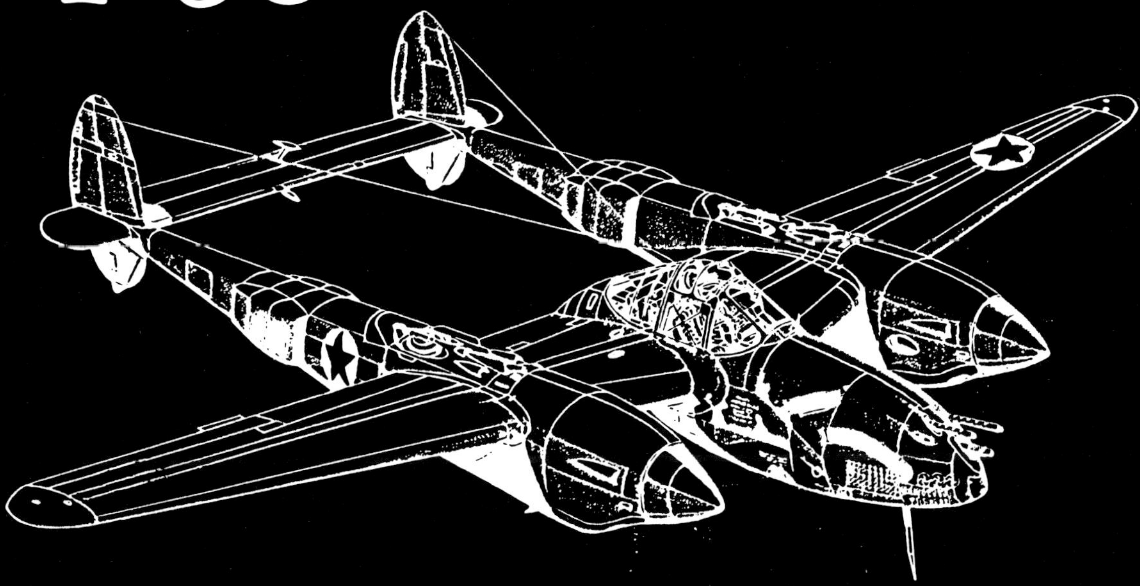

LIGHTNING

Publishers
Turner
PUBLISHING CO.

Publisher: Turner Publishing Company
Publishers of Military History
P.O. Box 3101
Paducah, KY 42002-3101

Writer and editor: Pamela Reynolds

Editorial staff: Gil Cefaratt
Former President
P-38 National Convention

Library of Congress
Catalog Card No. 89-050039
ISBN 0-938021-35-4
Printed in the U.S.A.

Limited Edition of 1,000 copies.
Books available from the publisher.

Table of Contents

DEDICATION

Kelly Johnson

There have been many squadron reunions, group reunions and squadron and group association reunions over the years and they are still taking place.

There have been memorials erected for Major Bong and Major McGuire.

In 1977 on the Fortieth Anniversary of the awarding of the contract, there was a Symposium sponsored by Warren Bodie and the Split-S-Society.

In 1987 on the Fiftieth Anniversary of the awarding of the contract, there was a National Convention held.

In 1989 there was a celebration for the First Flight Golden Anniversary, sponsored by Katherine Hulme and the Santa Maria Museum of Flight and Lockheed. All these events have taken place because of one airplane, a much liked airplane, one that brought her pilot's back on many occasions after being damaged, and one that took two fighter pilots to positions of Number One and Number Two Fighter Aces of America.

And this great airplane, the P-38 Lightning was produced by Lockheed Aircraft, but more importantly was designed by a brilliant engineer and designer. For making all this possible I dedicate this book to a gentleman, a family man, an engineer, a designer, and a friend—my boss, Kelly Johnson.

Gil Cefaratt

Gil Cefaratt

PRAYER

Father Al Williams

"Bless the Lord, O my soul! O Lord, my God, you are great indeed! You are clothed with majesty and glory robed in light as with a cloak. You have spread out the heavens like a tent-cloth, you have constructed your palace upon the waters. You make the clouds your chariot; you travel on the wings of the wind. You make the winds your messengers, and flaming fire your ministers." (Psalm 103: 1-4)

Lord of the high heavens, Creator of the universe, the young men who died for justice and for peace are always in our tender memory. They are in our hearts' love in an admixture of sorrow at their loss and of admiration at their bravery.

As their graceful airplanes cleaved the skies over so many diverse regions of a war-ravaged globe, they must have adverted to the feelings prayerfully set forth by the Psalmist we have quoted here.

In tearful poignancy we recall in particular those who flew to the purring rhythm of the airplanes described in this book. Their youth exulted in the control of the power and speed they commanded even as they flew into fearsome and unknown danger. And some, indeed, fell victim to the unyielding perils of enemy action or to the vagaries of inhospitable weather.

We ask you now and always, our Father in heaven, to give them eternal rest. May their eternity be one of peace, be one of never-ending and youthful joy, and finally, one glorious day, be of reunion forever with their families and their comrades in your everlasting Kingdom. Amen.

Father Al Williams

ACKNOWLEDGEMENTS

Gil Cefaratt

Since the Fortieth Anniversary Symposium in September 1977, I have met so many fine people in the P-38 Community and I cherish all the friendships. So many have contributed so much over the years and especially since the beginning of the history book. To you who have contributed, it is your book, and I am proud and pleased to have been a part of it.

I do have to acknowledge the following people: Scholer Bangs-aviation writer, Steve Blake-aviation historian, Warren Bodie-aviation writer/historian, Harry Brown-47PS, 15FG, 9FS 49FG, 431FS 475FG, Clair E. Buskirk-38FS 55FG, 338FS 55FG, Cal Butler-435FS 479FG, Oliver "Ollie" Cellini-1PG, Bob Carlton-49FS 14FG, Art Cole-49FS 14FG, Kevin Grantham-aviation historian, Bernice (John) Hallford-429FS 474FG, Marilyn (Bob) Hanson-474FGA, Fred C. Hayner-434FS 479FG, Walt Heck-94PS 1PG, 97FS 82FG, Bob Hildebrand-95FS 82FG, Jack Ilfrey-1FG, 20FG, Tony LeVier-Lockheed test pilot, Dave Lusk - aviation writer, Russell S. Miksch-9PRS 8PRG, Erik Miller-Lockheed aircraft photographer, Mary Lou Colbert Neale-WASP, Pilot Class 43-D, Dick Reinke-31PRS 1OPRG, Hal Rigney-94FS 1FG, Philip O. Robertson-1PG, 21PRS, B-25's, Wayne O. Sneddon-Lockheed tech. rep, Irv Styer-27FS 1FG, Patricia (Robert Symons), Symons Baldwin, Father Al Williams-54FS Chaplain, Milt Getker-Curtis Electric tech. rep. If I have overlooked anyone, please accept my apologies.

For the success of the P-38 National Convention of May 1987, I would like to thank all the supporting attendees and to the following: The P-38 National Convention Committee members, Advisory Committee members; Roy A. Anderson-Chairman, Executive Committee-Lockheed Corporation, R. Richard Heepe-Past President-Lockheed California Co., William P. "Bill" Lear, Jr.-Chairman-Lear, Inc. and Ed McMahon-Carson Tonight Show, Marvin "Lefty" Gardner, Dave Tallichet, Tony LeVier, Kelly Johnson, Chet Patterson, George Toumanoff, the late Pat McGeehan, Dick Baughman, Ward Beman, Dick DeGrey, Marge Bong Drucker, Ray Toliver, Jay Fiondella, the late Jimmy Mattern, former President Ronald Reagan.

Thanks to the following for the success of the P-38 First Flight Golden Anniversary in January 1989. The supporting attendees, Katherine Hulme and the Santa Maria Museum of Flight, Marvin "Lefty" Gardner, Tony LeVier, Nancy (Mrs. Kelly Johnson), Ward Beman, Dick DeGrey, Steve Hinton, Rex Barber, John Mitchell.

Gil Cefaratt

PUBLISHER'S MESSAGE

F5E Aircraft, 34PRS, 10PRG, 9AF, Chateaudun Orleans Sector, France, 19 August 1944. (Courtesy of Jack Quinn/Gil Cefaratt)

The Lockheed P-38 Lightning was introduced during WW II after many years of tedious designing and risky test flights. The P-38 served its men well through its versatility and adaptability to many situations. It served as a bomber escort, day/night fighter, precision bomber, long range photo reconnaissance plane, tank buster, and in numerous other capacities. The addition of belly tanks also made the Lightning the ideal plane for dangerous long range flights in the South Pacific and for flights from England to Germany.

Within this book is an historical text which gives an overview of the P-38 Lightning, the men who designed and tested the plane and the men and women pilots and ground crews who prepared the craft for combat. Also included are many special flight stories from the ground crews and pilots.

A special thank you is extended to Gil Cefaratt, former president of the P-38 National Convention, who not only made this publication a reality but whose meticulous attention to detail has made this book a treasured keepsake for all associated with the P-38. And of course, we must thank everyone who contributed biographies, photographs and special stories to help document the P-38 Lightning.

This volume is dedicated to the men who designed and test flew the Lightning and to the men and women who flew the P-38 during WW II. This book will serve to perpetuate the legacy of the P-38 Lightning.

Dave Turner

Dave Turner
President

P-38L-5-LO. Owned and flown by David Tallichet, now on display at McGuire Air Force Base, New Jersey. (Courtesy of Lockheed California/Gil Cefaratt)

P-38 Lightning
The Fork-Tailed Devil

"The power of the P-38 as a weapon of war was beyond all doubt. Its devastating, concentrated fire power, high speed, long range, and ability to turn up in - to the enemy - the damndest places had, according to captured German and Italian prisoners, endowed it with a terror inspiring legend and a grim pet name - The Fork-Tailed Devil."

It has been said that lightning never strikes twice in the same place, and that was certainly true for the P-38 "Lightning" fighter plane - because when the P-38 struck, you could rest assured that it would strike with deadly force and not have to strike a second time.

The name lightning embodied everything the P-38 could achieve - it was truly a "flash of light in the night," and nothing could stop it. It was sleek, fast, and beautiful. It was the whippet of the skies, as swift as the wind, and packed more sturdiness in its two engines than any other plane. The Japanese said it was two planes with one pilot - the Nazis called it the "fork-tailed devil." Some say it was one of the greatest fighter planes of World War II.

The P-38 was as versatile as it was beautiful, and its duties became more varied as the war progressed. Originally designed as an interceptor plane, the P-38 also served as a fighter-bomber, photo reconnaissance plane, droop snoot to accommodate a bombardier, low altitude troop support and it towed troop carrying gliders during the D-Day Invasion of Normandy.

Because of the ingenuity and perseverance of many test pilots the P-38 Lightning became one of the greatest fighter planes of World War II. The story of the P-38 is one of continual problems, solved by the ingenuity and creativity of those people who believed in the plane. The life of the P-38 did not only revolve around its war years; but also the many laborious years of the development and testing of the P-38 - years crucial for determining the viability of producing the plane for battle. The following is the story of the P-38, from its inception on paper, to the battles it fought and the personal stories and biographies of the pilots who flew it during World War II.

P-38 assembly, Langford Lodge, Northern Ireland, 1942 (Courtesy of Thomas J. Cleland)

A Shocking Idea A Twin-Engine Plane

Farsighted officers dreamt of an aircraft that would give the Air Corps that edge in creating the "will to win" amongst its pilots. During the 1930s, little money was provided for the development of dynamic aircraft capable of intensive aerial battles. But with the increasing threat of a Hitler-dominated Germany more money was allocated for the research and development of advanced aircrafts. Aircraft developers knew they needed an outstanding aircraft to revitalize a conservative industry. One group of designers envisioned an aircraft unlike any before designed - a plane with twin-engines capable of carrying heavy armament to high altitudes, having long range, and the capability of intercepting high-flying bombers. The requirement was given the name 'Specification X-608.'

After having envisioned such an aircraft, it was necessary to find a contractor. The contract went out to all aircraft manufacturers since the industry was economically depressed. After having experienced many ups and downs because of the poor world wide economy, the Lockheed Aircraft Corporation was very interested in securing the contract. Lockheed was well-known for designing and building high performance aircraft and they were looking for a lucrative government contract to provide stability and growth. "Specification X-608' was Lockheed's first venture into military production. Engineer Clarence "Kelly" Johnson, who had one of the most impressive careers in American aviation - he designed the U-2, F-104, and the SR-71 "Blackbird," - worked for Lockheed at this time and was looking for that one special plane, he found it in the 'Specification 608,' better know as the P-38.

Johnson, along with Lockheed engineer Hal Hibbard, worked diligently on 'Specification X-608,' which was designated as Model 22 by the factory, and realized it would require an abundant powerplant. As a result, the design became truly radical, for the one difference between it and all previously designed aircraft was the two engines needed to provide the necessary power for top speed performance and the ability to climb to high altitudes.

Johnson chose the V-12 Allison V-1710 powerplant since it could generate at least 1,000 horsepower and was very streamlined. The engine was designed as a liquid-cooled in-line or v-type engine. It was very compact and had a large horsepower output per unit of frontal area - it consumed fuel efficiently. In addition to the twin-engines, or twin-booms, the Model 22 featured a fuselage pod mounted on the wing between the booms, twin General Electric turbosuperchargers, tricycle landing gear, and the final radical element - an armament system located in the nose comprised of a large cannon with a battery of machine guns, which could accurately aim at an enemy with devastating results - giving the interceptor unparalleled fire power - truly lightning had struck.

While Lockheed's Johnson oversaw the design of the plane, the U.S. Army's Lt. Benjamin Kelsey made the final recommendations to the government in regard to whom to award the contract. Lt. Kelsey was very impressed with Model 22. On 23 June 1937 the government issued Air Corps Contract Number 9974 to the Lockheed Aircraft Corporation for the construction of one XP-38 which would carry the army serial number 37-457.

Lockheed officials realized that one aircraft would not mean the production of an entire fleet of planes, especially since the P-38 was so radically different. But Johnson and his co-workers did not give up hope, and as a result the United States Army Air Forces reaped the benefits of a high-flying twin-engine fighter of outstanding qualities. Work started in July of 1938 and progressed rapidly - the plane was physically complete by the end of 1938. The P-38 was unlike any other plane, to begin with it was huge, it

The control panel of a P-38. The many switches, knobs and dials, could be confusing when learning to fly the P-38., but the pilots grew to love the plane. (Courtesy of Gil Cefaratt)

weighed over 15,000 lbs, and had a wing span of 52 feet. It actually looked more like a bomber than an interceptor. Also unique to the P-38 was its metal covered control surfaces instead of fabric covered surfaces - metal provided for less damage when recovering from speed dives. The Lightning also represented the latest in aeronautical technology; flush riveting, polished metal, sleek installation of the twin Allison engines, some people compared it to a "custom-built European racing machine."

The final specifications of the plane included: twin engines, the ability to carry heavy armament to high altitudes, achieved via new turbosuperchargers - a new concept in the development stages, traveling long ranges without refueling, climbing quickly and intercepting high-flying bombers, and ground handling simplified by a tricycle undercarriage. The P-38 could also climb to 20,000 feet in six minutes and offered a top speed of 360 miles per hour at altitude. It had an average combat range of 600-700 miles. The P-38 had one 23mm Madsen cannon, some models had 37mm- and four Browning .50 caliber M2 air-cooled machine guns with 200 rounds per gun. The P-38 had an external bomb capacity of 3,200 lbs, and when on escort duty (not carrying bombs) it could carry a maximum fuel load and approach a range of 2,000 miles. It was not until 1941, after many test flights and technical changes that the P-38 went into final production. By the time of Pearl Harbor, the P-38 was in full production and remained so until 1945. The War Department awarded contracts totalling $52 million for combat fighters of the P-38E versions. The unit costs for the P-38 were as follows: 1939-1941, $134,284; 1942, $120, 407; 1943, $105,567; 1944, $97,147.

Lockheed built 9,924 P-38s for the United States and under a pooling contract Consolidated Vultee built 113 P-38s. During peak construction days, Lockheed built 15 P-38s a day. Each engine cost $16,000, the propellors cost $4,000 and the radio equipment alone cost $3,000. Lockheed also built the P-38 for Britain and France. France was impressed with the P-38 and ordered 500 in 1940. The Royal Air Force ordered the Lightning but it had to meet their specifications - they requested the turbosuperchargers removed, along with the counterrotating propellor. As a result, the P-38 Model 322-16 did not perform to the RAF's satisfaction and they cancelled their order.

Lockheed designed 18 versions of the P-38 which performed dozens of missions as: day/night fighters, bomber escorts, long-range photo reconnaissance planes, tank busters, railroad train and track destroyers, smoke screen layers, rocket launchers, torpedo carriers, aerial ambulances, high-altitude precision bombers, and fighter-bombers.

The 18 versions of the P-38 all served various purposes in different theaters of the war. For the China-Burma-India Theater, the P-38 was modified for Azon bombing.

Rocket-laden P-38 during factory testing. (Courtesy of Lockheed)

The camera compartment of a Photo Recon P-38. (Courtesy of Gil Cefaratt)

P-38 assembly line outside the Lockheed-Burbank factory. (Courtesy of Richard DeGrey)

A. W. "Tony" LeVier in front of a P-38. Inside the P-38 is Glen Fulkerson, LeVier's crew chief. (Courtesy of Gil Cefaratt)

Bombardier noses, also known as "droop snoots" were built into the P-38. The nose was made of plexiglass and held the appropriate equipment for a bombardier. In a 51 day period, 10 Lightnings were so equipped in the spring of 1945 at Langford Lodge, Ireland. Engineers installed the standard M9 bombsight, the Azon Bomb Directional Control and the automatic radio-bomb-release transmitting equipment. On such missions, the Droop Snoot P-38 would lead the Lightning formation equipped with 1,000 lb bombs. It was the droop snoot's job to give the orders to drop the bombs upon siting of the target. The bombs were usually dropped at high altitudes, around 20,000 feet, and was extremely effective. After the bombs were dropped, the P-38s could resume their fighter status and defend themselves.

In 1942, Lockheed established an aircraft modification center to handle the various versions of the P-38 to be utilized in the war. The center was located at Love Field, Dallas, TX. According to Leo Childs, project engineer for the photo reconnaissance planes, the center "provided a direct interface between various aircraft factories and the theaters of operations. It was a place where any production live model could be modified by design changes to best serve the needs of specific war zones. The operation took on somewhat of front line battle ground atmosphere with planes being flown directly from the factories or back from the war zones for modification." Key members of the project included: Captain "Ossie" Olsen, Walter Hazelbaker, and Don Miller under the direction of General George Goddard, all from Wright Field Photo Lab. Lockheed employees included: Bob Pote, chief engineer; Fred Legget, electrical engineer; Milt Middleton, production manager; and Hal Bruck, Jack Wilson and Jack Elden.

According to Childs, Lockheed converted 1,600 Lightnings into photo reconnaissance planes. This process was accomplished by removing the armament and replacing it with various aerial mapping cameras. Childs said, "Early models included the F-5G and F-5C which required no major design changes to the basic fighter nose but imposed considerable problems in mounting the cameras and indexing and photo interpretation. The F-5G required the replacement of the fighter nose with a completely new nose from the windshield forward but drastically reduced modification down time provided for critically needed longer focus length cameras, and eliminated the time consuming film indexing and photo interpretation."

Additional modification included increasing the radius. For missions over long ranges, as were common in the Mediterranean Theater and in the South Pacific Theater, the Lightning's radius was extended by adding internal and external wing tanks. By the Fall of 1944 the P-38 could carry a maximum fuel load. For the P-38J the maximum fuel load was 900 gallons. This fuel load was attained by adding a 165 gallon drop tank and a 300 gallon drop tank.

The Lightning was considered the most versatile of all World War II fighter aircraft with 10,000 ships delivered in six years of production.

Major Dick Bong seated in a P-38 with Lockheed test pilot Jimmy Mattern.

Engineering test pilot Herman "Fish" Salmon with engineering flight test manager Rudy Thoren. (Courtesy of Gil Cefaratt)

Triumph of the P-38 Test Pilots and Problems

"as airplane speeds and operating altitudes increase...consideration must be given to the effect of compressibility" C.L. Johnson from OF MEN and STARS, July 1957 Chapter V (Lockheed Aircraft Corp)

Plagued from the beginning, it is a wonder that the P-38 ever saw Europe. Fortunately, many dedicated engineers, pilots and government personnel believed in the P-38, persevered, and beat the odds that were seemingly stacked against them.

Engineers finished the XP-38, the first prototype, in December of 1938 and its first test flight was on 27 January 1939. The YP-38 flew on 16 September 1940, and the first extensive testing began on 26 June 1941. An early problem encountered by the test pilots of the P-38 was that the tightly-cowled Allison engines would have to be cooled and maximum streamlining would have to be obtained. In addition, an appropriate location for radiators and oil coolers would have to be found. When the XP-38 was ready to fly, it was transferred to March Field in Riverside, CA. It took several days to reassemble the plane, then Lt. Kelsey supervised the servicing prior to testing.

The first test of the P-38 consisted of a taxi-test, during which Kelsey found many problems, one of which almost prevented the P-38 from going into final production. During this taxi-run, Kelsey drove past the end of the runway, hitting a ditch. The plane was not damaged and the cause of this accident was attributed to insufficient braking power. The ingenuity of the Lockheed engineers resulted in a quick scavenger hunt through the Army's spare parts bin which produced a tank and pump which they assembled to fit the P-38. This tank and pump transported additional fluid to the brakes to increase braking pressure and cooling.

Having resolved the braking dilemma, Lt. Kelsey made the first test flight of the XP-

38 on 27 January 1939. After just a few minutes in the air, the plane began to flutter and the instrument panel became a blur. The flutter was so violent that the wings actually traveled three-feet up and down. After a very skillful landing by Lt. Kelsey, the cause for the flutter was immediately discovered - a lack of gap sealing and poor flap installation, which in turn caused three of four soft aluminum rods to break. The gaps were properly sealed, steel rods were added, and the damaged fins were repaired.

Even after two slightly disheartening test runs, the P-38 was ready for official army testing at Wright Field in Dayton, Ohio. Since the large plane required disassembly if transported by land, it was determined that flying it to Dayton would be easier. Lt. Kelsey flew the plane from Riverside to Amarillo, TX in two hours and 48 minutes, and thence from Amarillo to Dayton in two hours and 45 minutes. This was a cross-country speed record for the amount of time spent in the air and since the ground time was also calculated, the cross country speed record remained with Howard Hughes.

Having reached Dayton, it was determined that Kelsey could beat Hughes total hours in the air record (not counting ground time) if he left Dayton for Long Island, NY immediately. After briefly discussing this with General H. H. "Hap" Arnold, Kelsey continued on his journey. As Lt. Kelsey approached Mitchell Field at Long Island, New York, the P-38's carburetors began to pick up ice.

Ice would form on carburetors during certain weather conditions along with an idling engine, on this occasion ice formed in the carburetor throat therefore blocking off the fuel supply to the engine. The P-38 did have a carburetor heating system but the ice formed in the throat before the heating system had a chance to eliminate it. Quick thinking and years of experience with handling aircraft paid off as Lt. Kelsey brought the plane in for a landing, unfortunately, the plane had no power and the P-38 slammed to

Pilots of the 95th FS, 82nd FG based at Mines Field, CA during 1942, brushing up on their aircraft identification. From l to r are: S/Sgt. William M. Conner Jr., Captain Robert E. Kirtley, ?, S/Sgt Marshall E. Hyde, S/Sgt Charlie Langdon, S/Sgt Archie Malette, and John Litchfield. (Courtesy of Col. Robert Kirtley)

95th FS, 82nd FG on duty at Mines Field, CA in 1942. Back row l to r: S/Sgt Marshall E. Hyde, Archie F. Mallette, Virgil S. White. Middle row l to r: S/Sgt William W. Conner Jr., Claud E. Ford, and William J. Schildt. Front row, l to r: S/Sgt Harold G. Manning, Alex K. Hamric, Charles R. Langdon, and Samuel P. Bradshaw, Jr. (Courtesy of Col Robert Kirtley)

Piggy-back P-38-F. (Courtesy of Bill Hess)

the ground just short of the runway. Kelsey came out unscathed, but he was unable to save the plane.

Again, the makers of the P-38 were plagued with problems - this one possibly the worst - they had a revolutionary idea, but no physical evidence to support it. The Army decided to continue with the P-38 and on 27 April 1939 contracted with Lockheed for 13 service YP-38 aircraft. The immediacy with which the contract was negotiated can be attributed to Gen. Arnold who believed in the P-38, and even though the Army had no combat or performance records to study, the Army took Arnold's word that Lockheed could produce what they wanted.

The P-38 now got off to a fresh start, it had a new model number - 122, and the terms of production differed. In contrast to the speed with which the XPs were produced, the YPs moved slowly because of the changes to the technical drawings as well as the partially completed airframes. A few of the changes included: improved Allisons (the F-series - V-1710-27 and V-1710-29), in which the propellors turned outboard to reduce airflow turbulence over the tail surfaces, a change in the armament to the Colt Browning 37 mm weapon with 15 rounds, and the machine guns were two .50 calibers with 200 rounds per gun and two .30 caliber weapons with 500 rounds per gun. Additional changes included modifying the turbosuperchargers, radiators, and cockpits, expanding the cowlings to allow for better cooling, and enlarging the radiators.

Lockheed test pilot Eddie Martin often test fired the guns of the P-38 into the firing pit in front of the hangar where the ship was parked, right beside a small mountain of sand. He said, "I would push that P-38 back 8, 9, 10 inches; and you could sure feel the jolt. I got quite a thrill out of that, but always sort of wished that I could get one of them in the air and try the guns and see what the feeling was like." One day Martin did take a P-38 out on a test flight over the ocean on the way towards the islands off the California Coast. After he was clear of all people and boats he could then test fire the guns. Martin said, "You could see right where they (bullets) went straight ahead of you; it shot the water high into the air. Those bullets went right down in the water and no doubt killed a lot of fish."

Even though the YP looked basically the same as the XP it was a new aircraft and needed intensive testing to prove ready for military service. The flying hours were quickly built up by high caliber Lockheed test pilots whose work could be considered just as dangerous as the work of the fighter and bombers during the war. The P-38's advanced characteristics resulted in a high accident rate - as high as 12 a day. Air Force leaders decided to secure an experienced P-38 test pilot who could visit training schools and demonstrate the various capabilities of the plane. For that job, Lockheed recruited pilot James "Jimmy" Mattern, along with Milo Burcham, and Major Signa A. Gilkey. Additional test pilots included Marshall Headle and Ralph Virden, one of Lockheed's top test pilots, whose career lasted 13 years, accumulating 15,500 hours by November of 1941.

One of the most popular test maneuvers at this time was the power dive, in which the test pilot would take the plane to altitude, push the nose down, and then travel as fast as possible to see how the plane would handle and recover. This was a dangerous maneuver and revealed many problems with the P-38.

On 16 September 1940 the YP-38 was taken up for its first test flight by Marshall Headle. Additional major testing of the P-38, including many power dives, began in the summer of 1941 near Selfridge Field, MI. During a service test flight Major Signa Gilkey entered a power dive at high altitude and experienced severe buffeting at 320 miles per hour. Just as the P-38 had fluttered violently in 1939, again in 1941, as the Major's dive progressed the nose became heavy and the empennage shook violently. With skillful handling Major Gilkey successfully landed the plane.

After much wind tunnel testing, contract pilot Vance Bresse and Lockheed test pilots Virden and Mattern determined that the major cause of the buffeting was the disturbed airflow in the area of the wing-fuselage juncture. Wing fillets were installed to correct the problem. Unfortunately, this did not eliminate the compressibility problem of the P-38.

Testing continued in November of 1941 and the engineers wanted the test pilots to push the speed past 300 miles per hour at 30,000 feet. Mattern and Burcham refused, Virden agreed. In November at the main Lockheed factory 25,000 employees gathered to listen to Gen. Arnold, a supporter of the P-38 since its conception. Afterward, fly-by demonstrations performed for the crowd consisted of three Hudson coastal reconnaissance bombers with RAF roundels, a brand new Olive Drab P-38, flown by Virden, and a Vega Ventura bomber flown by Mattern.

The day proved to be ill-fated and revealed that the P-38 still had its problems. As Virden tried to recover from a power dive the tail came off - the plane started to zig-zag to earth, crashed into a house, and burst into flames. The occupant of the house survived, Virden did not. Later that day at a press conference, company president Robert Gross said, "Ralph Virden was a great pilot, but even greater, a man engaged in extremely dangerous work for his country." Some say this accident set the program back a year, others say "... the loss of the prototype in this case actually expedited the program."

The cause of Virden's crash was attributed to a structural failure: Ward Beman and Tony LeVier decided that a device was needed to warn the pilot of impending danger so he could pull out before it was too late. The two men decided that a dual manifold pressure gauge could be maneuvered to let the pilot know when he was too close to an ill-fated power dive. An engineer at Lockheed's Irish base recalibrated a dual gauge and the next morning LeVier test flew a P-38 with the new mechanism and the results were good. The gauge consisted of two disks, instead of needles, the bottom disk was painted white for 180 degrees and the top, only half a disk, was painted black. The closer the pilot came to danger, the larger the

Tents, supply crates and P-38s in various stages of repair on an airstrip somewhere in the Aleutian Islands. (Courtesy of Francis J. Pope)

band of white showed. It was an instrument easily seen which allowed the pilots time to come out of a power dive before it was too late.

Perhaps one of the most peculiar test performed with the P-38 occurred when Eddie Martin was asked to place a one pound sack of flour on the wing right next to the cabin and fly with the sack of flour on the plane. Martin said he performed some wingovers, banked the plane, turned the plane over and "that sack of flour just stayed right there, sliding back and forth right next to the cockpit, about three feet one way or the other; it never left the side of that fuselage." Whether or not this information was useful for wartime operations is not known, but it certainly was interesting trivial information to know about your plane.

Mattern, who accumulated 3000 hours in the P-38 - continued test flying of the P-38 and on 13 November 1942, introduced the "piggy-back" method of training to future P-38 pilots. The radio equipment in the pilot's compartment was removed, therefore providing just enough room for a second person to lay down with his knees in his stomach. Employing this method allowed a pilot to experience the P-38 before he actually took one to the air. Mattern visited all P-38 training bases and demonstrated every conceivable circumstance including "accelerated stalls, manipulation of all controls, single engine landings, and single engine performance." After three months of Mattern training of future pilots, the accident rate dropped considerably. Test pilot Martin said, "It was somewhat of a thrilling flight because you wondered how the guy laying back there in such a position could be enjoying it!"

Ward Beman, Chief of Aerodynamics at Lockheed from 1937 - 1947, played a key role in the development of many aspects of the the P-38. One serious defect the P-38 experienced in Europe was that the very cold weather over Germany caused hands and fingers to become unusable just when needed most. A special pair of heated gloves was developed in England to quickly remedy the situation.

The P-38 was more complicated to learn especially if the pilots were accustomed to flying small single-engine planes. Double controls (twin engines), a cockpit the size of a bathtub, 21 clock-like dials, three dozen switches, 22 levers, five cranks, two plungers, half a dozen thumb buttons and radio controls created a sizeable control board which could intimidate a new pilot without training. But once the pilots became accustomed to the power and speed of the Lightning they wouldn't trade it for any other plane. Mattern's piggy-back method of training certainly aided the new recruits and lessened the shock of such an advanced plane.

A training manual did not exist to assist new pilots on how to attain the long range and super performance that was available through the P-38. According to Beman, "To aid the new pilots, a small pictorial book with see-through fold over performance curves was designed. It was small enough to fit in a flight jacket pocket and was intended as a reference for guidance in emergency situations. It was important to know how best to return home if the plane had suffered damage but flight was still possible.

The 18 versions of the P-38 were most versatile and served the following duties: day/night fighters, bomber escort, long-range photo reconnaissance planes, tank busters, railroad train and track destroyers, smoke screen layers, rocket launchers, torpedo carriers, aerial ambulances, high altitude precision bombers, and fighter-bombers. The versatility of the P-38s contributed to its remarkable record in combat and played a great role in securing Allied air superiority in many theaters of operation.

"Strikes by Lightning" P-38 Operations during W W II

"*. . . any future war will inevitably open with great aerial activity far in advance of contact either upon land or sea, and that victory cannot but incline to that belligerent able to first achieve and later maintain the supremacy in the air.*" From a 1919 American aviation mission report upon return from Europe.

With the December 7, 1941 Japanese attack on Pearl Harbor, the United States first had to protect its two coasts from further attacks. One of the first duties of the P-38s was to provide stateside air defense. The P-38s joined the First Pursuit Group and flew combat patrol over the West Coast in case of a Japanese invasion. On 8 December 1941, the P-38s joined the First Air Force to began water reconnaissance to locate possible hostile forces which might approach the east and west coasts. Captain Newell O. Roberts said that this was his most dangerous flying because nobody knew how to fly the P-38s. He said that often the formation was so tight that the wing tips were only three feet apart.

The First Pursuit Groups later joined the 14th Pursuit Group at Riverside, CA. After a few weeks of patrolling the coast it became obvious that the Japanese were not going to attack, so the P-38s went directly to

Rainy weather was not uncommon in the Aleutians. Here a P-38 takes off for a mission to one of the many islands. (Courtesy of Francis J. Pope)

Attu would be retaken so the 54th was ordered to cease operations against Kiska and focus on Attu. The P-38 pilots maintained a constant daylight patrol over Attu to protect the U.S. armed forces from enemy air action and to strafe any objective holding up the ground forces. By 21 May 1943, the weather had cleared enough for 18 P-38s, under the direction of Col. John V. Hart, air liaison officer for the day, to hit villages and installations on Attu. On 23 May, Navy PBY sighted Mitsubishis west of Attu and P-38s found them and destroyed five enemy planes. Lt. Col. James R. Watt and Lt. Col. Harry C. Higgins are credited with one plane each in this mission and Lt. Frederick Moore is credited with three. Two P-38s were lost and seven enemy bombers were claimed to have been destroyed. Lt. John K. Geddes belly-landed in Massacre Bay and was rescued by a Kingfisher from the USS Idaho. Col. Watt radioed that he was hit, but was never seen again.

On 26 May 1943, six P-38s were credited with five Betty-type Japanese bombers. On 4 June 1943 the P-38s of the 54th Fighter Squadron flew many sorties, 30 of which reached Attu. Nineteen tons of bombs were dropped by the 54th. This ended the 11th's work in Attu, so it then shifted its attention to Kiska. The 11th had fought long and hard under unusually harsh conditions. With Attu behind them, the next target was Kiska - the target date set for 15 August.

Despite the low-hanging summer fog, the 11th dropped 270 tons of bombs in June. In July, weather improved slightly, P-38s, along with B-24s, B-25s, and P-40s dropped 55 tons of explosives. In August, eight P-38s attacked a target on North Head and Little Kiska. The air offensive reached its climax on 4 August - 134 sorties and 152 tons of explosives were dropped. Bad weather forced interruptions for the next four days.

The last tactical mission against Kiska was flown on 14 August 1943 from Amchitka. During this operation 1,098 sorties were flown and 202 tons of bomb were dropped on targets by the 54th. The 11th's responsibilities included bombing gun positions and airstrips on Kiska.

In September, 1943, the squadrons were back at Attu. In October the area of operation extended to Shemya, the official headquarters being at Amchitka. In November the 54th was entrenched in Alexia Point, Attu Island. The mission of the 54th had two purposes - to defend the area from attack by land, sea and air, and to prepare for long range missions to the Kuriles. The winter of 1943-44 was spent in training against the Kuriles. On 5 February, 16 P-38s and six B-25s covered the retirement of the North Pacific Force after it attacked the Kuriles.

The 11th AF encountered continued bad weather during March and April. The 11th dispatched ten missions, generally armed reconnaissance flights. During 1944, AAF installations were reduced, an action that affected all of the 11th's bases east of Adak (except Elmendorf): Yakurat, Annette, Fort Greeley, Naknek, Atka, all were reduced to the status of airdromes. Juneau, Cordova, Gulkana, McGrath, Port Heiden, and Ogliuga were designated as AAF airway stations. By the fall of 1944, the 11th was reduced from 17,000 to 11,000 men. Its headquarters were shifted to Adak from Elmendorf. The administrative work of XI AF Service Command was divided between the 11th's own HQ and the Alaska Air Depot. On 3 March, 11th Fighter and the 11th Bomber Commands HQ personnel were absorbed by the 343rd Fighter Group. On 3 August service command HQ was abolished. The next several months saw the construction of airfields in the Aleutians for future use.

The threat in the Western Aleutians was over. initiative in the North passed from Japan to the United States. The U.S. had a reduced force containing the Northern Islands from the enemy. Despite the needs elsewhere, Japan still kept more than 400 aircraft in the Hokkaido-Kuriles area in case of an invasion.

An invasion never occurred. The U.S. surface forces conducted light raids on the Aleutians, in addition to reconnaissance missions. Flights were long, weather was bad, and it was generally hard on the crews. Only 9 or 10 of the original 25 to 31 Lightning pilots survived the Aleutian Campaign - the pilots faced unusually harsh conditions - freezing sea, heavy fog, low overcasts and high winds and foul weather. The pilots of the 11th generally flew 1200-mile missions carrying two belly tanks and flying at 1400 RPM. The speed of 1400 RPM helped conserve fuel since the Lightnings would fly for seven to eight hours. It was also not uncommon for the Lightnings to return powered by one engine.

Gen. Arnold summed up the Aleutian

Lt. Col. Harold J. Rau, 20th FG C. O. and Col .Cy Wilson, 20th FG discuss a mission. 1944. (Courtesy of Jack Ilfrey)

Campaign aptly when he said, "This is one hell of an air force you have up in Alaska." In 1944 the crews might have retorted, "This is one hell of a war we have up here too."

Digging P-38s out of mud was not uncommon in the European Theater. (Courtesy of Marilyn Hanson / Gil Cefaratt)

P-38 pilots proudly display the tail section of a FW-190 destroyed on 25 May 1943. (Courtesy of Marilyn Hanson / Gil Cefaratt)

European Theater 8th Air Force

Even though the Lightnings did not see as much action in the European Theater as they did in other theaters, they certainly proved themselves as superior fighters. When assaulted by Focke-Wulf's the P-38s escorting Fortresses would refuse to break formation; as a result the enemy aircraft would withdraw and the American aircraft could return safely to base. The P-38s always fought in teams and stayed in formation. Any time a pilot would break formation or chase a decoy plane, the entire enemy formation would pounce on him. The escorting P-38s had about three minutes of shooting in them - highly concentrated firepower. The pilots usually held their fire until 100 yards from the enemy plane.

The 8th Air Force was activated on 28 January 1942 at Savannah Air Base, with Col. Asa N. Duncan as the Commanding General. It was originally designated as the air element of GYMNAST. GYMNAST was an early Allied plan for the seizure of Casablanca and the invasion of Northwest Africa. Feeling that his air force was ill-prepared to carry out GYMNAST, Gen Duncan requested that it be strengthened with several bomber and pursuit groups. On 9 April 1942, Headquarters Echelon was located at Bolling Field, Washington, D.C. in preparation for the 8th AF's move overseas. Elements of the 8th, consisting of bomber, fighter and base commands, along with the 15th Bomber Squadron, the 2nd Air Depot Group and a weather detachment, totaling about 1,800 men, sailed from Boston, MA for Britain; they arrived on 12 May. In May, General Carl Spaatz was assigned as the 8th AF commander and assumed command at Bolling Field.

Military commanders planned to integrate all U.S. air units in the United Kingdom into the 8th Air Force. The main objective of the AAF in the European Theater of Operations was to attain "air supremacy over Western Continental Europe" to prepare for any future invasions. In late June the Headquarters for the 8th AF was established at Bushy Park, a suburb of London. The 8th AF decided to use the P-38 not only as an escort plane, but also in ground attacks. The Lightnings participated in all day missions, escorting, strafing and attacking targets of opportunity whenever possible.

On 9 July, seven P-38s flew to Britain via the North Atlantic ferry route. This was the first time a single seater Allied aircraft was flown by this route. On 15 August 1942 a P-38 from the 27th Fighter Squadron along with a P-40 from the 33rd Fighter Squadron attacked a Condor, a German aircraft. The plane was destroyed and became the first German aircraft destroyed by a USAAF in World War II in the European Theater of Operations. The 27th Fighter Squadron was only in Greenland for a short while before leaving for England but while there, the squadron managed to considerably reduce the German threat. The German Focke-Wulfs discovered the Lightnings with their long range were dangerous.

The P-38s of the 27th FS, 1st Fighter Group, and the 14th Fighter Group arrived in England by the end of August.

To aid in Operation TORCH, the 8th AF sent all but one of its fighter groups to aid the 12th AF based in North Africa. TORCH was the Allied plan for landing in North and Northwest Africa in November of 1942. On 27 October 1942 the last two squadrons of the 14th Fighter Group (P-38s) left for North Africa, therefore completing the transfer of all fighting units to the 12th. In September of 1943 the 8th AF, having sent all P-38s to the 12th, began to re-equip with Lightnings. The 55th Fighter Group arrived in Nuthampstead in September. The 55th had the newly modified P-38 with an increased range, they now had the capability to escort bombers to Berlin and back to England, a distance of 1,300 miles. The first mission to utilize the newly modified P-38 was flown on 3 March 1944, bombers now had a plane that could protect them throughout an entire mission. The 20th Fighter Group, the 364th Fighter Group, and the 479th Fighter Group

Pilots relax in an underground shack, North Africa, 1942. (Courtesy of Emil Kielbasa)

Staff Sergeant Morris and Lt. A. G. Jackson check gun camera film used on a mission in North Africa, 1943. (Courtesy of Archie G. Jackson)

The Kraut Exterminators of the 96th and 95th Squadrons of the 82nd FG congratulate each other for their excellent record. (Courtesy of Charles E. Adams)

later joined the 55th - all equipped with P-38s.

It was determined that additional units sent to the 12th AF at the expense of the 8th would severely limit and hurt the air forces in the United Kingdom. In November of 1942, the Air Ministry decided that the 8th Fighter Command would support bombers as escorts on all attacks against the Continent. General Jimmy Doolittle, after having taken command of the 8th AF in mid-1944, decided to eliminate the P-38 from action with the 8th AF. Eventually the 8th AF was completely equipped with P-51s; the Lightnings and P-47s of the 8th AF were all transferred to the 9th AF. It was decided that the P-51 was more efficient and cost effective than the P-38 and by 14 July 1944, the remaining P-38 fighter group, the 55th, was completely transferred to the 9th.

European Theater 9th Air Force

While the 8th AF continued with its offensive against Germany, the objective of the 9th AF was the invasion of Normandy. What began as a small air force became within seven months one of the most powerful tactical air forces of World War II. During the invasion of Normandy it was the 9th's job to cover the amphibious landing in France and cooperate with the ground armies as they swept Germany. The 9th AF was a powerful branch of the AAF including every aircraft - long-range, heavy bombers, strategic, escort and tactical fighters.

Established in 1942 the 9th AF, commanded by General Lewis H. Brereton, replaced the United States Army, Middle Eastern Forces upon its elimination. The units that currently served with the 9th were transferred to the 12th AF. The personnel of the 9th prepared for its move from Egypt to England, by the end of November the bombers, fighters and service commands, along with various other companies, had completed the move. The 9th had its official start on 16 October 1943 with little more than a headquarters and a staff. Soon thereafter, the 8th AF transferred men and planes to build up the strength of the 9th in preparation for the D-Day Invasion.

The 9th had three P-38 groups - the 474th Fighter Group (operational on 25 April 1943), the 370th Fighter Group (operational on 1 May 1943) and the 367th Fighter Group (operational on 9 May 1943). Lightnings best performed in combat as fighter escort, the accompanied by other fighters escorted bombers on 1,178 effective sorties. These activities comprised the major share of the military effort of the 9th. By the end of March, the 9th had proven itself as a superior tactical air force and the fighters began attacking targets in France.

From June of 1942 until September of 1943 the 9th AF participated in two missions against Ploesti, Rumania oil fields, cut off supplies and reinforcements to the Axis African troops, supported British troops against Germans as they assaulted Egypt and later aided the North African Air Force in hitting Italy and Sicily. By the end of 1943, the 9th's service command found it necessary to modify several planes to improve the flow of planes to the fighter units. P-38s, in addition to other planes, were equipped with jettisonable tanks.

The responsibility of the 9th AF in the D-Day Invasion was one of the most critical of all the air forces. The fighter command was scheduled to provide escort for bombers, perform reconnaissance, and carry out offensive sweeps over France. The three P-38 groups, in cooperation with Lightnings from the 8th AF, would maintain daylight patrol over the invading troops. Around H-Hour, P-38s bombed enemy gun batteries and provided support for the ground forces. The Lightnings were utilized because they were more effective than the P-47 at low-altitude attacks and for maximum damage the military recommended the use of 2,000 pound bombs. On 25 April 1944, the P-38s of the 474th Fighter Group went on its first mission to patrol. But like the 8th AF, whose P-38s were transferred out by order of Gen Doolittle, the 9th AF's P-38 were also replaced. The two other Lightning groups, the 367th and the 370th, were replaced by P-51 Mustangs and P-47 Thunderbolts.

Overall, the P-38's role in the European Theater was short, but nonetheless, extremely important. Bomber escort and air cover over the invading troops on the Normandy Beachhead were roles the P-38 filled most effectively.

Iceland

An often forgotten and lonely area of war was Iceland. On 14 August 1942, the first aerial victory in the European Theater of Operations of the Allied Air Force was accomplished by Second Lt. Joseph D. Shaffer of the 33rd Fighter Squadron (a P-39 Squadron) and Second Lt. Elza E. Shahan of the 27th Fighter Squadron (a P-38 squadron) when they shot down a Focke-Wulf 200 off the coast of Iceland.

The Mediterranean Theater 12th Air Force, North Africa

"you couldn't learn to play an organ or a piano in eight months, and that damn P-38 had just about as many things to keep in tune" Eddie Martin-Lockheed Test Pilot

The Aleutian Campaign created a climactic environment to which many pilots were unaccustomed and which proved to be very hazardous. Likewise, the Northwest African Campaign proved to be one of the most daring, complex and hazardous undertakings of the Allied powers.

At Ste. Marie du Zit Airfield, Tunisia. L to R are: Bill Broome, Herbie Ross, Bob Hope, Fred Haupt and "Swede" Lindstrom. Bob Hope took a P-38 piggy-back ride. 1943. (Courtesy of Willian Broome)

America's interest in Northwest Africa went back to the collapse of Allied armies in France and the Low Countries during the summer of 1940. The Germans split France into two parts, but still maintained control over both. To the United States, Nazi-occupied Vichy-Africa posed as a threat to the Western Hemisphere. To England, it was a threat to the sea route through the Mediterranean, exposure of the sea route to U-boats and bombers and a threat to the air route across Central Africa to the Middle East.

In June, July and August of 1942, P-38s were flown by combat pilots from Lockheed in Burbank, CA to England and then to Africa. This was the first time fighter planes were flown non-stop across the ocean - this was accomplished by carrying an extra gasoline tank under the middle wing section. The first operations in North Africa were defensive to block the extension of the Axis powers. At the Arcadia Conference in Washington on 23 December 1941 - 14 January 1942 the plan GYMNAST was presented as an operation to land U.S. and British forces at Casablanca - an area of strategic importance in the Atlantic. GYMNAST, renamed TORCH in December of 1942, was advantageous because it was a countermove to a German entry into Spain; it sealed off and neutralized Dakar, forestalled Axis occupation of French North Africa; opened the Mediterranean to a limited degree; and secured bases for land and air operations against Italy for air attack on Germany if longer range bombers became available.

By 25 July 1942 the Combined Chiefs of Staff approved the command set-up for TORCH. General Dwight Eisenhower was the overall military commander for TORCH. Air Marshall Sir Arthur Tedder commanded the newly-formed Desert Air Force which included the new USAAF 9th AF. . A new air force was created to handle the vast amount of airpower needed for success in North Africa - the 12th AF - and it was built from the 8th AF in Britain. General Jimmy Doolittle took command of the 12th AF on 23 September. The 8th AF assigned large numbers of its own units to the new AF, including the P-38. The initial combat force comprised two P-38 fighter groups, two Spitfire fighter groups, two heavy bombardment groups, one light and three medium bombardment groups and one troop carrier group. The main duty of the P-38s was to protect Allied forces in Tunisia, comprised of six French battalions and Royal Air Force's reinforced paratroop battalions. On the eve of TORCH, unfortunately, there was no evidence of how the P-38 would perform against the Luftwaffe. Because of this lack of evidence and the misgivings some pilots had toward the P-38, Gen. Spaatz was eager to see the P-38 in action and the Lightning pilots received careful training prior to transfer to Africa.

The P-38s, because of their versatility and endurance, were used in a variety of ways. Because of this, Lightnings were not readily available for bomber escort. At the Casablanca conference, Gen. Arnold, a long time advocate of the P-38, practically demanded that all P-38s from England be sent to North Africa.

The men and equipment of the 14th Fighter Group of the 8th AF out of Britain, were moved, even though they didn't know where, while the planes were launched to Gibraltar. They were to eventually strike at Vichy-occupied North Africa. On 8 November 1942, the Allies hit the beaches of Casablanca, Safi, Fedhala, Port Lyautey and Oran, surprising the enemy and thus beginning Operation TORCH. On 11 November 1942, Gibraltar-based Lightnings flew their first mission and landed on the airfield at Tafaraoui; this was the same day the French surrendered. The P-38s of the 1st and 14th Fighter Groups of the 12th AF bore the brunt of the fighting in North Africa and were used mainly as early-morning weather reconnaissance of target areas. On 13 November, P-38s escorted C-47s carrying AA guns and aviation gasoline to Duzerville Airfield. P-38s of the 14th Fighter Group arrived from Britain for heavy bomber escort. Unfortunately many were damaged by the enemy on Maison Blanche Airfield.

On 20 November P-38s of the 48th Fighter Squadron, 14 Fighter Group were surprised by the Luftwaffe. The Lightning pilots endured the German hit and run raids and the harsh realities of living in the field - primitive conditions included living in tents, discarded fuel barrels or out in the open. The climate conflicted between hot and dry and wet. Because of these conditions, engine and other mechanical failures were not uncommon.

On 21 November P-38s of the 1st Fighter Group arrived at Nouvion to replace the 14th Fighter Group in escorting B-17s. The Lightnings of the 14th Fighter Group moved into Youks-les-Bains air field on 22 November. At the end of November, P-38s escorted bombing missions on Gabes, Djedeida and Bizerte. Other P-38s strafed Gabes Airfield, flew reconnaissance and shot down an Me-109 in aerial battle near Tunis.

December brought with it a variety of activities for the Lightning pilots. They flew reconnaissance and patrols over Tunisia and Sousse-Sfax-Gabes, escorted B-17s as they attacked Bizerte, and often had missions aborted because of heavy rains. In mid-December P-38s conducted sea patrol over the North Coast and the Gulf of Tunis and accompanied by P-40s flew widespread reconnaissance missions.

On 13 December, P-38s attacked several targets including vehicles north of Gabes and a schooner off Cape Dimasse. They continued their attacks on 14 December by destroying vessels off the north Tunisian coast, the road between Tunis and Bizerte, a train near Kerker and La Hencha, and trucks near Chaaba and El Djem. Throughout the rest of December P-38s had victories, claiming the destruction of German aircraft, two Ju-88s, FW-190s, vessels, locomotives, a tank depot, and motor vehicles. These attacks took place over Youks-les-Bains, Tunis, Sousse, Pont-du-Fahs and Tunisia. Activities increased near the end of December with the P-38s attacking docks at Sfax, flying missions to Tripoli and destroying several trucks. The Lightnings also flew reconnaissance over the Medenine-Tripoli area where they could monitor the disposition and supply of Rommel's forces.

In December, small forces of B-17s fiercely fought with FW-190s at Bizerte, the P-38s can be credited with helping the B-17 losses remain low, since enough Lightnings escorted the B-17s to divide the Luftwaffe's attention. Between 14-18 December, the P-38s of the 1st Fighter Group moved to the bomber station at Biskra under the control of the XII Bomber Command. This proved to be a wise move as the P-38s simplified the rendezvous and eliminated the need to coordinate each mission with a fighter command headquarters.

P-38 pilots took heavy losses and the 14th Fighter Group withdrew from combat by the end of January 1943. Of the original 54 pilots, 32 were lost in combat or accidents. The group claimed 62 enemy aircraft destroyed.

Operation TORCH progressed and the P-38s consolidated and increased firepower as new pilots and aircraft arrived. General Erwin Rommel moved his Panzer units against the Allies based in North Africa causing heavy losses. Germans received many of their supplies by air - and the Allies decided that this aerial supply must be broken. The P-38s of the 1st and 82nd Fighter Groups participated in Operation FLAX, a massive effort to attack and destroy German supply lines. The German transport fleets of JU-52 (Junkers) were easy aerial targets for the Allied fighters since the escorting ME 323 (Messerschmitts) were slow and unmaneuverable.

The first major encounter for FLAX was on 5 April 1943 when P-38s of the 1st and 82nd flew their usual bomber escort missions and encountered 65 JU-52s. On 11 April 1943, P-38s downed 32 Nazi cargo planes attempting to supply the crumbling Afrika Corps from Italy. The P-38s went through the fighter escort formation and hit the transports. Eleven Junkers were lost, along with their fighter escorts. In April a formation of P-38s successfully hit a convoy of Siebel ferries and a large barge. For the next several days P-38s, P-40s and British Spitfires attacked German aircraft formations. Allied aircraft were lost, but the Germans suffered more, losing hundreds of transports. The Luftwaffe was unable to protect the transport flights. The Allies strengthened their position in North Africa and gained control of the airspace by destroying the transports. On 13 May 1943, the Tunisian Campaign ended.

In September of 1943, Lightnings assaulted the grounds at Grazzanise; they did not find the appointed targets, but did hit many targets of opportunity. On 17 September, P-38s hit bridges, roads, railways and towns in Italy during 27 dive bombing missions. Such activities continued throughout September aiding the war effort in the Mediterranean Theater of Operations.

In October of 1943 a P-38 squadron was on detached service to Bengasi and Gambut to hit the Aegean shipping. Flying during a four day period, the P-38s claimed 17 planes destroyed. At this same time the 12th was to hit larger fields dropping thousands of frags and tons of general purpose bombs on Argos, Rhodes, Araxos, Crete, Greece and Athens/Tatoi. One of the most outstanding missions of October involved the assault against Wiener Neustadt. Lightnings escorted B-24s and B-17s on their assault on fighter aircraft plants, transports and barges. The bombers were attacked when over the Leghorn-Pontedera area and the P-38s successfully defended the bombers, claiming eight enemy planes destroyed. Also during October, P-38s strafed and bombed many targets of opportunity, roads, bridges, and towns.

In November of 1943, the 15th Air Force was activated and the units of the 12th Bomber Command transferred to the new air force. In December the 82nd Fighter Group's mission was to escort B-24s as they hit Udine in Northern Italy. On this mission, the P-38s experienced windshield icing conditions which interfered with their operation, however, enemy aircraft were destroyed by the P-38s.

While in North Africa, the P-38 proved itself as a weapon of war serving in a variety of ways - in high-altitude fighter sweeps, as a bomber escort, medium-altitude dog fighter, ground attack missions and reconnaissance flights. The P-38 was extremely valuable because of its 350-mile fighting radius. Allowing 10 minutes of combat and using an auxiliary tank the Lightning could reach assault beaches and remain flying above for

A ground crew member loads armament into the gun compartment of a P-38. (Courtesy of Francis J. Pope)

The ground crew of the 27th FS change an Allison 12 cylinder engine. Italy 1944.(Courtesy of Robert W. Share)

Right - The 15th Air Force's P-38s in a fighter formation over Italy. (U.S. Air Force photo, courtesy of Bill Hess)

Middle - The First Fighter Group takes off for a mission over Italy. (Courtesy of Bill Hess)

Below - Pilots of the 96th FS, 82 FG partying in Squadron Officers Club, Christmas Eve, 1943. The airfield was the former Italian Air Force Field at Leece, Italy. L to R are: Capt. Prann, Lt. Col. Coleman, bartender Lt. Dean Barlow, Lt. Dolezol and Lt. Thomas. Middle row - Lt. Berube, Capt. Suddeth, Capt. Perry, Lt. Houtz and Lt. Jones. Front Row - Lt. Augenour, Lt. Larkin, Lt. Chatmus, Lt. Downey, and Lt. Trammel. (Courtesy of Fred Selle)

Gen. Nathan Twining, CO, 15th AF fastens the Distinguished Unit Citation to the colors of the 1st FG, Italy 1944. (Courtesy of Robert W. Share)

Pilots of the 94th FS are briefed before a mission in Foggia Italy. (Courtesy of Robert F. Taylor, Jr.)

Lt. Charles E. Adams Jr., 95th FS, 82 FG, Foggia, Italy, the swastikas represent enemy planes shot down.(Courtesy of Charles E. Adams)

An intelligence officer conducts a debriefing after a mission with the 94th FS in Foggia, Italy. (Courtesy of Robert F. Taylor, Jr.)

P-38 pilots G. Milne, Hackbarth and I. N. Perry discuss the results of their latest escort mission on Europe. (Courtesy of Gregor Milne)

Lt. Col. Francis "Bucky" Harris C.O. 94th "Hat-in-the-Ring" FS, 1st FG, 15th AF and Capt. Everett Lindley, Ft leader of the 94th FS stand in front of Harris' P-38. Salsola, Italy, Feb. 1945. (Courtesy of E.S. Lindley)

an average of one hour. The conditions in North Africa were harsh and took their toll on both the planes and the pilots. The Lightning in Africa strafed, bombed and attacked Rommel's tanks with deadly accuracy and were able to land in mud or sand on their tricycle gear without flipping over. New areas of combat opened for the Lightning as it had proven itself in combat.

The Mediterranean Theater - NAAF, 15th Air Force Pantelleria, Sicily, and Italy

... but I have no Luftwaffe ...." Robert Ritter von Greim, Oberbefehlshaber der Luftwaffe, Berlin, April 1945.

The war in North Africa ended in May of 1943 and attention shifted to the islands of the Mediterranean. After the North African victory and before entering combat in the Mediterranean Theater though, the USAAF units re-equipped and built up strength in preparation for the invasion of the Mediterranean island of Sicily and Italy. Whereas in North Africa and the Aleutians, pilots had to deal with inclement weather, the pilots fighting in southern Europe were in an extremely dangerous zone of combat. It was not uncommon for pilots and crews to be badly beaten or killed by hostile civilians - rescue coming only when the Germans intervened to haul the downed fliers to a prison camp.

The Lightnings played a major role in the raids on German and Italian airfields - its heavy, concentrated armament was most effective. By the end of May the Allied Forces began the big push in the Bizerte-Tunis area and by 13 May the Allies were ready to hit Pantelleria, a Mediterranean island. To handle this new area of concentration, the Northwest African Air Force (NAAF) was created with the now legally non-existent 12th AF serving as the administrative organization for the American elements of the NAAF.

The objective of the new campaign was Sicily and the surrounding islands of Pantelleria and Lampedusa. The living conditions were harsh and the reverse of what the men experienced in North Africa - constant dust storms and dry weather caused numerous mechanical problems.

Pantelleria, a small island between Sicily and North Africa in the Mediterranean Sea, was the site of heavy enemy concentrations, one of the first targets of the Allies and one of their more ambitious undertakings. It was an outcropping of volcanic rock which had been strengthened by Italians since the 1920s. Pantelleria, and its neighboring island of Lampedusa, was of strategic importance and as long as each was held by the enemy, was a threat to the future Operation HUSKY. Operation HUSKY was the Allied plan for the conquest of Sicily, to be undertaken after the end of German resistance in Tunisia.

The airfield on Pantelleria was valuable to HUSKY landings and it was determined that heavy bombardment from the air would break down the morale, materiel, personnel and equipment for an early success.

The plan to capture Pantelleria was named Operation CORKSCREW and the date for the final assault was set for 11 June and would be launched from Tunisia. Operation CORKSCREW represented the first Allied attempt to conquer enemy territory essentially by air action. The P-38s were of vital importance to this operation because of their versatility.

The enemy had 900 operational combat planes and the Allies had 2,374, of which 1,276 were serviceable. The Allies had superiority and the real air offensive began on 18 May and lasted until 5 June (D Day minus 6). The offensive started with 50 medium and 50 fighter bomber sorties a day isolating Pantelleria by the end of the month. On 1 June P-38s and P-40s dropped 141 tons of bombs on the island and on 4 June dropped 200 tons to neutralize the coastal batteries. By 6 June the assault was full-fledged and the Germans and Italians began to respond.

From 8 May to 11 June, the North African Air Force flew 5,285 sorties, dropping 6,200 tons of bombs. Four aircraft were destroyed, 10 missing and 16 damaged. P-38s played a major role by bombing the island, escorting B-17s as they bombed Palermo, escorting B-25s as they bombed the dock and town area of Olbia, B-26s as they blasted Porto di Pantelleria and Marghana, and bombing the docks on the Gulf of Arcani, railroad bridges. Hundreds of bombing and strafing missions were carried out by the Lightnings of the 1st and 82nd.

To complement the attacks on Pantelleria (CORKSREW) and pave the way for Operation HUSKY, P-38s attacked zinc works at Iglesias, Sardinia and Carloforte harbor and bombed other targets on Sardinia - factories, hangars, oil dumps, railroad bridges and warehouses. The Lightnings continued bombing and strafing missions hitting Porto Ponte Romano, Chilivani, Cagliari, Guspini, Santa Caterina, Porto Torres harbor, Milo air field and Monserrato and Capoterra, all on Sardinia.

Pantelleria was twice given the chance to surrender, declining both times. On the night of 10 June the air forces attacked gun positions, and bombed and strafed parked aircraft. After several hours of medium bombing, a white flag appeared over Semaphore Hill; the military governor of Pantelleria asked to surrender and with that action the Allied forces shifted their attention to Lampedusa, which if successfully taken, would provide an air base to support the Sicilian campaign and the defense of the Sicilian narrows.

The highlight of Pantelleria was when 82nd pilot Lt. Col. Ernest C. Young, deputy Group commander, skip-bombed a 1,000 pound bomb directly into the open door of the Axis underground hangar.

Now, with Pantelleria taken by the Allies, attention shifted to Italy. On 25 August 1943 the 1st Fighter Group sent out 65 Lightnings to hit enemy airfields near Foggia, Italy. The P-38s flew almost undetected on the deck and claimed 200 German and Italian aircraft destroyed - only two P-38s were lost.

The 82nd won three Distinguished Unit Citations. The P-38s of the 82nd considerably weakened German air resistance through low level strafing attacks on Foggia airdrome and satellite airfields, under the leadership of Lt. Col. George M. McNichol. On 2 Sept. 1943, P-38s escorted 72 B-25 Mitchell bombers striking Cancello, Italy. They fought off 100 enemy fighters - 100 miles out to sea, shot down 24 of the attacking force and brought all of the bombers home safely. On 6 September 1943 the Lightnings went on patrolling, strafing, and bombing missions, covering the Allied Salerno landings.

In ten days, the Lightnings flew 96 missions and 990 sorties. September fighting continued to be intense with every available P-38 battling, the main targets being German troop and supply concentrations. On 18 September, it was estimated that approximately 300 enemy planes were destroyed or damaged after the raids on Foggia airfields.

Having moved to a base in Italy in October of 1943, the P-38s hit Balkan and Yugoslavian ground installations and ammunition dumps. By the end of 1943, 365 enemy aircraft were shot down - this number not including the enemy aircraft destroyed on the ground.

In January 1944 the 82nd had moved to an Italian base. They fought air battles over the Udine area in northeast Italy and later over the Lake Balaton region south of Budapest. The 82nd was now employing the latest type of P-38s with the extended radius.

Operation CORKSCREW proved the power of air bombardment in forcing the surrender and downfall of an area. The victories at Pantelleria and Lampedusa came before land assault troops could contribute because the Allied Air Forces literally bombed the two islands into submission.

Having paved the way for the Sicilian campaign by successfully taking Pantelleria and Lampedusa, the Allied Air Forces were now ready to begin Operation HUSKY. The base was in Tunisia.

The objectives of HUSKY were "to make secure the Allied line of communication in the Mediterranean; to divert Axis strength from the Russian front during the critical period; to intensify pressure on Italy; it also required Germany to extend military commitments into southern France and the Balkans, make Sardinia untenable and threaten Corsica, force the resignation of Mussolini on 25 July and lead to an Allied-Italian armistice on 3 September." Sicily was conquered in 38 days as a result of intensive air, ground, and sea power. The swift victory can be attributed to the superior men, planes, and ships as they maintained a constant and strong offensive against a skillful enemy.

Specifically during HUSKY, the P-38, as part of the NAAF, escorted heavy bombers on attacks at Olbia, bombed Milo, strafed radar installations in the east part of Sicily, strafed trains and other targets of opportunity southwest of Angitola, hit lighthouses in Southern Italy, attacked bridges in Catanzaro, hit trains, troop radar stations, and tunnels and escorted B-17s. The attack on Sicily took place on 10 July ending on 17 August of 1943.

On 22 October 1943, the Combined Chiefs of Staff approved a plan by Gen. Arnold and the Joint Chiefs of Staff to create a new air force in Italy - the 15th - elements coming from the 12th to be used in strategic bombing in Germany. The 15th was activated on 1 November with its headquarters located in Tunis, Gen. Doolittle was the Commanding General. The 15th AF was one of the most successful organizations in the Air Corps. It fought in Tunisia, Sardinia, Sicily, Italy, France, Germany, Austria, Czechoslovakia, the Balkans and the whole of southern Europe. The 1st, 14th and 82nd Fighter Groups were assigned to the 15th AF.

The P-38s of the 15th were to escort heavy bombers, B-26s, B-25s, B-17s, and carry out bombing and strafing missions over Austria, Greece, the Balkans, France and Italy. Lightnings also completed weather reconnaissance missions over the Adriatic Sea, Sofia, and Zara. Many of their December missions were restricted because of bad weather. The largest air force attack came on 29 March when 400 B-17s and B-24s, with over 100 P-38 escorts bombed Bolzano, Turin and Milan.

The final plan for HUSKY was approved on 13 May by the Combined Chiefs of Staff, and it called for eight simultaneous seaborne assaults along 100 miles of coastline - NAAF to provide the air support. The last three weeks of HUSKY the objective was the airdromes in southern and central Italy. P-38s and P-40s flew extensive fighter sweeps over Sardinia and the toe of Italy. The 1st, 14th and 82nd fighter groups handled operations against mainland targets, evacuation routes, in addition to bombing and strafing trains, tracks, motor transports, radar stations, bridges, and targets of opportunity. The Lightnings carried 500 pound and 1,000 pound bombs against little or no opposition. Overall the P-38s inflicted considerable damage, aiding in the success of Operation HUSKY.

Also during HUSKY, three groups of P-38s (from the Strategic Air Force) protected the ground troops - one group was specially trained for night flying so it could leave before dawn and return after dark. HUSKY, the final assault on Sicily, saw the first large scale airborne operations undertaken by the Allies in World War II.

In the latter half of 1944, the 15th AF was assigned the task of bombing the oil fields of Ploesti, Rumania. As a result, it was one of the most highly defended targets in Europe. The P-38s were employed to execute long range, low level surprise dive bombing missions. The target was the Americano Romano oil refinery and it was destroyed. According to Richard Willsie, "The Ploesti (oil fields) in Rumania became Europe's number one target during the latter half of 1944. Consequently, the German's defense of Ploesti was so intense it was rated as the most highly defended target in Europe. After the first raid on Ploesti in 1943 by B-24s at low level from bases in Africa, there was a lapse of raids on this target due to its distance from our air base. When the heavy bombers moved into Italy, the raid began again in earnest."

Willsie continued, "These heavy bombers were escorted with large numbers of fighters, P-38s, P-51s and P-47s, but due to the large numbers of enemy fighters protecting Ploesti, many B-17s and B-24s were shot down. Also, by the time the heavy bombers arrived in the target area, the targets would be covered by large smoke screen from smudge pots fired by the German's defense forces at the earliest indication of our air raid, hours before the bombers arrived."

"As a result the US Air Planners decided to employ P-38s on June 10 to execute a long range, low level surprise dive bombing mission. Results were the destruction of the Americano Romano oil refinery. The planners also decided that a more effective way to destroy the defending fighters guarding Ploesti would be to attack their airfields by a surprise low level fighter strafing raid, rather than limiting ourselves to air battles. The enemy's airdromes were located in an area we called the Foscani Airdrome Complex. These airfields were not within the range of our fighters based in Italy. This answer to this was to attack the airfields and then recover at friendly bases made available to us in Russia, specifically the Poltavo Air Base in the Ukraine," Willsie concluded.

The longest P-38 dive-bombing mission in the Mediterranean Theater of Operations, on 10 June 1944, was led by Lt. Col. William

A dozen 8th FS, 49th FG P-38s move along the taxi-way at Ligayen, Luzon for a napalm mission against a Japanese stronghold on Luzon. May 1945. (Courtesy of James P. Gallagher)

The 475th FG of the 5th AF returns after a mission. (Courtesy of Bill Hess)

The 475th FG over the Leyte Gulf. (Courtesy of Bill Hess)

P. Litton. Again the pilots flew at tree top level over 650 miles of hostile territory striking huge Romano-Americano Oil Refineries at Ploesti, Rumania, for this the unit received its third Distinguished Unit Citation.

Alphons J. Mikes remembers this mission well. "At the break of dawn, 10 June 1944, the 82nd Fighter Group took off from a dirt runway near Foggia, Italy on a sneak low level dive bombing attack on oil refineries at Ploesti, Rumania. Carrying a 330 gallon fuel drop tank, and a 1,000 pound bomb to a target over 600 miles away was taxing the P-38s to the limit." Mikes continued, "We were provided escort by the 1st Fighter Group. The mission was successful, however losses were severe. We were hit hard in target areas by many fighters and heavy ground fire. Many of the pilots returning to base had been airborne over seven hours," he concluded.

Charles E. Adams, of the 82nd Fighter Group also recalls the dive-bombing mission. "I can vividly remember the excitement that I felt when we found out that we were to dive-bomb the Refinery with our P-38s. We each carried two photos of the Refinery and surrounding area as well as a "flimsy" which gave courses, times, distances, etc. It is evident, from an inspection of the aerial photos, that the target area had been bracketed by B-17 and B-24 bombing efforts. We were confident that by dive-bombing with our P-38s we could accurately place our bombs in the critical refinery facility." He continued, "We liked the plan that called for flying at minimum altitude across Yugoslavia and "on the deck" across Rumania to hopefully avoid detection by enemy radar. As it turned out, we were probably detected because we encountered numerous enemy airplanes near Bucharest. At that point it was very difficult to stay on course and to avoid the temptation to shoot down a few before continuing on to the target." Adams concluded by saying, "After seven and a half hours in the cockpit (and gun fire that hit his right engine) - most of the time with both feet on one rudder pedal - I was completely exhausted and I welcomed help in getting out of the plane. I was sure proud of that fine Allison engine for the excellent performance on a 650 mile single engine flight."

Italy surrendered and the southern part of the peninsula was taken, all of this bringing to the Allies a certain amount of benefits. Additionally, the swift victory over Sicily through Operation HUSKY was owed to the superiority in men, planes and ships. P-38 pilots faced a constant and strong offensive against a strong and skillful enemy.

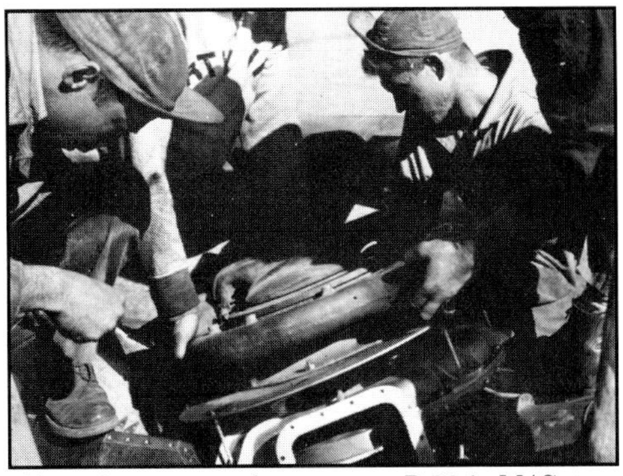

Changing a supercharger on a P-38 is M/Sgt Magann and M/Sgt Kobaly, Australia.

The 431st FS, taken at Hollandia campsite, June 1944. (Courtesy of Anthony Kupniewski)

Preparing engine for P-38 installation at DoboDura. (Courtesy of Anthony Kupniewski)

Engine change by Sgt. Greenberg,, S/Sgt. Meyer, and T/Sgt. Stidd. DoboDura.

Pacific Theater
South Pacific, 13th Air Force

Allied potential shone through in the South Pacific under able leadership and a sufficient armada of men, planes, guns and ships. It was determined that the South Pacific needed its own air force, which would be part of the U.S. Army Forces, South Pacific. On 5 December 1942, the 13th Air Force was created by order of General George Marshall. General Nathan Twining would

Operations shack, 433rd FS, DoboDura, New Guinea, 1944. (Courtesy of H. Cochran)

have his headquarters on Espiritu Santo and Admiral Aubrey Fitch would be ashore

Gen. Harmon faced difficulties in securing the needed personnel and units needed to adequately start the 13th AF. The 13th AF was officially started on 13 January 1943 with Gen. Twining as the commander. Regardless of what was written on paper, the newly-formed 13th AF still had very little with which to work and was reduced to borrowing and consistently requesting men, planes and supplies. Eventually, P-38s, P-39s, and P-40s were flown into Magenta Field, although most of the planes were landed at Noumea because of its central location.

For the next several months and during the beginning of the Guadalcanal Campaign, the 13th AF worked at a disadvantage from a lack of engineering and air base groups. Improvement was slow but sure, and by the spring of 1943, the 13th AF was on its way to building the strength it needed to successfully fight the Japanese.

Facing the Japanese at Guadalcanal, the 13th AF needed a plane that could bomb, strafe and harass the Japanese ground units. The P-38 was the ideal plane for the job. General Millard Harmon, Commanding General of the 13th Air Force of the SOPAC, noted the deficiencies in its air forces, explaining that the P-40 was not adequate. He needed a high altitude plane and the solution was the P-38, unfortunately none were available until November of 1943. Additional planes were needed in the defense of the southern Solomons and for the development of a plan and operations for the reduction of Rabaul. The main strength of the air forces in the SOPAC would be in its two squadrons of P-38s. The powerful twin-engines allowed the Lightnings to negotiate the long distances between island bases. Harmon also asked for three P-38 squadrons to be dispatched to New Caledonia so major strategic points could be held against the enemy.

Admiral Robert L. Ghormley was the Commander of the SOPAC; established under him was the joint United States - New Zealand Naval Command. On 20 May 1942, Admiral John McCain assumed command of the air force in the SOPAC. Brig. Gen. Twining was the Commanding General of the 13th Air Force in the South Pacific. The Pacific Theater commanders continually requested P-38s, citing the strength in such a versatile plane. Gen. Arnold, even though admiring the P-38, praised and recommended the P-40s, saying that the Atlantic needed the P-38 just as much as did the Pacific. In North Africa, the P-38 was needed to cross the Atlantic or fly from the United Kingdom to Oran or Casablanca. It was believed that withdrawal of the P-38 from Africa would mean the abandonment of the African campaign completely.

In December, Gen. Harmon went to Guadalcanal and found that conditions had improved. The P-38s provided excellent service with a minimum amount of maintenance. Gen. Harmon singled out the P-38s for special praise because they provided cover for bombers, served as excellent reconnaissance planes, and had the potential to be a second bomber. The P-38s performed well in the worst conditions - low ceilings, limited visibility and driving rains. Harmon simply could not get enough Lightnings nor could he praise them enough.

At this time the 13th AF included the 18th Fighter Group, the 347th Fighter Group, the 6th Night Fighter Squadron, and the 4th Photo Reconnaissance Group. Henderson Field, Guadalcanal served as the main base of action for the Lightnings. On 18 November 1942 the 13th went on its first bomber escort mission.

The Shortland-Poporang area off of southern Bougainville was considered the principle Japanese base in the Solomons. Fighter Command ordered a morning attack strike force of sixteen P-38s and Corsairs. Several of these planes succumbed to engine trouble and only six went out. Captain Thomas G. Lanphier Jr. of the 70th Fighter Squadron, 347th Fighter Group led the group at 50 feet altitude and attacked enemy destroyers and many float planes. The only casualty was the P-38 flown by Lt. Rex T. Barber of the 339th Fighter Squadron, 347th Fighter Group who flew too close to a warship and lost three feet of his wing tip - skillful handling brought the plane in for a normal landing.

The Japanese had an excellent target in Guadalcanal since it was stockpiled with fuel, ammunition, supplies and equipment, Lunga shore sported cargo vessels and many warships were docked at Tulagi. The Japanese did attack at the first of the month. Forty-two fighters, six of them P-38s, engaged the Japanese in an intensive air battle during which the Japanese lost 20 aircraft, the Allies lost six. The Japanese again attacked on April 7, sending out 160 dive bombers. Fighter Command sent out every available fighter whose main fighting occurred near Tulagi off the Russell Islands. Flying at the highest altitude were the P-38s led by Capt. Lanphier. The fighters claimed 39 enemy aircraft downed. Seven allied planes were lost and one pilot, Maj. Walden

Williams of the 70th Fighter Squadron died.

Next, on 18 April 1943, the 13th AF successfully carried out a surprise attack against the Japanese. After having broken the Japanese secret code, the Americans had determined that Japanese Admiral Isoroku Yamamoto and other high ranking Japanese officials would fly to Ballale airfield in Betty Bombers, protected by Zeroes. Eighteen P-38s, eight from the 12th Fighter Squadron, 18th Fighter Group, eight from the 339th Fighter Squadron 347th Fighter Group, and two from the 70th Fighter Squadron 347th Fighter Group, were assigned to this important mission. Two P-38s aborted after take-off. Four of them were designated under the attack section with Capt. Lanphier of the 70th while 14 pilots provided cover under the command of overall flight leader Maj. John W. Mitchell, formerly of the 20th Pursuit Group, now Commanding Officer of the 339th Fighter Squadron.

P-38Fs and P-38Gs prepared for the long range flight with a mixture of 165 and 310 gallon drop tanks on the underwing pylons of the planes. They left Henderson Field on Guadalcanal on 18 April for the 500-mile flight to Shortland Island. The enemy was intercepted as scheduled and the Japanese took heavy losses - losing one of their greatest commanders - Isoroku Yamamoto. Mitchell and the members of the attack section received the Navy Cross. The USAF gives half credit each to Rex T. Barber and Thomas G. Lanphier. The P-38s of the 13th rounded out their activities in May by bombing air fields at Munda, strafing barges at Bambatana and by accompanying Navy Marine airplanes as they pounded Gatere and Pelipelpi Bay.

Throughout May the P-38s continued striking Munda Air Field hitting gun emplacements, runway and revetments. On the third, P-38s, P-40s and Navy dive bombers and fighters attacked buildings and AA positions in Rekata Bay. P-38s thoroughly strafed Nyanga and AA positions at Vila. The end of May saw 34 P-38s, P-39s and P-40s along with 62 Navy and Marine fighters and six Royal New Zealand Air Force P-40s intercept 20 planes over Russell Islands and Tulagi area. They claimed 16 enemy planes shot down.

The assault on Munda continued in June. The 13th AF expanded its area of assault to include Buin and the Luti Bay area of Choiseul Island. In early June the P-38 pilots attacked the shipping area of Buin damaging a destroyer and two smaller vessels. They claimed 15 Japanese planes down. By mid-June the P-38s of the 13th AF were strafing barges northeast of Ganongga and the Kahili area, as well as the Malaita Islands.

Just as they had proven themselves battle strong in other theaters of operations, the Lightnings did the same in the South Pacific and Brig. Gen. Twining wanted to form a full P-38 group. Gen Harmon passed along the request for 75 aircraft plus a reserve of 50 percent to be received by 1 Sep-

Pilots of the 431st FS listen to Capt. Willie Hanning as he briefs them for a mission to escort B-17s. 12 Mile Strip, Port Moresby, New Guinea, 5 Sept. 1943. (Courtesy of Frank A. Nichols)

Original pilots of the 431st FS, 12 Mile Strip, Port Moresby, New Guinea. (Courtesy of Frank A. Nichols)

From l to r are: Fred Dick, James Keck, and Dick Ganchan. Lingayen, Luzon, 1945. (Courtesy of James P. Gallagher)

Capt. L. T. Daniels and crew chief T/Sgt. W. M. Powell n 1944 on Biak Island, South West Pacific. (Courtesy of L. T. Daniels)

P-38 formation over the Philippines. (Courtesy of Bill Hess)

tember. Generals Twining and Harmon recognized the importance of adequate fighter coverage for all amphibious landings. Also, with P-38 fighter cover, fighter command would be better able to isolate the battlefield. General Arnold reviewed the situation and decided that the available Lightnings should remain in North Africa.

On 27 July, Gen. Ray Owens became the Commanding General of the 13th Air Force. On this same day, eight P-38s and 70 Navy and Marine aircraft again attacked Munda area hitting positions on Bibolo Hill and targets at Gurasai, Munda Point and Munda Air Field. The 13th also began strafing and bombing Ballale in late July. The same activities continued in August as the Lightnings escorted B-25s as they bombed Vila and Buki harbors. They also extended their area of operation to include the Central Solomons.

On 20 September 1943, Lt. Henry Meigs II, flying a P-38 broke all local records by shooting down two Betty Bombers over Guadalcanal within 60 seconds.

In October the P-38s continued to bomb and strafe Kahili as well as Buka I Air Field. The 5th Air Force was to aid Admiral William F. Halsey's venture against Bougainville. General George C. Kenney set aside B-25s and P-38s which already had begun attacking Rabaul Air Field. Now the attacks would be coordinated with Halsey and the 13th AF. It was decided that the P-38s and the Corsairs would defend the Russells. The mission was slated for execution on 27 October 1943.

The Lightnings covered the landing of New Zealand troops on Stirling and Mono Islands. The P-38s also played a major role in the U.S. Marine landing on Bougainville, flying at 20,000 to 25,000 feet, then sweeping over the waters northwest of the Solomons. The Lightnings of the 18th Fighter Group flying at 23,000 feet met enemy fighters and Bettys, claiming seven Zeros shot down. The Marines landed on Bougainville with the objective of eliminating the last Japanese stronghold southeast of Rabaul. In a surprise attack on 26 October eight P-38s strafed Buka-Bonis, killing approximately 200 laborers. These activities laid the groundwork for the landings on Empress Augusta Bay on the west coast of Bougainville.

In November the 13th AF continued its concentrated effort on Bougainville and the surrounding islands. Encountering only weak resistance, 58 P-38s escorted 27 B-24s as they dropped 81 tons of bombs on Rabaul. On 1 November eight P-38s of the 18th Fighter Group flying at 23,000 feet, claimed seven planes. The second day of fighting was no less harassing for the Japanese. Bad weather had prevented Kenney's 5th AF B-25s and P-38s from an assault on 2 November but the weather soon cleared and the 5th carried out an attack on the shipping at Rabaul - 70 P-38s escorted 75 B-25s in attacking Rabaul Air Field. Three destroyers were sunk, 12 aircraft on the ground were destroyed and 68 were shot down. On 5 November 58 P-38s and 27 B-24s from the 5th AF pounded the wharf with 81 tons of bombs.

In the Guadalcanal campaign the P-38s escorted B-26s because the Lightning was the only plane that could stay with the bomber at full range. Striking against Kahili and Buin, the P-38s furnished high cover to the targets while the P-40s covered withdrawal of the bombers.

By 9 February the enemy resistance in Guadalcanal ended. The Allied Air Force took on a secondary role despite the major role played by the P-38s and her pilots. As the pilots had heard before, the Pacific was a minor theater, even though just as many lives were lost and lost for the same reasons as every other man fighting in other seemingly more important theaters.

Pacific Theater
Southwest Pacific
5th Air Force

". . . they fly above our positions as if they owned the sky." December 1942, Japanese diary found in New Guinea.

". . .we have the Jap air force whipped." January 1943, General Whitehead, following P-38 assault near Cape Endaiadere.

The Southwest Pacific generally conjures images of tropical islands and relaxation - it was anything but that for the the men of the Allied Air Forces. And their mission in the Southwest Pacific (SWPAC) was even more stressful because they were considered low priority on the military totem pole, as the pilots had heard too many times before, the war in Europe came first. SWPAC included the Philippines, Borneo, Celebes, New Guinea and Australia. Life in the Southwest Pacific consisted of primitive living conditions, a lack of recreational activities, unfamiliar rations, war weariness, inexperience and inadequate training of some of the more recently arrived units, a stagnant promotion list, and a lack of proper hospitalization. It rained day after day, and the entire area became a quagmire. Trucks mired in it. Jeeps became immobilized. The short distance to the fighter strip took an indeterminable length of time.

General Douglas MacArthur commanded the Southwest Pacific and Gen. George H. Brett was the commanding general of the Allied AF - the unit bases were in north and east Australia with advanced facilities in the Port Moresby area. Tensions were high between Gen. Brett and other military leaders and to the benefit of the air forces in the SWPAC General George Kenney assumed command on 4 August 1942. Gen. Kenney was a sharp, able leader who directed a close-knit disciplined army of pilots.

Kenney faced a formidable job. The men were tired and the planes were the worst for wear. He reorganized, reassigned, and ordered more planes and equipment. By the end of August, Kenney was able to announce the formation of the 5th AF, whose main objective would be to thwart the Japanese air power to allow naval and ground troops to function on the islands. The 5th AF was official on 3 September and was named in honor of MacArthur's fighter and bomber commands in the Philippines. The 5th AF came under the control of the Allied Air Force (SWPAC) which was created in Australia to control USAAF, RAAF, and Dutch elements.

Kenney was a strong supporter of the P-38. The Lightning was an excellent choice for the Pacific because its twin engines allowed it to travel great distances. Another definite advantage of the Lightning was that it could lose one of its engines and still have one with which to fly home. Kenney requested P-38s for the Southwest Pacific and by the end of 1942 several Lightnings arrived. In September of 1942, Brisbane had 25 P-38s, several of which had mechanical problems that had to be repaired before they would be ready for any missions.

Kenney pleaded with Arnold to send more P-38s. Because of Japanese tactics, U.S. bombers needed fighter cover (P-38s) to fly as Kenney said, "To engage the Zero coverage up top-side while the P-39s and P-40s took on the bomber." Kenney received 60 P-38s by October, but none were combat ready - they had leaky fuel tanks, superchargers in need of repair or replacement, and dilapidated water coolers and invertors. Also the armament required adjustment. Another 25 arrived shortly thereafter but needed feeds for the guns. Fortunately for Kenney and his 5th AF, 16 Lightnings of the 49th Fighter Group were flown up to Port Moresby in late October.

Even though the Japanese Zero was a better dogfighter, the Lightning was faster, could outclimb, outdive, perform better at high altitudes and carry more armament than the Zero. The P-38 also had features the Zero did not have - a self-sealing fuel tank and armor plating to protect the pilot.

The first mission flown by the P-38s in the SWPAC was in December of 1942. A pilot named Captain Robert Faurot, of the 35th Fighter Squadron, 8th Fighter Group, took off in his Lightning carrying two 500 pound bombs under its wings. The mission's objective was to make holes in the runway at Lae. Captain Faurot dived on a Zero, and realizing his bombs would hinder a potential tangle with the Zero, dropped both of them. Upon hitting the water the resulting splash hit the Japanese plane and caused it to crash into the water. This was the first Japanese plane brought down by a P-38 in air combat in the SWPAC. For this Faurot was awarded an Air Medal.

The Japanese air strength in the Southwest Pacific was tremendous - they were capable of launching large-scale attacks against the Allied ground forces at any time. The Japanese had a string of air bases stretching up the New Guinea Coast - and to the east at Rabaul, New Britain, they had amassed even greater strength. Australia was the feather in the Japanese cap for it served as the ideal base for any future conquests.

The islands between the United States and the South Pacific were crucially important to the U. S. so the line of communications would be secure and so the U. S. could more easily support the naval, air, and amphibious operations against the Japanese. The Allies had to monitor the enemy's progress toward Australia and destroy enemy troops and supply ships, aircraft and bases in eastern Malaysia, New Guinea and the Bismarck and Solomon Islands. William Bishop, who was in a photo reconnaissance squadron, said his P-38 was "equipped to take mapping pictures with trimetragon cameras. That is three camera operations simultaneously, two scanning right and left and one straight down." He continued by saying that photo reconnaissance squadrons generally operated alone in P-38s. "We had no armament, no fighter escort and weapons at all. Our job was to photograph, period."

It was determined that destroying as many of the Japanese strongholds as possible was necessary to gain control of the air and the 5th's greatest strength for that task was in its 80 Lightnings. Fighter escorts were necessary and the P-38 had the range necessary to escort bombers to and from Rabaul and Wewak. They also served as photo reconnaissance planes to provide needed information on Japanese activities on many of the islands.

At this time, though, the 5th AF only had three P-38 squadrons: the 80th Fighter Squadron of the 8th Fighter Group, the 39th Squadron of the 35th Group, and the 9th Squadron of the 49th Group. The 475th Fighter Group was activated as a P-38 outfit in Australia on 14 May 1943. The first base for the P-38s was Jackson Field, Port Moresby in New Guinea under the command of Captain George Prentice. A firm believer in the capabilities of the P-38, Capt. Prentice tried continually to acquire more P-38s, but to no avail. A limited supply of P-38s did eventually arrive and they formed the 8th Fighter Group, behind which followed the 49th.

By late December, the P-38s were ready for flight in major combat missions over New Guinea. December of 1942 brought with it the Papuan Campaign. The patrol over the islands was strengthened by the P-38s flying cover. On 27 December 1942 approximately 12 P-38s led by Captain Thomas J. Lynch made their first interception by diving on a Japanese formation of more than 20 fighters and seven dive bombers near Cape Endaiadere. One P-38 was shot down at Dobodura, New Guinea where they claimed nine enemy fighters and two dive bombers shot down, other sources claim 15 Japanese planes were downed. By January of 1943 the Papuan Campaign was drawing to a close - it was one of the greatest victories of Kenney's men - having secured Allied air control over Papua. General MacArthur said,

"To the American 5th Air Force and the Royal Australian Air Force no commendation could be too great. Their outstanding efforts in combat, supply, and transportation over both land and sea constituted the key-stone upon which the arch of the campaign was erected. They have set up new horizons for air conduct of the war." Page 128, Volume IV, The Army Air Forces in World War II.

General Kenney certainly agreed and responded by saying they had learned a great deal from the Papuan Campaign and would perform better in the next campaign.

Soon after the ending of the Papuan Campaign the Battle of Bismarck Sea was well under way. While the sea battles were fought, Zeros and P-38s fought battles in the air. It was the job of the 5th AF to neutralize Japanese airfields in Rabaul, Kavieng, Buka, and Buin. On the south-central coast of New Britain a large convoy headed for Lae to reinforce it. Allied bombers, escorted by fighters, had a field day.

On 6 January 1943, Lt. Richard I. Bong claimed three planes downed, and other P-38s accounted for 13 enemy planes downed. But the convoy did get through and did reinforce Lae. At one point the Japanese Zeros had to decide whether to attack B-25s, B-17s, Havocs, or the Beaufighters. They chose the B-17s. One B-17, flown by Lt. Woodrow W. Moore, was severely damaged and seven of the crewman managed to abandon the plane. Captain Faurot in his P-38 and with his two wingmen, Lt. Hoyt A.

The 459th FS returning to base at Rhumkapalong, India after a dive-bombing mission over Burma. (Courtesy of Leonard Boyd)

Cyclone Davis, Jay Robbins, Gen. Whitehead and Gen. Wursmith.

Debriefing for a mission, 459th FG, Chittagong, India, 1944.

A P-38 flies over the Hump in the CBI Theater. (Courtesy of F. Durant)

created a well-coordinated air supply, service and maintenance route.

The three countries had been linked for the purposes of American military strategy since February 1942, when Lt. Gen. Joseph W. Stilwell had been sent out to command the CBI Theater of Operations. CBI became predominantly an air theater, with just cause, the vast expanses of land and mountains would have been nearly impossible for land troops to maneuver.

10th Air Force India/Burma

The 10th Air Force, whose main area of operation was the India/Burma area, was activated at Patterson Field, Fairfield, Ohio on 12 February 1942. India served as the main base for all Allied Air Force operations and for aid to China in 1942, so each one served as Allied bases. Bombay and Karachi, India were the only ports free of enemy attack in 1942. On 17 February Col. Harry A Halverson became the Commanding Officer. In March, General Lewis H. Brereton arrived in India to formally take command of the 10th AF. On 8 March, the 10th AF began its move from Patterson Field, Ohio to India - heavy bombers and P-40s arrived and by the end of March had begun assaults on warehouses and docks at Rangoon. On 16 March 10th AF HQ made its move from the United States to New Delhi, India.

Activities relaxed for a while and on 26 June, General Earl R. Naiden became the Commanding General of the 10th AF, succeeding Gen Brereton. In addition to Gen. Brereton leaving, Gen. Elmer E. Adler and several key subordinates left, as did the crews of the 9th Bomber Squadron - the 10th AF was practically without men and planes.

Often the trip to your theater of operation was uncomfortable and not without difficulties. Philip D. Robertson, a retired USAF Colonel, was a member of the 9th Photo Reconnaissance Squadron during World War II and had to make a trip from Florida to Karachi, India, a distance of approximately 12,000 miles.

He said, "The trip to India was made in groups of three or four at a time. The route was from Miami to Puerto Rico to Trinidad to Georgetown, British Guiana, to Belem and then Natal, Brazil. We flew in Air Corp C-53s with Pan Am crews. No night flying, so we RON-ed at each place except Georgetown. The South Atlantic crossing was in one of the great Pan Am Boeing 314 Clippers nonstop to Fisherman's Lake at Freetown, Sierra Leone. It was bucket seats in the Gooney Bird from there to Karachi. Stops were at Accra, Kanok, Fort Lamy, El Fasher, Khartoum, Asmara, Aden, and Masira Island. The whole trip took ten days. Many interesting places, but a real pain in the sit-down."

Robertson continued by saying, "Operating conditions in the CBI during 1942 and much of 1943 were very difficult. Our maps were 1 to 2,000,000 scale, black and white, with vast areas of the mountain country between India and Burma, northern Burma and southwest China marked "UNEXPLORED." The black lines represented anything from a road or river to the mapmakers imagination. Places and pilotage features were frequently some distance from where the map showed them. One of our big tasks was to provide the aerial photography so that decent maps of the whole area could be made. We began getting the new maps around the middle of 1943. In the meantime, we found some 15,000 foot peaks in an area where 8,000 or 10,000 was thought to be the maximum. In late 1942 and early 1943, we had some pilots land in rice paddies and one in a river bed because they couldn't find an airfield. There were no navigation aids in eastern India at that time."

The 9th Photo Reconnaissance Squadron received F-5s in June of 1943. Robertson said, "They (F-5s) were ferried over by the South Atlantic route using 310 gallon ferry tanks. This increase in fuel capacity enabled us to greatly increase the length and area of coverage of our missions. With the F-4 and staging through RAF bases at Agartala and Chittagong, with a post mission refueling stop at the same places, we couldn't go much beyond Rangoon." Robertson continued, "With the F-5, missions of seven or eight hours and more became pretty common. On 5 September 1943, I flew a mission which lasted eight hours and 55 minutes. It was cited by Headquarters, 10th AF as 'one of the longest photographic reconnaissance missions on record.' It was the longest flown in the CBI up until I left the theater in February, 1944. If we had been given some training in or understanding about good cruise control procedures, the flight time would have been a piece of cake, as it was, I was down to a bit less than 15 gallons of fuel when I landed."

While the 9th Photo Reconnaissance Squadron was busy preparing the materials needed by the P-38 pilots to successfully carry out bombing, strafing and escorting missions, the 10th AF was determining what the military priority in India would be. One of the major concerns of the 10th AF was the protection of the air supply route to China, which General Naiden fully expected the Japanese to try to sever. The 10th AF presented a constant struggle for Naiden and his staff, the 10th had limited supplies and eight squadrons without proper equipment. Yet again, command changed hands from Gen. Naiden to Gen. Clayton L. Bissell who assumed command on 18 August 1943.

The 10th's P-40s and bombers continued assaults on Rangoon, Myitnge, and Loiwing and other areas throughout the rest of the year - the P-38s had not yet entered action in this theater. The brunt of the fighting was on the shoulders of the pilots of the P-40s, who did need relief, but had to wait until summer.

Photo reconnaissance squadrons were based in the India-Burma sector and were vital to the success of the war effort there. The 9th Photo Reconnaissance Squadron, based in Karachi, India flew F-4s, P-38s modified for photo reconnaissance missions. The F-4s were ferried from Dallas to Brazil, Ascension Island, Ghana, Khartoum to Aden. From Karachi to Agra, India thence to Chakulia, India. Photo missions began in December of 1942. Russ Miksch, who was with the 9th Photo Reconnaissance Squadron from 1942 - 1944, said, "Three pilots and planes were on continuous assignment to China. The remaining pilots and planes operated over Burma from Chakulia and later Pandavasevar, to provide photo reconnaissance for the 10th AF headquarters at Barrackpera, near Calcutta. In addition to the F-4s, later supplemented by F-5s, most of the pilots also operated B-25s modified for photo reconnaissance. Much of the time in 1943, a detachment of from one to three planes (P-38s or B-25s, or both) operated from Dinjan (Assam Province) over North Burma."

On 20 August 1943, the Allied Air Forces, India-Burma Sector of the CBI Theater was activated at New Delhi with Gen. George E. Stratemeyer as the commander. It was comprised of the 10th AF, China Burma India Air Service Command (Provisional), China Burma India Training Unit (Provisional) and several lesser units. On the previous day, Gen. Howard C. Davidson became the Commanding General of the 10th AF.

It was well into the summer of 1943 before the 10th AF received the relief it needed in the area of new planes, men and equipment to relieve the already war weary troops. The 311th Fighter-Bomber Group arrived, under the leadership of Col. Harry R. Melton. One of its squadrons was the 459th Fighter Squadron of P-38s. They had recently been deactivated from the Mediterranean Theater. The 459th was assigned to the 80th Fighter Group and was stationed in Kurmitola in Bengal where it operated as escort for the B-25 squadrons. The 459th was the first squadron of P-38s to fly in India. The previously used P-40s were supplemented by P-38s, P-40Ns, P-51As, and A-36s - all of which had superior altitudes.

On 11 March 1944, the 459th surprised the Japanese early in the morning when then Captain Luehring, " . . . led 12 Lightnings in as the Nip fighters were taking off to intercept. The unfortunate Japanese were hardly off the ground as the Lightnings swept in and picked them off one by one as they attempted to gain altitude." (459th Fighter Squadron history) In his report to Intelligence on the mission, Lt. Garland, of the 459th, reinforced everyone's belief in the durability of the P-38. He said, " . . . during the mission I received a number of hits on my plane. The hydraulic system was shot out, elevator damaged, rudder trim shot out, hole in left prop, one hit in left wing, and a few hits in the gondola and right boom. The aircraft returned safely . . . ."

On 13 March 1944, the 459th was separated from the 80th Fighter Group and placed under the direct control of General Davidson's 10th AF and under the tactical direction of Air Marshall Sir John Baldwin's Third Tactical Air Force.

The Lightnings provided for heavy bombers the long-range escorts they needed to carry out long-range missions. The real struggle with the Japanese Air Force was met by the 459th and the P-40s up in the Assam Valley and by the English with their Beaufighter, Mosquito and Hurricane aircrafts.

Dan Schmittner, a Lockheed Tech Rep, was with the 459th in its beginning weeks to assist the crews and mechanics in familiarizing themselves with the plane. To prepare the aircraft and crew for immediate combat specialized maintenance personnel was needed. Wayne O. Sneddon said, "This caused some confusion as most of the maintenance personnel was used to the concept that the crew chief did everything on big aircraft. It ended up with special crews for electrical work, prep work, engine work, oxygen servicing, etc. With specialized crews the aircraft and personnel were ready for serious combat in less than three months."

November brought with it some of the first activities for the P-38s of the 10th AF. On 27 November they escorted B-24s as they struck at Insein were they met fierce resistance from the Japanese. Six fighters and three heavy bombers were downed by the Japanese - the Americans claimed 13 enemy fighters downed.

On 1 December all available P-38s and P-51s were to escort B-25s on a bombing mission to Insein. The P-51s were unable to refuel in time to meet the rendezvous. Only the P-38s escorted the bombers. A large number of Japanese fighters were encountered over Insein and the Americans had high losses.

January 1944 brought with it a variety of activities for the P-38s. On 18 January P-38s hit air fields and satellites at Meiktila. On 20 January 16 P-38s escorted five B-25s as they knocked out a railroad bridge, pipeline and railroad tracks in the Natalin area South of Burma. Similar activities on 23 January found the Lightnings escorting B-25s as they knocked out bridges at Myittha and damaged a bridge at Samon. They also damaged several trucks and railroad cars. Other areas hit included gasoline plants at Yenangyaung causing fires.

While the P-38s and their pilots did experience many successes, the men often experienced difficulties, the first of which was a shortage of planes. The P-38s were often sent to Europe since the war in Europe usually came first, because it was seemingly more important. Sneddon said, "At one time the squadron (459th) was down to four aircraft. To keep the Japanese from knowing how short we were of aircraft, the planes were dispatched on three missions a day ranging from Northern Burma down to Rangoon in South Burma. Most of these four plane missions were very successful"

Bombing and strafing missions continued into February and the P-38s of the 10th attacked bridges, encampments and buildings along the Prome-Taungup road damaging a bridge and road, also claiming a tugboat sunk at Akyab. On 9 February P-38s along with P-51s, B-25s, P-40s and A-36s carried out 75 attack sorties against Seinnyinba, bivouac and supply areas in Mogaung-Loilaw, bivouac and buildings at Pyindaw, camps at Kumnyen and near Maingkwan and Shingban, and storage areas on Mongaung-Kamaing road near Sawnghka. A bridge over Namtawng River near Maingkwan was also destroyed.

The P-38s were very valuable as they were superior high-altitude planes for bomber top cover. And most significantly, P-38s were needed for escort during all daylight missions to Rangoon.

Technically, the P-38 engines were extremely reliable in the Burma-India area. Sneddon said, "It was discovered that the turbine settings were causing low power detonation and engine failure. The test for this condition was simple - while in cruise at altitude we placed the props in fixed pitch (Curtiss electric) - left the throttle alone, brought the mixture from auto-rich to auto-lean - the normal drop was 60 rounds per minute. Anything over that indicated the turbine pressure on the carburetor was causing it to lean out excessively." Sneddon continued, "It was determined that both the input and output arms configuration had to be changed to control the deck pressure on the carburetor. All planes arriving in the CBI were so modified before going out to combat." When the P-38 units in the Southwest Pacific learned of this modification they requested someone to modify their planes.

Air Marshall Baldwin was pleased with the activities of the Allied Air Forces in Burma-India. In a report he said, ". . . one squadron of USAAF P-38s alone, operating from Chittagong, destroyed 96 enemy aircraft in this way within six weeks of coming into operation in 224 Group." The combined efforts of heavy and medium bombers, in addition to the valuable fighters, led to success in a harsh and unrelenting theater of operations.

14th Air Force, China

"But for the 14th Air Force, 'we could have gone anywhere we wished.'" Postwar testimony of a ranking Japanese official in China. The Army Air Forces in World War II, Volume V)

The 14th AF was activated at Kunming with General Claire L. Chennault as the Commanding General on 10 March 1943. The 308th Bombardment Group (heavy) along with four fighter groups, a medium bomber squadron and a photo reconnaissance detachment (the 8th Photo Reconnaissance Group, P-38s) comprised the entire 14th AF. Bad weather, fuel shortages and a definite need for new equipment created unfavorable fighting conditions for the 14th AF. In July, six P-38s arrived and were immediately sent to the east, their arrival was needed, but their greater fuel consumption made them somewhat less desirable.

In 1942 China only had a single base unit which was set up in Kunming. The Kunming base, along with a Chinese factory, was the only air service available for Chennault and his air force.

On 21 August nine P-38s encountered 12 Zeros south of Changsha, they claimed three shot down. On 24 August P-38s escorted B-24s as they bombed air fields at Hankow and Wuchang. They claimed 24 enemy interceptors downed. On 30 August 10 P-38s of the 14th were on an armed reconnaissance mission from Sinti to Yoyang to Sienning when they strafed and bombed several targets of opportunity. The Lightnings continued bombing and strafing activities on 31 August, hitting a dike near Co Bi barracks. Four P-38s dive bombed Yoyang railroad yards and Sinti warehouses.

In September P-38s dive bombed and strafed barracks at Yangsin and Pho Lu, demolishing three buildings and two locomotives. On 9 September, Lightnings bombed Whapoa docks, hit Yangtze River shipping near Chiuchiang and strafed targets of opportunity in that general area. By 12 September the 14 P-38 combat units had increased their area of combat to include the Hong Kong area, hitting the Yangtze River traffic at Chiuchiang and severely damaging two vessels. Near Kunming P-38s intercepted enemy planes and claimed 17 shot down.

In October of 1943, P-38s escorted heavy and medium bombers as they hit barracks at Bayard and the Chiuchiang dock area. The support services of the Lightnings continued throughout November and on 25 November 16 P-38s escorted B-25s as they attacked air fields at Shinchiku - 32 enemy aircraft were claimed destroyed in the air and on the ground.

In attacking the Formosa Straits, the P-38s lead the way and knocked out the enemy. They claimed 15 airborne planes destroyed.

On 12 December 1943 Japanese bombers hit the west side of Hengyang airfield - P-38s and P-40s intercepted them and claimed 20 enemy planes shot down. P-38s strafed a troop train near Changanyi and attacked 25 sampans on the Yangtze River above Huangtang Lake. On 23 December P-38s dive bombed and strafed Huan Shan Kou.

On 22 January 1944 P-38s of the 14th hit bridges at Shektan and Sheklung. On 10 February, P-38s continued strafing the Chiuchiang area. The next day, 12 P-38s knocked out one bridge and damaged another at Skelung. In February, 24 P-38s, P-51s and P-40s intercepted 25 fighters near Suichwan, claiming seven shot down. On 24 February 10 P-38s on a sweep of the Yangtze River from Chihchow to Pengtse claimed four river steamers, five tugs, five motor launches and 15 sampans destroyed. On 29 February 12 P-38s and P-51s sank a large motor launch northeast of Anking, strafed barracks and three tugs in Teian area, and hit barracks northwest of Nanchang, Yangsin.

November of 1944 was characterized by continuous bombing and strafing of targets of opportunity throughout southwest and southeast China. The Lightnings hit rivers, roads, rail, traffic, buildings in Thailand, Burma, S. China, and N. Indochina.

China was considered to be outside the main theater of combat. The Japanese tried too late to attack and conquer China. As a result the 14th Air Force did not really have the opportunity to exhibit its full battle strength against the enemy in the China Sea. But like in every other theater of operation, the P-38 proved itself in combat in China, showing that it and its pilots were extremely versatile and able to fight on any front.

The P-38 pilots fought in every theater of war, from the harsh, cold weather of the Aleutians to the hot, steamy jungles of the South Pacific and the dry, dusty areas of North Africa. Numerous modifications adapted the plane for the different climates, making the P-38 one of the most versatile and sought-after planes of World War II. Unfortunately, the glorious plane saw an inglorious end as many were simply destroyed after the war instead of being ferried home. Some people believe it was the cheapest way to reduce war inventory of combat planes. A bulldozer was used to break the booms between the tail and fuselage, engines were removed and the wreckage was sold for scrap. Several Lightnings remain and are well-taken care of by the people who understand the contribution of the plane and the debts its pilots paid.

After 40 years, memories are hazy, but accomplishments are not. The battles fought and citations awarded comprise a litany of a job well done. Earl C. Hedlund best summed up the P-38 and her pilots saying, "The old P-38 always brought me back and I loved it. World War II was a long, tough engagement for those of us in the air. It was most challenging and demanding of both men and machines - the P-38 was just a "great" airplane which will be long remembered as one of our most effective fighters of World War II."

Sources Consulted

Airwar. Edward Jablonski. (Doubleday and Company, Inc. Garden City, New York, 1971)

Air Power November 1973, Vol. 3, No. 6.

Man-made Lightning Warren M. Bodie.

The Army Air Forces in World War II, Combat Chronology 1941-1945. Compiled by Kit C. Carter and Robert Mueller. Albert F. Simpson Historical Research Center, Air University and Office of Air Force History. Headquarters USAF, 1973.

The Army Air Forces in World War II. Edited by W. F. Craven and J. L. Cate. The University of Chicago Press, 1949.

USAAF Fighters of World War Two. Michael O'Leary. (Blandford Press, Poole, New York, Syndey, 1986)

Additional information compiled from stories submitted by: Wayne O. Sneddon, Philip Robertson, Eddie Martin, Richard Willsie, Alphons J. Mikes, Earl C. Hedlund, and Ward Beman. Also gathered information from the following submitted group histories and articles: 82nd Fighter Group History, 2nd Anniversary Overseas Report on the activities of the Twin Dragon Squadron (459th Fighter Squadron), 54th Fighter Group History, Journal - Aviation Historical Society, Volume 16, Number 1, Spring 1971, Jack Purdy "Borrowed Time Pilot."

Left: In an informal ceremony, Capt Dick Kiick of the 432nd FS receives the DFC and Air Medal from Gen. Frederic H. Switch, Jr. CO of the 5th FC. (Courtesy of Richard Kiick)

Right: The ground crew of a P-38 squadron stand in front of a P-38 at Chittagong, India, 1944. (Courtesy of H. H. Sealy)

Below: P-38s of the 459th FS are readied for a mission over Burma. Note the orderly layout of belly tanks. Rhumkapalong, India-Burma border, April 1945. (Courtesy of Leonard Boyd)

David C. Tallichet's P-38 at 474 FG Association 1975 Reunion at Minneapolis, MN. (Courtesy of Ron Ennis)

P-38 NATIONAL CONVENTION

by Gil Cefaratt

The P-38 National Convention had its beginning in November 1985 when Lockheed employee Gil Cefaratt put a notice in the Lockheed newspaper announcing the first P-38 National Convention to be held in 1986 and was looking for committee members. The first people to respond were Lockheed employees; Joe Kuhn (P-38 Fighter Pilot-401FS 370FG), John Lewis (P-38 Fighter Pilot 18FS 343FG), Frank Miron (P-38 Fighter Pilot 485FS 343FG), Don Johnston, John Taylor, Barbara Christie, Dave Skilling, Frank Lane, Lisa Skinner and Beverly Hahne, Ken Holdren (P-38 Fighter Pilot 475FG).

Other interested people to join the committee were: Millard Coffman (P-38 Fighter Pilot 12FS 18FG), Art Cole (P-38 Fighter Pilot 48FS 14FG), Dr. Al Denman-Fan, Monroe Deutsch (28PRS), Ralph Eden-Fan, Richard Fine (485FS 370FG), Bob Hildebrandt (P-38 Fighter Pilot 95FS 82FG), Stumpy Hollinger (P-38 Lockheed Tech Rep-retired), John Hood (P-38 Fighter Pilot 475FG), Lloyd Levine (P-38 Fighter Pilot 434FG) Doug McArthur (P-38 Fighter Pilot 77FS 20FG), Harold Madison (P-38 Fighter Pilot 431FS 475FG), Chuck Mauss (P-38 Fighter Pilot 37FS 14FG), Mary Lou Colbert Neal (P-38 Ferry Pilot-WASP), Ray Neale (Lockheed Test Pilot - retired), Joe Onesty (P-38 Fighter Pilot 48FS 14FG), Richard Ostronik (P-38 Fighter Pilot 82FG), Thomas H. McGuire (P-38 Instructor Pilot 434AAFBU), Ed Peters (P-38 Fighter Pilot 38FS 55FG), Seymour Prell (P-38 Fighter Pilot 431FS 475FG), Walt Schranz (P-38 Fighter Pilot 1FG), Bert Shepard (P-38 Fighter Pilot 38FS 55FG), John Stege (P-38 Fighter Pilot 48FS 14FG), Dick Willsie (P-38 Fighter Pilot 96FS 82FG), Sandy Willford (P-38 Fighter Pilot 9FS 49FG), Jack Walker (P-38 Fighter Pilot 82FG).

Others assisting the committee members were; Bill Caldwell (P-38 Fighter Pilot 35FS 8FG), Charles Dahlin (P-38 Fighter Pilot 19FS 318FG), Hiawatha Estes (P-38 Fighter Pilot 37FS 55FG), Alan Hale, Jr. (Actor-fan), Bill Ham (P-38 Fighter Pilot 401FS 370FG), Ken Nunamaker (37FS 14FG), Monie Pease (99ers-fan), Irvine Styer (P-38 Fighter Pilot 27FS 1FG).

Always present were the wives to assist; Nancy Kuhn, Ann Madison, Fanette Deutsch, Naomi Hale, Mrs. Holdren, Shirley Johnston, Irene Fine, Marge Miron, Jan Coffman, Abby Skilling and Mrs. Eden.

It was decided to have the Convention in 1987 for the 50th Anniversary of the contract for the P-38. In the beginning I thought that to have the P-38 Convention, we must have three things: first a P-38, we had two and one of them flew-Lefty Gardner's, the other one on static was David Tallichet's, (that has since been sold to Doug Arnold from England). There was another P-38 nearby on static, belonging to The Planes of Fame Museum. Second we had to have the Famous Hollywood Palladium for our Banquet, we ended up with a nice Banquet room at the Universal Sheraton. And third, we had to have the Famous Tex Beneke Band, and we did.

The day at Chino Airport to see P-38s included a tour of The Planes of Fame Museum and a Super Country BBQ catered by The Castaway's Restaurant (Specialty Restaurants -David Tallichet).

Program Day was a busy one, with the program beginning at 0900. The first speaker was Dick DeGrey - Lockheed Engineer who talked about Lockheed History from the beginning through P-38 production. Dick was loaned out to the P-38 program for three months, but that ended five years later. Dick has since retired after 50 years at Lockheed.

The next speaker was George Toumanoff (flew in from New York with his wife). George was a Flight Test Engineer on the P-38, and he spoke on various problems, but specifically on the Spin Test. He claims to be the only living grandfather who has been spun in a P-38 as a passenger. George is retired from Lockheed.

The third speaker was Pat McGeehan, a 1940s radio personality, who at one time had filled in for the famous P-38 Test Pilot Milo Burcham in an Army Air Force Training film when Milo had sprained an ankle.

The fourth speaker was Jay Fiondella-Explorer/Restauranteur, who spoke about his expedition to hunt and find The Lost Squadron: The six P-38s and two B-17s buried in Greenland. Jay said The Lost Squadron is still flying under 200 feet of ice. Since Jay was there, an architect from Georgia with an expedition has been to Greenland trying to get to The Lost Squadron and they made a report at the Golden Anniversary of the P-38 program in Santa Maria in January 1989. The fifth speaker was Chet Patterson (P-38 Fighter Pilot 55FG) who showed a (lump in the throat) type slide show on the history of the 55FG, very well done. Chet also spoke on some of the missions he was on and told his flights that when you break, break for England.

The sixth speaker was Richard E. Baughman, Chief of Public Affairs at the USAF Museum WPAFB, now retired. Dick had a great slide and speech presentation on the AF Museum. The last speaker was Tony LeVier Famous Air Racer/Test Pilot who had a panel of speakers. They were; Ward Beman-Lockheed Engineer (retired), Jim White-Lockheed Production Test Pilot (retired), Dick DeGrey -Lockheed Engineer (retired from the Skunk Works), and George Toumanoff-Lockheed Engineering Flight Test Engineer (retired). The panel spoke on some of the problems during production and in operation and on some of the fixes for the problems. One story I remember about LeVier was when he had been sent to England because this one particular Group was having problems with their airplanes. This one particular night Levier was having trouble getting the Pilot's attention, so he headed for the flight line to his assigned P-38 and went aloft. At altitude, he nosed over and headed directly for the Officer's Club and pulled up close to the club, LeVier said he got their attention then.

The Banquet period was a busy one. The Banquet was opened with The Flag Salute, led by Dick DeGrey, then a clip from the movie "This Is The Army," with Kate Smith singing God Bless America. Introduction of Committee Members as a Group. Introduction of Marge Bong Drucker, a Gracious Lady, Introduction of Carl Bong who had left early. Reading of telegram from Nancy Johnson (Mrs. Kelly Johnson). Introduction of Bill Lear, Jr. one of the four Advisory Committee Members, and Bill gave us a nice speech. Bill bought the last P-38 sold to a civilian and had a nice thing to say about Kelly Johnson, "If a young kid of 18 who had never flown a P-38 before could get in a P-38 and fly it and then land it, then it had to be the way it was designed."

Ray Toliver (P-38 USAAF Test Pilot/Author) showed slides of Aces and introduced the Aces in attendance. The guest speaker was Bob Stevens (P-38 Fighter Pilot/Cartoonist) famous for his "There I Was" flying/airplane cartoons. Bob is a witty speaker and had us laughing when he told about times he had to use the relief tube while holding the end of one in his hand. I read the telegram sent to me from President Ronald Reagan for everyone gathered.

A guest at the Banquet was a former Miss Lighting in 1944, Barbara Scott Berish and her husband Dan who had been a Naval Pilot. While he was in Guam he got to fly a P-38, anyone lose one?

Also, in the beginning I knew that if the P-38 National Convention did have a place in history, there were two other things that should happen; first a P-38 Association should be formed and a P-38 History Book should be published. Well, the P-38 National Association was formed and the P-38 History Book came to being.

L to R: Stumpy Hollinger-Lockheed Tech. Rep., Jackie Brundage Gardner-the Lightning Lady, Col. R.E. Willsie-96FS, 82 FG, WASP Mary Lou Colbert Neale, John T. Taylor and Gil Cefaratt -Lockheed. Yankee Air Corps, Chino, CA. P-38L, 1980.

I was pleased that the P-38 National Association was formed; it was the only time it could have been done. Robert D. Hanson, of the 474FG Association tried years ago but didn't have the opportunity we had, and as I stated at the Convention "I would like the Association dedicated in his memory. General Officers at the Convention were; Maj./Gen. Leo Dusard 347FG, Maj./Gen. Harold "Hal" Gray, Jr. 475FG, Lt. /Gen. Earl C. Hedlund 474FG and Gen. Seth J. McKee 370FG, and M/Gen. Russ Gustke 20 FG.

WASPS in attendance were: Kay Gott Chaffee, Nancy Batson Crews, Iris Cummings Critchell, Lauretta Beaty Foy and Mary Lou Colbert Neale. Unfortunately Betty Huyler Gillies became ill and was not able to attend.

Tech Reps in attendance were: Milton J. "Milt" Getker, Curtiss Electric-Propeller Division, 475FG, 39FS 35FG; Robert J. "Bob" Baker - Lockheed Aircraft; Thomas J. Cleland - Lockheed Aircraft; and Wayne O. Sneddon - Lockheed Aircraft 459FS 33FG, 80FG; and Dick "Stumpy" Hollinger - Lockheed Aircraft 82FG.

The P-38 National Convention was a tremendous success and I thank all the committee members for their time, dedication and loyalty and support I thank all the attendees for their interest and support.

P-38 ASSOCIATION

by Joe Kuhn

The P-38 National Association was formed on May 16, 1987, at a business meeting which convened following the National Convention which was held in the L.A. area. A slate of officers was elected by unanimous vote of the convention attendees present at the meeting:

Joseph A. Kuhn, President, Millard F. Coffman, Vice President, Claude Murray, Secretary, Lloyd Levine, Treasurer.

At this same time Board of Directors was elected as follows: Jack Ilfrey, Ed Vassar, Harold N. "Pete" Madison, Mary Lou Neale, Art Cole, Jack Walker, Thomas F. Pugh, William A. Ham, Jack W. Mullan, and Gerald A. Greene (now deceased).

Board meetings are held once or twice a year at the discretion of the officers, and conventions are to be held every two to four years. At the first board meeting, assistant officers were appointed to handle the large workload.

Frank Miron was appointed 2nd vice-president and membership chairman, Don Johnston was appointed asst. treasurer, and Mary Lou Neale was appointed asst. secretary.

The mission of the Association is to act as a headquarters or central clearing house for all P-38 and F-5 related matters. A newsletter, "Lightning Strikes," published three times a year with Joe Kuhn as editor and Mary Lou Neale Asst. Editor, publishes stories of the P-38 and the people who flew, maintained, designed, and built the plane. One of the goals of this organization is to establish a place, room or building, to house all the memorabilia being gathered by the P-38 National Association.

At the present there are 1,190 members in the Association.

One of the many women who was part of the WASPs, Jane Emerson Carter is pictured with her P-38. Dec. 1944. (Courtesy of Gil Cefaratt)

WAFS/WASPS 1942-1944

by Mary Lou Colbert Neale

When the Army Air Corps faced the fact of its acute pilot shortage following Pearl Harbor, the decision was made to employ qualified women pilots as civilians for ferrying duties. The 28 who reported in to New Castle Airport, DE in September 1942 had an average number of hours of 1,162. This group started delivery of light planes two months later and was under the leadership of Nancy Love. At the same time Jacqueline Cochran began interviewing women pilots with a minimum of 300 hours and a commercial license to go to Texas for Aviation Cadet training. The first class of 23 graduated from Ellington Field in April and joined the original group, now deployed to four Ferrying bases — Wilmington, Dallas, Long Beach and Romulus. They were called WAFS (Women Auxiliary Ferrying Squadron). After two subsequent classes joined the WAFS, there was a demand for pilots for other duties, such as target towing, administrative and utility flying, instructing, and for various missions, so the name was changed to WASPs (Womens Airforce Service Pilots) in August, 1943.

The WASPs distinguished themselves during the 28 months they were in service, flying 60 million miles in every type of plane used by the Army Air Force from primary trainers to the Super Fortress (B-29). Although 38 were killed in line of duty, their accident rate was lower than the male ferrying pilots, according to the testimony given Congress by Barry Goldwater, 1977. The Ferrying Division of WAFS/WASPs, about 300 strong, delivered three-fifths of all the pursuits by September 1944. These included the P-39, P-40, P-47, P-51, P-61 and P-63, and of course the P-38. However, only about a dozen of the ladies were fortunate enough to ferry the P-38s, as that airplane was not part of the Palm Springs or Brownsville Pursuit School curriculum. Five of those who did, attended the P-38 National Convention and still had stars in their eyes, remembering the thrill of flying that twin-engine beauty.

The five were: Nancy Batson Crews, Mary Lou Colbert Neale, Kay Gott Chaffee, Lauretta Beaty Foy, and Barbara Erickson London.

WASPs at the P-38 National Convention in 1987. L to R: Nancy Batson Crews, Barbara Erickson London, Mary Lou Colbert Neale, and Kay Gott Chaffey. (Courtesy of S. Eisner)

Mary Lou Colbert Neale.

Miss Jackie Brundage, the Lightning Lady of the 96 FS, 82 FG, and Richard E. Willsie at Lockheed Aircraft in 1944. (Courtesy of Gil Cefaratt)

THE LIGHTNING LADY OF WORLD WAR II

compiled from newspaper articles

Jackie Brundage, chosen by a U.S. Army Air Force P-38 fighter squadron as its "Lightning Lady" during WW II from among pinup photos of Lockheed women who helped build the famed aircraft, was reunited with retired U.S. Air Force Col. Richard Willsie, at the cockpit of a Lockheed P-38 Lightning in a pose similar to that in the 1945 photo. In the 1945 Lockheed Star photo, Dick Willsie, 25, is given a warm welcome to Lockheed-Burbank by 22-year old Brundage at a P-38 aircraft. The search for the Lightning Lady was undertaken by the former 96th Fighter Squadron members who wanted her to be the special guest at their reunion. The search took four months and Jackie (Brundage) Gardner said, " I think it's wonderful that they remembered me. I never dreamed that after 42 years anything like this would happen."

It was former squadron Sgt. Dick Ligenfelter of Boise, ID, who thought it would be great to have the Lightning Lady at the reunion. It was through a newspaper article that Gardner found out about the search. "It will be fun," she said of the reunion. "The pinup was wonderful. It was a good morale builder for them and for me."

Gardner now lives in Rancho Cucamonga, 40 miles east of downtown Los Angeles, and still has the snapshot of herself in a black dress that became the pinup. But she said she'd like to get one of the posters if they're still around. Gardner's picture was the feature item on the enlisted men's clubhouse walls and was painted on the nose of the squadron commander's plane.

Forty-two years later, Brundage and Willsie meet again at the Planes of Fame Museum in Chino, CA. 1986. (Courtesy of Gil Cefaratt)

"I think every one of them will want to see her," said Maj. Richard Willsie, the unit commander. "If you've never seen a pin-up in flesh and blood, you do want to see her eventually, I think."

In 1944, Jackie Brundage was a 21-year old clerk and assembly worker at Lockheed Aircraft in Burbank, which built the P-38 Lightning fighter planes the squadron flew. She became the Lightning Lady and the squadron's exclusive pinup in a contest. Squadron members, based in Foggia, Italy, got the pictures of 38 women from the Burbank plant and voted on the winner.

Ligenfelter said that meeting the former pinup girl after all these years probably, "Will be about the same thing as seeing an old buddy you haven't seen for 42 years." "She's marvelous...she's beautiful. Certainly, I'm thrilled. Lord knows, the gang did all right when it picked Jackie from the rest of the beauties. That wasn't an easy job...but I'm satisfied with the results!"said Maj. Dick Willsie, squadron commander of the Slugging Jackrabbits, upon their first meeting with Jackie Brundage, Lightning Lady winner.

KELLY'S DREAM A REALITY BUT NOT WITHOUT SOME GRAVE CONSEQUENCES

by Tony LeVier

The very first P-38 was flown by Captain Ben Kelsey of the U.S. Army Air Corp and, without sufficient testing, was dispatched on a U.S. transcontinental speed record attempt, from which it crashed short of its destination on Long Island in New York. He had set a blazing new record, and Lockheed lost the world's most advanced military fighter interceptor. Kelsey had such raves for the P-38, the Army ordered a second P-38 to be built, and Lockheed was launched into the most exciting program for an aircraft of that time period — 1940-1945 — practically the whole period of WW II.

The second P-38, No. 2 prototype, was maintained at Burbank with only a limited number of test pilots ever to fly it. The late Marshall Headle, Chief Pilot of Lockheed at the time, flew it first and later passed it on to the late Milo Burcham, Kelly Johnson's first Chief Experimental Test Pilot when Johnson formed his own Flight Test Organization, later to become known as the "Skunk Works."

As the second P-38 was expanding its incredibly vast envelope of performance in speed and altitude a very strange behavior arose at high altitude wherein the P-38, at very high air speed, started to buffet, followed by a very strong, nose-down pitching moment. Recovery was not immediate which, luckily, so impressed Burcham of the gravity of the situation, that he hesitated to venture further into exploring this strange phenomenon. Johnson and his engineers dug into the problem with no appreciable success. The only immediate improvement was changing the angle of incident of the horizontal stabilizer to a more negative angle, "nose up" condition. It was minor in effect of the principal cause to be learned later.

About this time, late 1941 or early 1942 (records are available), the Army Air Corp got its first P-38s and I think assigned them to the First Fighter Group stationed at Selfridge Field, MI.

One day, Captain Gilkey took a P-38 up for a high altitude flight only to return in ghastly fright of what happened to him in a dive. Unaware of this new, high-speed phenomenon, Gilkey peeled off into a somewhat steeper dive than Lockheed's test pilot, Burcham, had ever ventured to go and for very good reasons. Gilkey's P-38 had sustained some damage, as his recovery was down much lower where the stress on the P-38 became critical.

This incident created great concern for all parties, with Lockheed getting considerable demands from the Government to solve the problem. About this time, another Lockheed pilot entered the picture — Ralph (Killer) Virden, an ex-United Air Lines pilot, wrestler and strongman with lots of shear guts, took over to help Burcham in the exploration of high speed flight.

Lockheed Engineering Flight Test P-38 S/N 5266, the only completely two place P-38 with full controls in rear seat. (Courtesy of Gil Cefaratt)

At this same time the news of the P-38 problems abounded with every so-called engineer and pilot expounding on what was causing the problems. The military came out to Burbank to discuss the P-38 problems and to witness a dive and high speed pass down the runway at what is now Factory B-1 (the original factory). It was just about noon, and most everyone was waiting for Virden to come roaring over the factory when suddenly we heard the horrendous roar of his propellers off in the distance toward the City of Burbank, followed by a terrible over-speed noise of the props and a "thud" when the aircraft hit the ground.

The accident was a new low point for Lockheed and the P-38. I think the date must have been early 1942, as I had not yet joined Johnson's Flight Test Division. There were other incidents of uncontrollable dives reported which gave the Army Air Corp cause to believe Lockheed didn't have the skill to cope with the problem. Johnson had a good understanding, but try as he did to convince everyone it was a high speed problem, he was continually rebuffed by NACA, (now NASA) stating that there was no problem of the P-38 that couldn't be solved in their 125 mph wind tunnel.

I can tell you right here and now that the situation was grave for the P-38. The other problems of development were momentous, to say the least —the Allison engine being a major source. Johnson was quick to move ahead on all the problems with the high speed trouble of the P-38 being "top priority." He next proposed to Gross that he be allowed to hire his own pilots, so that he could train them for the very special task of experimental test piloting.

Johnson already had Milo Burcham and Joe Towle on indefinite loan and needed two more. Jim White and I were chosen. White had had excellent flight training with a stint at TWA. I was 29 years old at the time with a wide range of flying, including high speed air racing, testing, airline piloting and, just before joining Lockheed, was Chief Research Test Pilot for "Boss Ket" of General Motors.

We were a happy group of four. Burcham was just a wonderful guy and boss. His boss, Thoren, was tops too, and his boss "the old goat" Johnson was one in a million. We were flying everything from dawn to dusk, and my first real love for an aircraft, the P-38, became my main interest.

We entered 1943 with America deeply involved in WWII with a demand for the super P-38 greater than ever. The Allison engine had been a constant headache, of which I was assigned a principal part to do engine development test flying. I became a pretty good expert on that engine. I had also experienced this high speed phenomenon and told Burcham, Thoren and Johnson of my experience. I had also expressed a willingness to participate in the exploration of the P-38 high speed problems if they might need another pilot. The chance to do so was not long in coming. Johnson had made every possible effort to get the P-38 into a high speed wind tunnel,

Lockheed test pilot Jimmy Mattern and USAAF Major Dick Bong. (Courtesy of Gil Cefaratt)

and finally accomplished it through the help of General H.H. (Hap) Arnold, Chief of the U.S. Army Air Force.

In this high speed tunnel, Johnson and Ward Beman, one of his engineers, found what they were looking for. True to belief, it was indeed a high speed problem of gross proportion. The P-38 and its high altitude and speed was ever so close, even in level flight, of entering the new world of "supersonic flight." The P-38 had been dipping into what is now called the "transonic realm." The wings and other parts of aircraft perturbrances were being subjected to airflow of sonic to supersonic speeds. The Schliern photographs of the wind tunnel model showed very sudden change in pressure distribution about the P-38 wing which resulted in three major and detrimental effects.

Formation of a supersonic enclosure about the wing produced: 1). loss of lift (due to shock stall); 2). Aft movement of wing center of pressure (causing nose to tuck down); and 3). Severe tail buffeting in wake of turbulent airflow down stream of the shock wave on the wing; also a decrease in down wash upon stabilizer.

In essence, the P-38 was acting something like being in a normal slow speed stall but, instead of speeding up to recover, it now must "slow up." But how can you slow up with the aircraft in a steep dive from which you have lost control? That was the $64 question.

The term "mach number" arrives. Pilots called it "match" number until they learned it was named after Earnst Mach, a German physicist who had predicted years before that high velocity projectile, "bullets, etc.," would misbehave if they went too fast.

Johnson devised a special device called a compressibility dive flap. It would fit under the wing along the main spar. It was found to partially solve each of the aforementioned detrimental factors. 1. It restored some loss of lift. 2. It produced a nose-up pitching movement. 3. It increased the drag which reduced the mach number to below critical, which was approximately mach = 0.70, or 70% of the speed of sound at the test temp.

In March of 1943 Johnson invited me to join with my boss, Burcham, to conduct dive tests of this new (first ever in aviation) compressibility dive flap.

The purpose of two pilots doing the test was to assure that there was always a pilot ready to do the test in case of illness or in case of death.

Burcham and I were given a free hand to conduct the dive tests and decided to start our dives from 30,000 feet. Burcham made the first dive at an approximate 20 degree nose down angle and landed with good results. I made the second dive to confirm Burcham's first dive and made a second dive slightly steeper with good results. We were making what I consider very good progress. The data was excellent (the first ever), and Johnson was confident of our success when, suddenly, appeared Col. Cass Hough, dubbed "the world's fastest human."

It seems as though the Army Air Force was not happy with the way we were conducting the dive test which, in essence, had already been accomplished by Col. Cass Hough's vertical dive from 43,000 feet a few months earlier. His mission was to explain to us how he did it. There were eight men in the meeting — Kelly Johnson, Rudy Thoren, Milo Burcham, Joe Towle, Jim White and I of Lockheed, and Col. Cass Hough and Col. Ben Kelsey of the Army Air Force.

Col. Hough spent the next 30 minutes explaining in great detail how he had made this fantastic dive straight down from 43,000 feet and pulled out at about 12,000 feet, as I remembered him telling us in the meeting. At the conclusion, we Lockheed pilots ambled back to our office. On the way, I made a comment to Burcham something like, "For gosh sake, Milo, is there some secret thing we don't know about diving the P-38?" He turned on me and gave me hell for venturing the thought!

That afternoon, Kelsey (God bless his soul!) made a statement to the effect that he would make the next dive and get the thing over with once and for all. The next day at 1:00 p.m., Kelsey took the P-38 dive test aircraft up to approximately 40,000 feet and put it into a dive. We were waiting with great anticipation for his successful dive recovery news when, suddenly, we were beseeched with phone calls from people having seen what appeared to be a P-38 spinning down minus a tail. A pilot was seen floating down by parachute.

Well, we lost one good P-38 needlessly. For some unexplained reason, Kelsey was unable to extend the dive flaps and, in desperation, pulled so hard the whole tail parted company. We were dejected, to say the least, but were given orders to prepare a second P-38 as quickly as possible for further tests.

In June 1943 our second P-38 dive test aircraft was ready to go and, from our previous data that Burcham and I furnished Johnson, he had selected two instead of four separate dive flaps and made them production models, fully expecting success.

Burcham and I had to start all over again from 30,000 feet and were just getting started when orders from Army Air Force Headquarters in Dayton, OH demanded we increase the gross weight of the P-38 by 2,000 pounds and start our dives from 35,000 feet.

Burcham and I sort of gulped and sat down to figure out what to do next. I was in this up to my neck and thinking of every imaginable way to improve our ability to dive the P-38 accurately. We had no special instruments, such as a mach indicator, as it was not yet invented. We were both complaining of this lack when it struck me that some sort of dive angle indicator could be very helpful, from which I suggested turning the plane's directional gyro 90 degrees clockwise in the instrument panel. It took about 30 minutes to do and it worked perfectly the entire test.

Burcham and I started diving in increments of 1,000 feet higher at a 30 degree dive angle until we reached 35,000 feet. At this level we started our dives. The procedure was using rated engine power in level flight to stable speed at which point we set our dive angle indicator at zero, put our dive flaps out, reduced power and pushed forward to the desired angle. This went on for several days, working up in five-degree increments of dive angle. We were plotting our results after each flight, and "things were looking good." Burcham came down from his last dive of 55 degrees which is damn steep and exclaimed, "It looks like we have reached terminal speed, or close to it."

I drew the next dive, which was a duplication of Burcham's 55 degree dive which resulted to be the same as his. I returned to 35,000 feet, set her up for my next dive to be 60 degrees. All set up, top speed dive angle set at 0, flaps out, power back, I pushed over to 60 degree dive angle in about 2,000 feet loss of altitude. Maintaining the 60 degree dive angle was the most important to me, but I was scanning rapidly my speed and altitude to note anything at all that might be different from other dives when all of a sudden the P-38 started to buffet and nosed down steeper.

This was the first time we had experienced "tuck" with the new dive flaps, and my first inclination was to get out of the dive right then, if possible. I hesitated at 60 degrees and decided to hang in there and, as long as I could hold a 60 degree angle, things should get better. Our dive pull out altitude had been established to be 20,000 feet from the very start. I rode her down to 20,000 feet and then started pulling in ernest. The buffeting became excessively heavy, and I was now pulling harder than I had ever before. At 19,000 feet my dive angle was 59 degrees; I had only gained "one lousy degree" in 1,000 feet. My mind was going a mile a minute — at this rate I wouldn't make it. I pulled harder, the plane got rougher, and at 18,000 feet the dive angle was 57 degrees. Hey! It's coming around, but slowly! I pulled as hard as I dared! At 17,000 feet I'm at 53 degree dive angle, and I can now start feeling those glorious "Gs." She's starting to really bite in now, and I'm gaining more degrees each 1,000 feet.

Finally I reached my maximum "G" and buffeting at about 12,000 feet, at which time all six "red" overstress warning lights on top of my instrument panel lit up! Now I contemplated the aircraft coming apart.

I eased off the "G" and zoomed up a couple of 1,000 feet, slowed the P-38 down to about 130 Vi. and limped home thinking, "I'm coming unglued." I was over Muroc Dry Lake (now Edwards AFB) and it took a while.

There was elation at Lockheed that day.

P-38 production line at the Burbank, CA factory. (Courtesy of Gil Cefaratt)

Lockheed test pilot Eddie Martin sits in a P-38. (Courtesy of J.A. Liebeck)

THE LIFE AND TIMES OF A TEST PILOT

by Eddie Martin
Lockheed Test Pilot

I had to drive 47 miles to get to Lockheed, the farthest of any of the test pilots. I was practically always the first one up there in the morning, usually ahead of the janitor, and I was usually about the last one to leave. We were supposed to work six days a week, sometimes seven, and get one day a week off. It didn't work out that way because there was a lot of activity for a real long time and I was having so much fun I really didn't care whether I had a day off or not. I stayed with Lockheed for two and a half years and every day was just like a picnic. I loved what I was doing. I just couldn't wait to get there in the morning and I hated to come home at night.

When I first went to Lockheed, I had my eye out on a P-38, but I didn't say anything about it. I took my turn in flying whatever they wanted me to fly: the Hudson Bomber, the PV-1, the Lodestar. Within a month or two, I started testing the P-38. Lockheed was building a lot of P-38s, and I got more of those than any other type of ship.

One day I tested eight P-38s. Eight P-38s in one days is a record, I'm sure of that. One was a production flight; first production flights took about three hours

and 30 minutes. Then there were a couple of acceptance flights, which didn't take very long—there were only a few of us that were acceptance pilots. And I think I had one delivery to Long Beach. During my 2-1/2 years with Lockheed, I tested over 250 P-38s.

Altogether, Lockheed had over 60 test pilots, but some of them were scattered all over the country. For example, Tony LeVier was sent over to Europe for almost six months at one time. He was there testing planes for the Army because the Army didn't know how to fly everything like Tony and the rest of us did. For example, towards the end of the war, a P-38 went down next to where I was plowing in my field. The boy who was flying it had been enlisted in the Army only eight months before, and he was on his first flight with that P-38. It was his "get-acquainted" flight with the instruments and with the handling of the ship, and the next day he was going to join the force combing the ocean for submarines with that P-38. And that was his job, and here they were sending that poor little boy out—eight months from the time he was enlisted. You couldn't learn to play an organ or a piano in eight months, and that damn P-38 had just about as many things to keep in tune. I loved that P-38 but I'll tell you, after you have flown it for a couple of years it was sure a lot different than the first time you flew it. Any ship is, where you're not completely familiar with all the instruments. I knew that boy was in trouble for a long way off-he was scared to death, and he did everything wrong. He went down right in front of my tractor, buried himself and the ship 16 feet into the ground.

A "MAJOR" ENCOUNTER

The P-38 was a wonderful airplane: it was such a clean airplane that if you put it in a dive and held it too long, you'd pick up too much speed and it would go into compressibility. If you let it get away from you in compressibility, there was no way under the heavens you could get it out—it would usually go on into the ground. To slow the ship up, they put on dive flaps. That slowed the ship to a certain extent, it enabled us to dive another 20 or 30 mph faster before pulling out (before going into compressibility).

A Major came back from overseas and said that all the talk about compressibility with a P-38 was alot of "bushwa," that Lockheed pilots were just afraid to dive the P-38. So this Major was put in charge of the P-38s flying out of Burbank. He decided, and he had the authority, to take a P-38 out to prove to the pilots that you could dive a P-38 as much as you wanted and it wouldn't get into compressibility and get away from you. He took a ship out over Calabasas, at the far west end beyond the San Fernando Valley, and went up to probably 15,000 feet or better and put the P-38 in a dive. And he let it dive for quite a bit and sure enough, he got into compressibility. There was nothing he could do about it, although he was trying, but that P-38 shook itself so badly that it shook a wing off. The wing come floating down like a leaf—it floated one way and then the other, kind of skipping through the sky. It landed out there in the hills without much damage, but it had torn right off that P-38, and I could hardly see how that could happen because the P-38 was so strong. The Major was able to get out of the ship, which was a hard thing to do in a P-38, so they claim, when the ship was all together. It was said that with a ship as fast as the P-38, you'd hit the tail assembly before you were clear of the ship and it would hurt you, at the very least hang the chute with your harness in the tail section. The Major did get out alright and got down, but he broke an ankle. He didn't have much to say about diving after that; he was still on the Lockheed field for a month or so until his leg got better but he didn't give us any more guff about being afraid to dive the P-38.

DOUBLE LOADED

Every 25th P-38 was made into a camera ship. To do this, they had to take off the armaments—all the guns and ammunition. To replace the weight that had been removed, lead shot—sacked in canvas bags, was put in under the cowling. Then the cowling was sealed. This procedure was done by the night crew. A new boy was hired on this particular night crew, and his job was to remove the cowling and put in the amount of shot that was proper, seal it all up again, and sign it off. Well apparently, they didn't give him proper instructions, because the ship that he went to work on that night had already been signed off by the previous crew. He put another load of shot in that P-38, the one I drew the following morning, and there was no way for me or for anybody else to inspect that part of the ship before we flew it. We had to trust the paperwork that it had been done.

That morning we had a very strong northwest wind up at Lockheed. During this wind condition, we were requested to use the short runway. The P-38 that was double loaded on shot was my first trip out that morning. I inspected the ship and got ready to fly and taxied clear down to the south end to use the short runway, which headed just about straight into the wind out northwest across the San Fernando Valley. This put you out over the desert area the minute you got off the runway and right over the revetments built on the field to put airplanes behind in case the war ever got us over here. I got rolling with that P-38—I had checked everything on it, all the tabs and the whole works was just perfect. I got down about 2/3 of the way on that runway—too far down to abort the flight, and

P38s ready for modification to F-5 photo planes at Lockheed Modification Center at Love Field Dallas, TX. (Courtesy of Leo Childs)

A P-38 F5C prototype ready of initial flight test. (Courtesy of Leo Childs)

I had about 110 mph or so and was about ready to lift off the ground. The end of the runway was coming up awfully fast, but I just couldn't get the nose up. Inherently, I dropped my arm and turned a tab—I thought maybe the tabs were rigged wrong. I rolled the tab clear back to 16 degrees—that wouldn't get the airplane off the ground, but it made it easier to control, because that control was clear back in my stomach and I was hanging on with one hand as I pushed the rest of the power on. The P-38 would pull 68 inches or so if you had to use it, just about enough to blow the head of the cylinders. I knew that, but I had to get off the runway, otherwise I'd roll right out into the desert and crash and burn; it was a rough territory.

So with the throttles wide open, I just barely got the nose and the wheel off the ground, close to the place where Milo Burcham had been killed with a jet plane. I'll bet I didn't have two feet of runway left, and I was sure thankful for each one of them. I was too busy to phone home and ask any questions, I was trying to get the ship into the air, but the darn thing just wouldn't climb. I got a few feet at a time, but it took practically all the power I had, which was too much to be using, to get the airplane to climb. My big aim was to get up to 4,000 feet, if possible, before I made any fuss about it and the if I had to get out of the ship I could. The theory was that you couldn't jump out of a P-38, but I had made up my mind that I would make every effort to get out rather than go down and pile up in a fire with it.

I got clear out over the northwest part of the San Fernando Valley and I was getting a little altitude; I didn't have my four thousand feet yet, but it was climbing slowly under full power. I headed to the right over towards Newhall; if I could get up over Newhall, I could call for an emergency and go back to Burbank.

I was almost at four thousand feet just south of Newhall, when I made the first call in to Burbank to tell them about the trouble I was having. I was going to try to get back to Burbank and told them to give me immediate clearance and keep everybody else out until I got in. By the time I got over Newhall I had my 4,000 feet and I pulled the throttles down a little and tried to drop the ship, but the minute I did that the nose started to go clear down under, and I knew that it would only be a matter of a few seconds until it was clear out of control and headed for the ground.

So the only way I could get that airplane down was to close the throttles and let it drop a little bit and catch it with the throttles and the elevators and pick it up on the level again. And I went all the way down from Newhall to Burbank, just tipping it over and catching it with power, just like a rollercoaster. I'd drop that thing maybe a couple or three hundred feet at a time or a little bit more if possible, and catch it with power to get the nose back up. I came into Burbank and dropped it just before I got on the end of the runway and picked it up again with the power and let it settle right onto the end of the long runway. I made a pretty decent landing out of it. I let it roll to the far end of the runway—I didn't try to come in on the short runway, I was going too fast, and I didn't want to have to maneuver, just land it straight even though the wind was out of the northwest and still real strong.

I pulled up to the same pit that I had taken it out of, and there must have been at least 20 mechanics around that thing in the first five minutes trying to find out what was the trouble. They talked to me and I told them what happened—I didn't know what caused it. In just a few minutes they had all the cowling off that P-38 and they were all frantically trying to figure what happened. It probably took them 15 minutes to run into the fact that it had two loads of shot buried in the front end under that cowling.

There was never a word said to me after that about that incident, nobody seemed to care to know anything about it. Not once did Lockheed have anybody come and try to talk to me and find out just what the results were and why and how. It showed that there was a war on, and that factories didn't really care about learning how an airplane handled in a situation they didn't expect.

A "DEAR MOM" FROM FOGGIA

Fred L. Montgomery
82nd Fighter Group
96th Squadron

Recently, while doing the sad but necesary chore of going through some of my late mother's personal effects, I found a box crammed with most of the letters I wrote her during my brief career in the USAF. After the first small feeling of guilt at reading someone else's mail, I sorted them by postmark date and started reading. Disregarding the blarney, exaggerations, white lies and gripes I came up with a great picture of an average Air Force G.I., from his enlistment in November 1941, through all the fun days of the Big War, to his honorable discharge in June 1945. Well, I had a lot of good laughs and some real throat-lumpers over incidents that I would truly like to rerun with buddies who have since dropped out of the formation.

At the risk of boring hell out of some of you, but hoping to bring a twinge of nostalgia to the elite brotherhood of retired P-38 Crew Chiefs and their hard-working Assistants, vintage 1942-45, I submit hereunder an unedited portion of a letter I wrote on Thanksgiving evening, November 23, 1944. Hope some of you enjoy it as much as I did, especially those who shared the experience on those cold, muddy mornings at Foggia, II. Just for laughs I'll title it:

HERE'S MUD IN YOUR EYE, CHIEF!

"You say you want to know what my average day is like, Mom? Well, for instance, yesterday a.m. the whistle blew at 4:45,

A P-38 in North Africa with bombs mounted and belly tanks ready. (Courtesy of Bill Hess)

and I fell out in the mud and the pitch dark to preflight my ship. Dawn comes about 5:30 these days, and the sun, on those rare days when it does, rises about 6:00. If you think it's not a thrill to run up a '38 in total darkness on a muddy or snowy morning, you must try it sometime. The only way you can tell if she's sliding is by the relative position of your wing lights with those of the ships around you. When you've got one engine turning up 2300 rpms and pulling 30 inches manifold pressure...you're holding the brakes so hard your spine is putting dents in the armor plate in back of you...you're watching all your instruments, especially the tachometer while you check your mags...then all you've got to do besides is watch those red and green beacons on either side of your ship.

Of course, if you've got extra-wide vision and are real quick, you can admire the cherry glow of your supercharger decks and the foot or so of pretty blue flame shooting out of your exhausts, but you have to look over your shoulder slightly so it's not advisable...Before you check your mags you have to run her up to 2600 or 2700 to clear out your plugs, and during those brief seconds you have eyes for nothing but the black void around your ship and an occasional fleeting glance at your tach...By the time you've checked one engine, your legs are so tired that you have to set the parking brakes and let her idle while you rest up for the other one.

If both engines check out O.K., then you're plenty happy, but if either one cuts out on mag check, that's life's darkest moment. There you are, half a mile from Tech. Supply, no spare plugs on hand, no assistant, no light, no nothin! You want to beat holes in the dirty bird with a hammer, but you don't. You either start tearing cowling off, holding the flashlight in your mouth, and praying that your Flight Chief will come along in the jeep and go fetch you some plugs, or you go on a mad dash from one neighbor to another in search of some they'll part with. Anyway, you get them somehow and change a minimum of twelve, burning hell out of your hands in the process.

By now a dirty grey light is beginning to filter in over the eastern hills. You jump back in and run up the sick engine again, meanwhile muttering assorted mechanic's prayers. Hooray, she's O.K. this time! You hit the mud again, slap on the cowling, grab two rags (if you have any!), one wet with gas and the other with water, and a chamois...climb back up and clean out the mud from the cockpit floor, wipe the instruments, windshield, canopy, side windows and mirror. Check the oil in both engines, the hydraulic reservoir and gas tanks...the latter have been filled after yesterday's mission but must now be "topped off" when the gas truck comes around.

That finishes topside, so you slide down again, prime the nose strut shimmy dampener, drain supercharger balance lines, make out your daily form...and, in your spare time, give the whole ship a visual inspection, being darn sure you have removed the covers from the pitot tube (air speed) and the relief tube (you only forget this one once!). It's now just about time for the pilots to arrive from Briefing, and your assistant, having eaten early chow, takes over.

You are free to shuffle through an acre of mud to your shack, grab the M-1 mess gear and go get your powdered eggs, oatmeal, bread and jam and good old G.I. coffee. You eat, wash your "skillet," return to the shack, brush the moldy molars, take care of Nature's demands, and it's time for school, 8:00 a.m. to 11:30...(note: At that particular time crew chiefs were attending a six-day Allison-Lockheed refresher course taught by G.I. tech instructors, who were excellent.)

School is out just in time for you to grab early chow and return to the line to relieve your "boy" (with no disrespect whatsoever most assistant c.c.'s were known as boys. Some of them may have been older than their chiefs and probably knew just about as much.). He has been "sweating out" your ship ever since takeoff. It's a hard and fast rule that there must be a man "on the line" for each plane while it is flying, in case of an "early return" (probably the biggest bugaboo in a crew chief's life)...So now you take up the cold vigil. When your worrybird at last comes home, the pilot signs the form and tells you what's wrong with her today. After he's gone in to Interrogation, you run her up and check mags and general operation. This is known as the afterflight check and is not as extensive as the morning preflight, but it must be done, even though the pilot reports the ship O.K. (a rare occasion and one for great rejoicing).

Next, if she checks out O.K., you begin your Daily Inspection, which takes about an hour. Then wash down landing gear, doors and booms, clean the mud out of the wheel wells, hang belly tanks for tomorrow's mission. While waiting for the gas truck to fill them, you enter flight time, cumulative time on aircraft and engines, gas and oil consumption, mechanical work done, etc., on your daily form. After you've gassed up, locked surface controls, closed and covered cockpit, relief and pilot tubes, it's chow time, so you hit the road for home.

But wait; you're not through yet, Mr. Crew Chief! You still have some pencil pushing to do, so you stop by the Engineering shack and transfer pertinent data from your daily form to your permanent maintenance record. Then to the Operations office to turn in the daily form. Oh, yes, and while you were at Engineering you forgot to check the Tech. Order compliance chart and bulletin board to see if there are any new T.O.'s or AF memos which you must read and initial. But, by now chow is darn near over, it's almost dark and colder than a gold digger's passion, so you say "Manana" and slog home for the bright hour of the day - chow, mail, and The Sack, the heavenly, lumpy old sack of straw, your Beautyrest!! You'd like a couple of belts before you turn in, but you know you haven't the willpower of a gnat and if you go to the Club, you'll "spin in" four or five hours before the whistle blows you out of the sack again. Lately you realize you're not as young as you were a couple of years ago, and that Preflight described above just ain't the time and place for a hangover.

So you take pen in hand to write your Mom a nice, long, cheerful letter. But what keeps running through your mind instead?...That left booster pump sure sounded funny tonite; hope it doesn't go out tomorrow...Did I safety that oil drain?...Wonder if there'll be another early work call in the morning?...Finally, promising yourself faithfully that you'll write home tomorrow night, you tear up the nearly blank page and hit the sack.

But comes 11:30 p.m. and a none-too-gentle rousing from dreamland by the nice guy known as the C.O., who regrets to say there is to be a surprise bombing party for some of Hitler's boys tomorrow, and you must therefore go out and drop those belly tanks you struggled to hang a few hours before. You cuss a little but pull on your sheepskins, grab your flashlight and take off. If you're real lucky, the Flight Chief comes by and takes you and your sleepy mates out in the Jeep. Upon arrival, you find that one belly tank has already been gassed up, so you sit around shivering and dis-cuss-ing Army life with the guard on whose post your revetment lies, until the gas truck arrives to pump out the tank. Then, after you drop your empty belly tanks and remove the standpipes, you give the armorers a hand in bombing-up the ship or go help one of your buddies drain and drop his tanks...And so to bed again until the whistle blows reveille sometime between four and six a.m., and the routine starts all over again, with slight variations.

The most "amusing" of those variations is the Early Mission-Standby-Stand Down-Training Flight, which goes something like this: The whistle blows around 5:00 a.m., and you go out and pull one of those nerve-racking preflights in the dark (and it's usually raining or snowing on one of these days). When everything is checked out and the worrybird is ready and eager to take flight, you go in, grab chow and rush back out to shoot the breeze with your pilot before takeoff. But, comes takeoff time and no pilot. So you huddle under the wing, somewhat sheltered from the wind and rain, and await their Military Majesties' will with that profane patience learned in three years of Army life...Eventually the Great Secret leaks out and comes down the grapevine to you: The weather has closed in at the target, and we are now on Standby status in case it clears up. And there you remain, "standing by" in your warm nest under Bertha's wing, until 11:00 o'clock, when you send your boy in to eat early chow and then come back to relieve you. When on alert or standby, there must always be a man at the ship to tuck the pilot in if they decide to fly.

You finally get chow yourself and return

Officers of the 459th FS in front of Officers Club in Dud Kundi, India, 1945. (Courtesy of Nuehring)

Pilots of the 18th FS, Attu, the Aleutians. (Courtesy of Robert B. Minute)

to the ship. By now the sun may be shining, and the Chief comes around and tells you that the "Brains" have finally chucked the whole thing and the status is now Stand Down. This means that you are free to work on your ship, which undoubtedly needs it after having flown steadily for four days...probably has a 50-or 100-hour inspection long overdue. So this news of genuine Stand Down is music to the ears of the crew, who, if uninitiated, immediately pounce on the weary bird and strip off her shining plumage, laying bare her oily innards to their ministrations. Poor eager innocents!

No sooner have the nuts and bolts begun to fly, and a grin of contentment has replaced the scowl on your greasy puss than there arrives on the scene...with much honking of Jeep horns and bellowing by the Flight Chief...that shining example of young American manhood, the Eighth Wonder, "The New Pilot." He is willing — no eager — to do you the honor of flying your aircraft on his First Training Mission. Nor will he take no for an answer, though you deprecate your poor, unworthy ship till you're blue in the face.

With a sick smile on your face and murder in your heart, you slap her back together, button her up and try to instruct the loo-tenant in the idiocyncracies of your baby. In your present mood it's a bit difficult to squeeze out the proper degree of official respect, but you get through it somehow without earning a court martial. Then, a last-minute climb back to the cockpit to tell him his left engine is flooded and will probably start if he'll turn off the damn booster pump. (He almost blows you off the

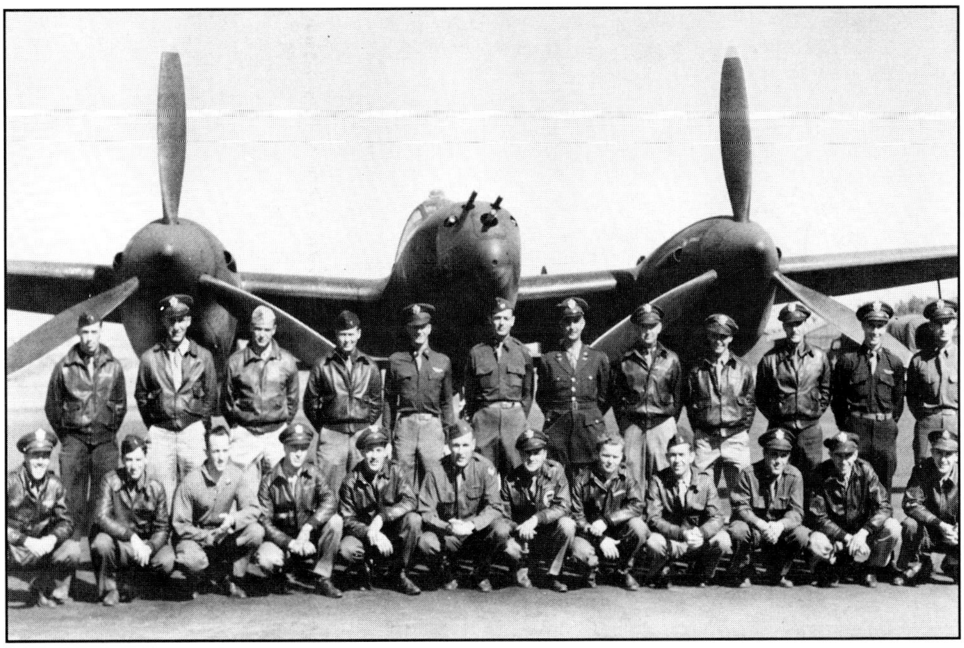

U.S. Naval Air Station, San Diego, CA. 16 Feb. 1943 (Courtesy of Gil Cefaratt)

wing when it starts.) Finally, you wave him out of the revetment and sink down upon your tool box, a beaten man. The only advantage of a muddy day is the "Hot Shot" can only blow a few gobs of mud, instead of a ton of dirt, in your face as he blasts out at full throttle.

Now all you have to do is sweat out his takeoff and landing...and he probably has the most fascinating new wrinkle on landings that he's dying to try out. (The wrinkle will undoubtedly show up in Bertha's nose section, just forward of the strut, when he proudly taxies back in)...After you've noted all the things he found wrong with the ship — all of which you told him about in "briefing" so he wouldn't write them up — patted him on the back and sent him on his merry way, secure in the knowledge that he's the hottest jockey in the USAF...then you just have time to run up, clean up, pull a quick Daily, make out your forms, etc...Providing, of course, that you don't have to change a nose strut as a result of his trick landing some fifteen feet from terra firma. "Sure is a rough runway," he said!

And so you log another entry into your book titled "Crew Chief Daze."

A GERMAN MUD BATH

by Gene W. Sidwell

One mission I will always remember was next to my last. It was on April 22, 1945, the war was winding down and everyone knew it. It put on extra pressure not to get yourself killed. The mission was to dive bomb and strafe in the Po Valley in northern Italy. We took off in the mid-morning from Foggia, Italy and flew up the Adriatic Sea. When we left Foggia the sky was clear but when we made landfall somewhere west of Venice it was overcast and the ceiling was down to less than 1000 feet. Since it was socked in too much for dive bombing the squadron split up into elements looking for targets of opportunity (an element is two airplanes, a leader and a wingman). After we had flown in on the deck for about 20 minutes I noticed we were being shot at as we passed a farm house. I called my wingman on the radio and suggested we go back and drop our bombs on it. We each had two 500 lb. bombs. Because the ceiling was so low there was no way to pull up and turn around to make a straight run but instead we made a circle just above the trees. We could keep the target in sight by watching the tracers coming at us, (every 6th round is a tracer bullet) as we approached I noticed there was a large irrigation pond beside the house and you could see the German machine guns shooting from behind some sand bags in the yard. When we dropped our bombs we were in a slight turn and the bombs missed the machine gun nest and hit the pond instead. When I looked back there was a huge sheet of mud just ready to engulf the gunners and the entire house. I would like to have seen those Germans a few seconds later, it was probably pretty funny but I figured we had better get out of there while we were ahead.

When I think back on it, at that stage of the war we really didn't want to kill anybody anyway - a mud bath was plenty good enough.

YO-YO

by Arthur Cole
48FS, 14FG

"Up and at 'em, Cole," was the yell that came out of the African pre-dawn darkness through the thin walls of my army pup tent. As I hastily dressed in the freezing blackness, I could hear the air crew getting the two P-38s out of bed and warmed up. Mark, my element leader, and I were on the early morning, alert shift and after a gourmet breakfast of hardtack and peanut butter washed down with a canteen cup of warm water, we were sitting (semi-alertly) at the end of the runway on watch. Two squadrons of the 14th Fighter Group and one company of British paratroopers were stationed in a tiny valley in southeast Algeria anchoring the Allied flank of the Allied invasion of North Africa. Our mission was to keep a look-out for the lead elements of Rommel's retreat from the British Eighth Army's attack in the Middle East.

Two hours after Mark and I had assumed our watchdog position, French Intelligence came through with our mission - a German armored column was proceeding along the coast. Blood and engines pounding, we took off for the 40-minute flight to the Gulf of Gabes. Sure enough, there in the distance was a column of 1 1/2 ton trucks, armored vehicles, and light tanks crawling behind the last transport in line; I gave Mark enough room ahead of me so that any rounds bouncing off the ground wouldn't hole him and I took the left side while he took the right. As soon as Mark opened up the vehicles halted and men began to dive for the ditches on either side of the road.

Mark finished his run and pulled up for home and shortly thereafter, I started to follow; but then I noticed that the first vehicle in line (and the last to stop) was untouched, and I knew that it must be the command car. I wanted that car! I pushed the nose down for one quick squirt and pressed the firing button. As I hauled back on the control wheel, WHAM!, came an awesome jolt that I have previously experienced only on crash landings - that airplane must have bounced 100 feet in the air. Dazed, I realized that the craft was shaking as though it had delirium tremens, and I finally understood that the right engine was going to vibrate right out of the nacelle, or else the wing was going to come off. Chopping the throttle and feathering the prop reduced the vibration, but the bird still flew some sort of drunken path. All I could think of was how overjoyed those shot-up Germans would be if I came down anywhere near what remained of their beat-up column. Trimming finally permitted me to return to Youks-Les-Bains, but what a mess that beautiful aircraft had become.

Apparently, I had literally ricocheted off the command car with it passing under my right wing just next to the right engine nacelle. The prop blades had been neatly docked or custom curled, the air scoops had been cleanly wiped off, the coolant shutter on the outside of the boom was flattened, and the horizontal stabilizer outside the rudder had disappeared.

Despite these on-the-job modifications, my P-38 had brought me safely home once again, but I'll bet it took one giant, pancake turner to get that car off the pavement.

A SWIFT CLEAN EXIT

by Len Boyd, 459th FS, 10th AF, CBI

It was a few minutes before 8pm on 25 March 1945. The night was dark and clear, a large orange bomber's moon moved slowly off the horizon. Bright stars blinked down on Rhumkapalong, India, the new home for the 459th Fighter Squadron, known in the CBI as the Twin Dragon Squadron.

Situated on the knuckle of that slender finger of land that joins India and Burma and points to the northern entrance of the Arakan coast, "Rhumka" faces to the west. The Bay of Bengal, across which rested this night, the city of Calcutta, hunkered down in its day's sweat, but all aglow, no longer afraid of bomber attack. To the north and south, the RAF airfields at Cox's Bazaar and Akyab had already launched their night patrols. The brooding Chin Hills, which formed the bomb line, loomed unseen in the distance. Further southeast into enemy territory lurked the serpentine Irrawaddy River, coiling like a monstrous Indian python the length of Burma, its beady eyes staring into the passes of China while, 1425 miles away to the south, its tail thrashed in the waters off Rangoon.

With the end of the Imphal, Kohima and Myitkyina campaigns, the threat to India from the Japanese Army Forces was over. The many battles to keep the supply routes open into China had kept China in the war, pinning down one third of the Jap Army. The British were now air-lifting record loads of auxiliary supplies to their rapidly advancing troops in Burma.

In the Arakan, British General Christison's 15th Corps had retaken Akyab from the Jap 28th Army under General Sakurai and the Ramree Islands quickly fell on 9 February 1945. Mandalay had just fallen to General Slim's 14th Army with General Kimura's Burma Area Army in full retreat. The bitter hand-to-hand struggle for Meiktila had begun with General Honda's two crack divisions opposing the surprise flanking thrust of the 17th Indian Division of British General Messervy's 4th Army Corps. Meiktila, surrounded by vital airfields, was the logistical nerve center for the central Japanese Burma campaign, a prize to be taken before the May monsoon began.

The 459th Fighter Squadron was now engaged in support of this desperate campaign. The war in the air was practically over. The last pilot lost in action was Lt. Conrad Tolladay, killed in late January in a fighter sweep over the Jap airfield at Heho, east of Meiktila. A few days later, Captain Henry Mahler, flying his new P-38L "Dragon Lady," and his wingman, Lt. William Mouser, shot down two Oscars attacking the British invasion of the Ramree Islands. Now, operating solely under the 3rd Tactical Air Force of Major-General Stratemeyers Eastern Air Command, the "Twin Dragons" were daily active in the dangerous task of low level attacks on enemy troops, bridges, railyards and road and river traffic.

The fight in central Burma raged on, but here, at this forward combat station on this March night, a peaceful setting was belied only by the guarded areas where lay dispersed the 25 P-38s, armed and ready for the morning mission. A quick jeep ride away were the airmens quarters flanked by the basha huts that housed the fighter pilots. The poker players were placing their bets, others were writing letters. Faintly heard from the makeshift Officers Club, the recorded voice of Sinatra hung in the air..."sorry for you, she has no sister," while inside, a few pilots were knocking back a Carews smooth booze with grapefruit juice.

As I headed for the photo tent, I could see Lt. James Pashley in his basha at his favorite past-time of building model airplanes. I entered my tent and switched on the light. To keep out any light leaks, I tied the tent flaps tightly and placed a 4 x 8 sheet of plywood against the flap entrance.

I was in the midst of preparing the developing fluids when I first heard it: the alien throb of a plane, fairly high, the sound rapidly receding as it flew on. It was no Beaufighter, that I was sure. Jokingly, I called out to Jim, "there goes Washing Machine Charlie" and he briefly replied "Yeah, I hear him." About 15 minutes went by. I was running film through the soup when it happened. It was sudden, it was violent and alarming. The still night air cracked with the explosive sound only created by throttles and prop governors being slammed to the wall. Colt 45 in hand, I was running hard across a rice paddy as the Jap bomber pilot roared out of his silent gliding dive. The explosions and the quick glance backward to see the sky light up and fires begin added and extra step. I became aware of others running near me: Lts. Bob Guthrie, Bill Bohn, Ray Christian, Gene Hinsvark, and others were also doing the 100 yard dash in Olympic time. The Jap Lily bomber, both engines growling and very low at 300 feet, suddenly appeared at two o'clock to me, and transfixed itself right in the middle of the rising moon, like a huge black center in a giant orange bulls-eye. It was a foolish thing to do but I fired my pistol at that bomber. Other shots rang out, adding to the panic. Somehow, we failed to shoot each other in the dark.

Rushing back to the bombed area, we found the Mess Hall ablaze, the medical shack and several bashas burning out of control. Two civilian Indian workers were dead, one bomb beheaded. The engine still running, a Staff Sergeant sat in his Jeep in a catatonic state, his face freckled with black dots of shrapnel. Voices cried here and there for medical attention. Our medical supplies in flames, the problem arose of evacuating over 20 wounded airmen. Having just a dirt airfield without lighting, we now missed our former base at Chittagong with its hard runway and full facilities. The Squadron Commander, Ace Major Hampton Boggs, and the Engineering Officer, Captain James King, decided to fly the wounded out. In a B-25, off a dirt field illuminated only by Jeep headlights, Boggs and King made a daring takeoff and flew the injured to care and safety.

The following morning, we found out why the Lily had not inflicted more damage. The Jap pilot had done a remarkable job of finding us in the dark and planting his load on target with complete surprise, but his anti-personnel bombs, with their windmill fusing, flew out of their cannister too near the ground and a lot of them did not arm for explosion. While some of us were digging the unexploded bombs out of the soft earth, others were very busy digging foxholes.

I went back to the photo tent for the first time since it all began. The tent flaps were still secure, the 4 x 8 board firmly in place on the inside. Walking around the tent, I could find no evidence of forced exit. It was as exciting a night as any of my combat missions, but today, 42 years later, I still do not know how in the hell I got out of that photo tent.

DICING WITH DEATH

by Capt. Jack Wassom, Army Air Corp

Prior to May of 1944, photo reconnaissance planes flew at altitudes of 20,000 to 35,000 feet over German-occupied Europe. The high altitudes lent an element of safety from enemy anti-aircraft guns. But with the Normandy invasion imminent, the strategic war planners needed clear, detailed pictures of the positions of enemy beach obstructions and gun emplacements. Many Allied invasion troops' lives depended upon this information.

When pilots from the 31st Reconnaissance Squadron of the 10th Group suggested flying at 10 to 15 feet levels to capture accuracy and details on film, other reconnaissance groups in England stated it would be certain suicide. These beaches were known to be the most heavily defended anti-aircraft areas on the European Front.

One 31st pilot, however, was determined to try it.

Lieutenant Albert Lanker from Petaluma, California, was that pilot. Louie, as everyone knew him, was ruggedly handsome with piercing dark eyes and a nose that must have been broken in some bygone skirmish. His chest, arms, and neck were very well developed. With his sleeves rolled up, as they always seemed to be, his physique would have impressed any college football coach. Louie possessed a combination of daring determination; and a love for his twin-tailed P-38 Lockheed Lightning. He named his P-38 "Outlaw" in reference to the movie of the same name starring Jane Russell. With her famous voluptuous pose painted on the nose of his plane, Outlaw seemed to be fitting title for Louie, Jane, and the Lightning.

To increase speed, P-38s in photo reconnaissance were stripped of their armament. Guns and protective armor plating were usually replaced with one twelve inch continuous exposure camera in the nose and two 24 inch cameras in the armament compartment. By World War II flying standards, Outlaw was a very fast ship.

Louie continually persisted to be assigned to a low-level flying mission. His clearance came on May 6, 1944. He took off from the base at Chalgrove, England, started across the channel at Dungeness, and gradually lowered until he was flying so near the whitecaps that German radar could not pick him up. After hitting the continent at Berksurmer, his assigned point, he firewalled as he turned around a sandbar. With forward camera and both left and right obliques running, the Outlaw headed down the beach. At LeTreport, Louie headed home with about 20 miles of low-level beach pictures on his camera. Photos later showed that sandbar to be a gun emplacement. Louie said his most vivid memory of that mission was how the Germans working on landing obstructions scattered. They were so startled, the only one he saw shooting at him was standing on the cliffs.

It was a joyous day for the 31st, since Lt. Lanker proved lower-level flying missions were possible. "Dicing with death" is what Louie called it. From that day on, all future low-level missions were called dicing. Several pilots of the 10th Group lost their lives dicing, but the beaches were completely and accurately photographed.

Along with Captain Robert Holbury and Lieutenant Rufus Woody of the 31st, Louie flew dicing missions through some of the most critical areas between Cherbourgh and Calais, France. The Royal Air Force originally took credit for a widely published photo reconnaissance picture actually taken by Louie on one of his dicing missions. But he was duly recognized when the Air Corp's General Weyland came to the base to decorate Louie, Woody, and Holbury in honor of the significant roles these men played in the battle for Europe.

Lieutenants Thair Best, Russ Ward, and I, the novices in the squadron, were impressed with these men's records. Louie became a big brother to us. At night, he sat on our bunks in our Nissen hut barracks at Chalgrove, telling of his methods of navigating and evading enemy fighters and flack. He told of flying back from missions at tree-top level. When asked how he found the base, he replied, "You've got to develop your own radar. You'd better get to work on it." His procedures were not always orthodox, but then neither was Louie.

In war time, rules and regulations were relaxed, but Louie bent and stretched them even further. He knew he was not a conformist. Even with his outstanding flying record, he did not expect promotion beyond First Lieutenant. His knowledge prompted him to call himself our Underground Flight Leader.

As the Allies began to break out of the Normandy Beachhead, word came from headquarters that the 31st was soon to move to a landing strip called A-17 near Carentan, France. By the time some of the engineering and ground troops got there, our squadron was ordered to occupy the German airbase at Rennes that had been captured by Patton's 3rd Army.

Instead of Nissen huts, our homes were now GI 6-man tents.

Unsure of where the Germans would be that first night at Rennes, Louie chose a tent he considered positioned in a strategic spot. Looking like a freedom fighter with his 45 revolver strapped across his chest and a razor sharp trench knife on his hip, he made the announcement, "Anything that moves in the tent tonight, I'll deal with in hand-to-hand combat." Louie's build backed up his warning, so no one argued. The fear that pervaded that evening was more of Louie than of the Germans.

With Patton breaking out and starting to roll, 31st pilots were fast logging flying time shooting reconnaissance pictures for the 9th Air Force and the 3rd Army. Louie sustained his love for dicing even when the assignment was "Bomb Damage Assessment from 20,000 feet." On the way back, Louie always diced in to see what the country and people looked like at close range. He told of how he once slipped between two rows of frauleins working in a potato patch. During interrogation, he never explained the small arms fire holes that pierced the top side of his plane.

After the squadron moved to a new landing strip in France, Louie decided to take a look at some property his family had deeded in Bern, Switzerland. It was well known that the Swiss were neutral and entering their country was forbidden.

On a gray October day in 1944, Louie received the assignment to fly a Bomb Damage Assessment mission in the Strasbourg area. He left for this mission using his self-patented Outlaw takeoff. He set the throttle to 45 inches manifold pressure, held it just off the ground until his air speed reached 300 miles per hour, then headed straight up as far as his two Allison engines would pull him. Unlike other pilots who dared this method of take-off, Louie never stalled out. His Underground Flight knew he would fly two sorties that day. Although Bern officials must have seen an evasive Lightning far inside the Swiss border, 31st Intelligence never did thoroughly interrogate Louie about this incident. But Louie's comment, "Boy, I know where I'm going to live after the war!" gave us the only clue we needed. What Louie did not know was that in a few short months another kind of dicing mission would end his dreams.

When the squadron moved to the temporary field A-64, in St. Dizier, France, it was the closest base to enemy lines on the continent. The 10th Group shared it with the 354th Fighter Group, who by that time had an impressive list of flying aces to back up their arguments about who was actually winning the war. Louie envied them and confided that many times he had wished to have "just a couple little old 50 calibers sticking out of Outlaw's nose."

The weather started to get colder that October. Penetrating deep into Germany for the coverage of targets, 31st pilots flew some days when even the ducks were grounded. If a most urgent priority target came in, someone tried to fly the mission. Heavy reliance was placed on the Allied forward radar men whose navigational assistance saved the lives of numerous reconnaissance pilots that winter.

Late in November, the 31st moved to Jarny, France, near the forts at Metz. Patton's forces had bypassed some of the Germans, leaving them still holed up defending their thick-walled concrete fortifications. Louie often mentioned his great admiration for General Patton and what a great crew of tank drivers he had. When Patton's 4th Armored Division pulled back from the Saar River, they remained in our area overnight before heading straight north into heavy fighting. While talking with the men from the 4th, Louie shouted, "Hey, how about taking us on a test drive tomorrow?" The Patton troops answered, "Drive one yourself!" And drive one Louie did! He was as daring on land as in the sky, as he drove the tank through ditches, over fences, jolting over ground swells, and rumbling down the road again. Sticking his head through the porthole, he really looked the part.

By late 1944, the 31st Squadron was flying all types of reconnaissance depending upon the nature and size of the enemy's offensive. Meanwhile Patton's 3rd Army was blasting away on the south flank of the German Army. The 101st Airborne was holding out at Bastogne. The 101st radioed the 31st and requested accurate photography of all territory surrounding them. We complied. Captain Woody shot the first group of pictures. He had to dice in at tree-top level and drop the pictures to the 101st in a modified aircraft fuel tank while the German tanks were firing 88mms at him point-blank.

The 101st radioed for more of the same the next day. Veteran Captain Wolcott flew the coverage. Louie volunteered to dice his way to the 101st area to lay Wolcott's pictures into General McAulife's lap.

But the dice rolled wrong for Louie that day. This time the Germans scored a direct hit on Outlaw. Louie gave his life flying the gambling style he liked best. He always knew the odds were greatest at dicing with death, but when he scored, his feeling of accomplishments were worth it. Because he was brave enough to risk the odds, Louie's wins were many but his life was short. He will never be forgotten by those who rolled high and low through the skies with him.

FLYING UNDER THE STORM

by James H. Costley

What I remember the most was having to fly under those huge storms that originate in the South China seas. We had three options when a storm appeared: (a) fly around it if possible (b) climb over the top (c) fly under it if low on gas as it was the only chance we had to make it back to base.

When flying under a storm all planes tucked in as tight a formation as possible so as not to lose visual contact. The bigger the storm the tougher it was because the further you flew in under the storm, the darker it got and the harder it rained. It was impossible to see ahead, so whoever led the mission flew just above the waves. Instruments were useless and visibility was down to around 50 to 70 feet above the surface of the sea.

If there had been any ships sailing in line of our flight a collision would have occurred and all planes lost because those who avoided collision and pulled up into the storm would also crash because the up and down drafts would have spilled the gyro compasses and those planes would surely have been lost as well.

Any pilot who ever flew such a mission must remember the feeling while under the storm, when it started to grow less dark, the rain let up and suddenly you were back into the bright sunshine again.

I can't speak for the others but I know I always said, "Thank you God for helping us out."

I WAS SHOT DOWN BY A P-38

by Hal Rigney

The P-38 with black crosses was observed by several of the pilots on this particular mission. I was the "tail-end Charlie" of our flight. Since we were so greatly outnumbered and fully engaged with the enemy, it was everyone for himself. It was impossible to maintain any semblance of a formation.

I noticed a P-38 behind me. I assumed that it was one of our aircrafts and was joining up for mutual protection. However, it seemed to lag back for a while failing to close or form up with me. About the same time, as I dismissed him from my thoughts, I was startled by hits on my aircraft. I immediately broke left to see where the shots had originated. There was no aircraft but the P-38. I suspected the firing was accidental. There have been times when pilots in their excitement have hit their gun button instead of the microphone button. Both are located on the wheel.

As I slowed down the lone P-38 shot by me. It was directed back to my own aircraft. As I smelled smoke, I observed a fire under the cowling of my port engine. Unable to extinguish the fire and recognizing the gravity of my situation, I called for escort home, or at least to Sicily. I was informed that none was available as disengagement was almost impossible.

By this time the fire was becoming more intense. I decided to make a run for home. I rolled over, feathered the engine and dove from 17,000 to the deck trying to blow the fire out. Needless to say it didn't work. The fire had spread to the wing. I realized I was in a precarious position, too much fire, too far from home base. I made the dreaded decision-to bail out. I jettisoned the canopy, unfastened my safety belt and pulled the nose up for additional altitude.

With the canopy gone flame and smoke were sucked into the cockpit. Visibility was so restricted that I was unable to determine my altitude. Fearing there was insufficient height and safety I abandoned the aircraft. I shoved the control column forward. The next thing I knew I hit the water. Since I had failed to refasten my safety belt I was abruptly propelled or catapulted out of the aircraft and into the water some 50 miles off the Isle of Capri.

Due to my fire, descent and preoccupation with my emergency I failed to notice but learned later, that 5 ME-109s were using me for target practice. All I can say, "Thank God they were lousy shots."

During my travels behind enemy lines, and after Italy had capitulated, I had the occasion to talk to some Italian fighter pilots. I mentioned to them that I was shot down by what I believed to be a P-38 with German markings. They confirmed that a P-38 in fact was captured and used by one of their pilots on several missions against USAAF fighters and bombers.

HEHO HI

by Ben Beach, 459th FS

On 11 March, 1944, Lord Louis' Montbatten South East Asia Command directed the 459th Fighters Squadron (The Twin Dragons) to sweep the Japanese Air Base at Heho in Central Burma. The Heho area was known to the world as the place where local Burmese priestess kissed the heads of striking cobras.

The Heho base was paralleled by two sets of low lying hills. It was loaded with aircraft and heavily defended. The Dragon's tactics were to be high speed runs executed with total surprise.

Leading eight aircraft, Captain V.D. Leuhring, the Squadron C. O., and pilots observed and fired upon, both aircraft on the ground and attempting to take off. A 180 degree turn was executed by the Twin Dragons and the eight aircraft streaked back through to attack again. On one of the passes, 2nd Lt. Ben Beach, a totally untried pilot with only three missions to his credit, observed a series of very black puffs with fiery scarlet centers pacing at his level just out of range at 10 o'clock. This served to keep him tightly glued to his leader. As Leuhring pulled up off the strip, Beach caged his guns. An action that would never be repeated again.

20th FG Droop Snoot. King's Cliffe, England. (Courtesy of Jack Ilfrey)

459th FS, Dud Kundi, India, 1945. (Courtesy of Nuehring)

The next instant signaled a call "Bogie, 2 o'clock high." The C.O. immediately fire-walled his throttle and began to leave.

This left Beach looking off at what appeared to be an AT-6 with no thought of what he was facing. The oncoming aircraft turned, seeming lazily, directly into him. As the oncoming attacker suddenly became very large, smoke appeared at the sides of the approaching aircraft. The thought exploded, "He's trying to kill me." The knee jerk reaction of Beach was, "If I'm going, he's going with me." It was head on. At the very last moment, Beach uncaged the guns and a short burst was fired. Edging his aircraft up slightly and looking down over the left wing, during a flick of a eyelash, he met the eyes of the Japanese pilot passing 12 to 15 feet below. Looking back as he fire-walled his aircraft, he observed the aircraft spiraling down with smoke.

Back at the base he learned he took one hole into the intercooler. He began to realize he was a survivor.

THE POSSUM SQUADRON'S RISKY RESCUE

by John A. Berry
433rd FS

Satans Angels were instructed to destroy a heavily guarded bridge at Cavite, Luzon. The 433rd Possum Sqdn. sent two flights of four each P-38s. The first flight was led by Lt. Jack Purdy, the second flight by Capt. Charles Howe. They took off each carrying one belly tank and one 500 lb. HE bomb. The orders were to bomb the bridge and strafe the machine gun nests surrounding the bridge area.

Arriving at the bridge, Lt. Purdy's flight lined up in trail, strafed the area and bombed the bridge as they passed over. Capt. Howes flight followed immediately amid a hail of gunfire from the bridge area. All of the bombs were on target.

Lt. John Berry was Tail End Charlie on Capt. Howe's flight. After dropping his bomb, he remained low over the trees checking out the area when he spotted a P-38 smoking heavily with one engine out, approx. 500 feet high and losing altitude. Failing to get any response from the crippled P-38 he called his flight leader and reported the incident. He was ordered to get back in formation which at that time was about ten miles away. Lt. Berry decided to stay with the crippled A/C and assist any way possible. At this time the P38 was about 100 ft above trees looking for a clearing to crash land which he found about a mile further; a small clearing with two rounded 50 foot knolls in the center. With a lot of luck and outstanding skill the pilot stalled the P-38 through the tops of the trees and bellied in on the upside of the nearest knoll, bounced in the air and landed on the downside of the other knoll about 600 feet away.

At this point Berry noticed the A/C no. and recognized it was Jack Purdy. Berry circled tightly and saw Purdy slowly get out of the cockpit, walk out on the wing, pause, go back to the cockpit, get the form I-A and got off the aircraft which had six foot flames coming from the right engine. Suddenly Berry noticed a Japanese Jeep full of soldiers coming into the clearing. He strafed them knocking the jeep over.

Coming around again he saw what looked to be an armored car with about 20 soldiers following heading for the crashed P-38.

He made two passes at them, strafing the car which caught fire and scattered the soldiers. Looking ahead, he noticed a Japanese barracks about a mile from the crash scene. He strafed the barracks and turned back to the crash site; more Japanese were almost to the A/C. He strafed them and ran out of ammo.

Meanwhile, Purdy was walking up a little trail with his parachute over his shoulder as if he were headed for the ready room at the base, completely unaware of the activity around him. About ten Filipinos waving white pieces of cloth came out of a ravine, grabbed Purdy and rushed back into the ravine. Berry, with no ammo made two more passes at the Japanese to hold them off and gave the rescuers time to reach safety. Berry then returned to the base.

Lt. Purdy arrived at the base about 17 days later by the Philippine underground. His head had been badly lacerated upon impact and was severely shaken. He vaguely remembered what happened after the crash. After taking a little time to recuperate, he was rarin' to go on the next mission. There were a lot of close knit pilots in the 433rd Possum SQDN and they always came through.

Sitting: Supp, Monyelle, J. Larson, Bloom, Alford, Hein, Edwards, Corey, Kilgore, Barrett, Daggett, and Berrier. Kneeling: Lt. Ordway, Bower, Garmon, Rees, Leone, R. Larson, Cloer, Callen, Miller, Slifer, Fox, Elwood, Loizeaux, Sampson, Hornbuckle, Anderson, and Bowen. Standing: Capt. Stephens, Lt. Reamy, Capt. Smith, Lt. Thomas, 1st. Lt. Rinn, 2nd Lt. Vickery, Strong, Meier, Modzelski, Estill, Norwich, Geresi, Farnham, Hamilton, Liston, Hogland, Mayhew, Banks, and Maj. Zwanzig.

THE FORGOTTEN BASTARDS OF ICELAND

by Wayne O. Sneddon
50th FS

My first foreign assignment as a Tech Rep for Lockheed was to Iceland to cover the 50th Fighter Squadron and the North Atlantic Ferry Route. Due to the cold the entire tread of the tires flew off (on the landing at Goose Bay). New tires were installed in a heated hangar.

I departed from Presque Isle, Maine on

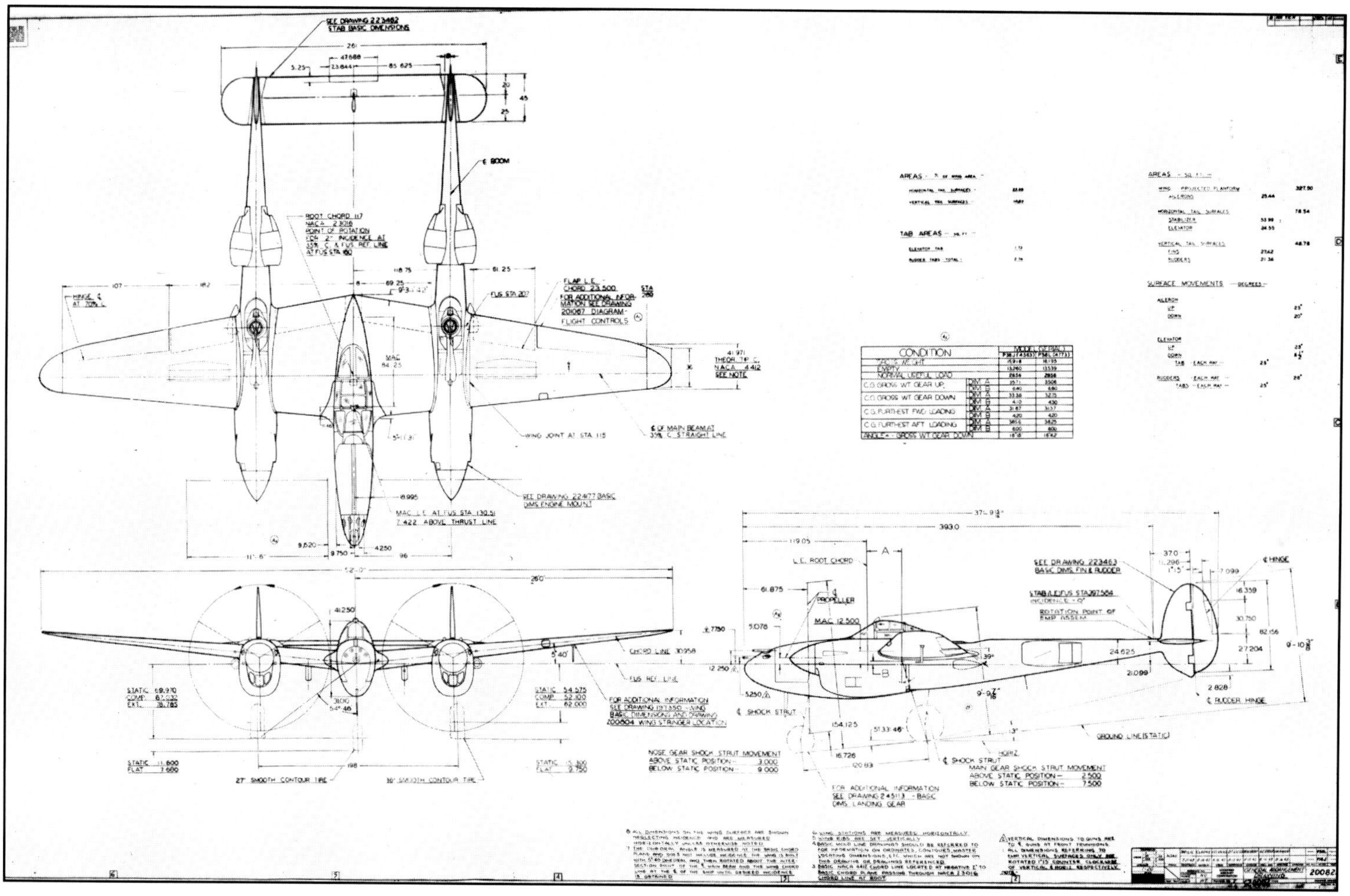

Engineering drawings of P-38 specifications..

November 19, 1942 on a C-87 with the crew that, on the return flight, became the story of the book "Island In The Sky," by Ernest K. Gann. Basically it was a problem of a new radio station in Detroit (if my memory serves me correctly) and it was using the same frequency of the approach ADF of the Goose Bay Airport. The ADF reading of Detroit caused the airplane to pass by Goose and end up on a lake in central northeastern Canada.

The flight across Greenland was about seven hours in a whiteout and the outside temperature was below 50 degrees as the gauge only read to 50 below, and the gauge was pegged. Even with all the clothes we had on we had to keep moving within the length of our oxygen masks and keep the oxygen bag under our clothes to keep it from freezing.

The landing at Iceland was a surprise as it was about 45 degrees Fahrenheit and the grass was green on November 21. The 50th Fighter Squadron was the last squadron of a group of 38s being ferried across to England. The 48th Fighter Squadron was the first to reach Iceland and it was held there until the 49th arrived. The 49th was held there and the 48th went on to England. When the 50th arrived, the 49th then went on to England and the 50th became the Iceland P-38 Squadron as part of the 33rd Group. The 33rd Group had a squadron each of P-38s, P-39s and P-40s.

The first German airplane shot down in the war was shot down in Iceland by a P-39 and a P-38. The first contact was made by the P-39 and then the P-38 finished the German plane off. A second P-38 arrived and ordered the first P-38 off the target but the German airplane blew up before the second P-38 could fire.

From a Tech Reps viewpoint some of the interesting points in Iceland were:

A. Learning to place 55 gallon barrels loaded with rocks, ice etc. and placed in the ice to anchor it and then tying a rope to the nose wheel of the P-38. In the high winds (80-130 mph) the airplane would teeter around the barrel as the wind changed directions. The props would rotate hour after hour in these high winds. Only one airplane took a short cut and hit the barrel in the entire winter I was there. It was not uncommon in the morning to see the airplanes 180 from their original parked position of the night before as they circled around the anchor barrel.

B. We all awakened one morning to find all the airplanes and Quonset huts in one to four feet of ice and water. A volcanic eruption under a glacier had caused the airport to be covered with ice and water. We had orders to get the entire squadron out of this ice in one day. The airplanes were jacked up or pulled out of the ice. The brakes all had to be heated to get the wheels to rotate and then the airplane ferried to solid ice where it took off to another field. One of the planes had to abort on the ice. The pilot used his engine to cause the airplane to rotate on the ice until he was going backwards down the ice. He then used both engines to stop the airplane. A very interesting day getting all the aircraft out of the ice.

C. As a Tech Rep one of my assignments was to help get the B-24 and two B-25s ready for the Bernt Balchen (famous Norwegian pilot) mission to eliminate the German weather station in Northern Greenland. This station was misleading several of the ferry flights by giving false weather reports and heading information. The mission was a complete success.

D. A Tech Reps job often took on an aspect that was not very technical in respect to the aircraft. As an example...very little flying took place during the darkness of winter. As a result a football player who normally had no problem pushing rudder pedals had lost leg strength sitting around during the winter. In the spring he wrote up "stiff rudders" on all the aircraft he flew. After checking one aircraft thoroughly and finding nothing, and noting that every aircraft this person flew had the same complaint the pilot was ordered by the flight medical officer to run around the airport twice a day for two weeks. The pilot was amazed to find that this was the correction for the stiff rudders.

E. One of my interesting experiences of this assignment was the purchase of a typewriter. When I left Burbank, the company had no typewriter available for me. They authorized me to purchase a typewriter if I could find one. After some time in Iceland an army group was being moved to England and put all the Officers Club furnishings up for auction. They had a new Royal typewriter for which I put in a bid of $126, hoping no one else would bid over $125. About three weeks later the 50th Fighter Commanding Officer drove into the revetment where I was working on an airplane and asked me what I had bought that took three 6x6 trucks to deliver.

It turned out that my bid was the only one received, and I got the ENTIRE furnishings of the Officers Club. Needless to say, the 50th Fighter Squadron got a new Officers Club—and I got my typewriter. The list of material acquired is still the topic of conversation in some circles at Burbank. Incidentally, the typewriter ended up in Lockheed's London Office after the war.

F. Three days before Christmas in 1942, 12 Army personnel, including Paul J. Pelteir of the 50th Fighter and I were asked to go out on the south coast of Iceland and check the condition of a P-38 that had floated ashore on the midsection of a boat that had been attacked at sea. Going out was no problem as the weather was good. The investigation showed the airplane to be in good condition. Just as we started to return a violent blizzard hit us and we spent the next three days out in the blizzard with no food. Finally the storm let up sufficiently for us to get directions from an Iceland sheep ranch, and, following a power line we finally got back to a road that could be recognized as such, and we returned to the squadron. Quite a memorable Christmas.

G. Just in case you wonder why the men of the 50th (and some other personnel assigned to Iceland) don't like ground spinach or red beets, there is a simple explanation. It seems that all canned goods in that area turned out to be one of the above regardless of whether the container was marked corn, beans, etc. This problem went on for several months.

H. Another incident (non-technical) that I thought interesting was that about four o'clock one afternoon while I was in town I asked an Icelandic girl for a dinner date that evening. I walked about 3/4 mile back to my quonset hut to get dressed for the event before I could get ready there was a knock on the door—it was an intelligence officer. I was informed that the girl was sympathetic to the German cause. I was briefed on what to say and what not to say. The date took place as scheduled and everything went as planned.

On returning to the quonset hut I did not have time to get ready for bed before the Intelligence Officer was there telling me that the girl had sent a message in code to the Germans and that they had intercepted the message and decoded it—the information I gave was planned. All this took place in less than 45 minutes after I left my date. The behind the door fighting during the war was always intriguing.

I. In the middle of winter when only a couple of hours could be considered daylight we had a plane undershoot the runway caused by misjudgment between the coarse lava rock approach and the fine lava rock of the runway. The accident occurred a little after 3 p.m. and I classed it as a night landing. The airplane was demolished, but the pilot was not hurt even though the nose section broke off and preceeded the airplane down the runway. It took several letters to convince everyone in Burbank and Dayton that a 3 p.m. landing in winter could be classed as a night landing as far north as Ireland.

J. As we approached Greenland on the return flight from Iceland on one of the new DC-4 aircraft a reflection of the western sun on the ice caused a very beautiful cross of red light in the clouds. Unfortunately there was not one camera among

Low level pass over an English base. (Courtesy of Bill Hess)

An F-5-B photo reconnaissance plane. (Courtesy of Bill Hess)

Jay T. Robbins and his P-38.

the numerous personnel on board the aircraft. It was the most eventful sunset I have ever seen. The coast of Greenland and the ice cap with the cross above was certainly something to remember.

K. There are many more incidents that could be related, of course. Some of the aircraft I was called on to work on or advise about were the Hudson bombers, C-30s, C-87s, B-26s, B-25s, and B-24s.

The 50th Fighter Squadron was affectionately known as the F.B.I. Squadron. F.B.I. standing for the Forgotten Bastards of Iceland.

While I was there five German aircraft were shot down by the pilots in the P-39 and P-38 squadrons.

On September 1, 1943 a Major Keeler arrived in Iceland with orders for me to proceed to India-Burma via Dayton and the Lockheed factory in Burbank.

P-38S ESCORT GEN. EISENHOWER

by Archie G. Jackson
Lt. Col., USAF, Retired

On July 29, 1943 four P-38 pilots, two from the First Fighter Gp. and two from the 14th Fighter Gp. were briefed for a July 30 and 31 escort of Gen. Eisenhower to Palermo, Sicily and to Malta.

Take off was 9:00 AM July 30 from La Marsa air field near Tunis, Tunisia. General Ike's plane was a C-53.

The landing at Palermo was without event. The pilots also wanted to go into the city but General Patton's orders were no one in town without a steel helmet, so it was wait in the shade under the P-38s until Ike returned.

The take-off from Palermo was about 3 PM and to escort Ike to Syracuse, Sicily near Gen. Montgomery and then to Malta where Ike was to confer with Gen. Alexander. No enemy aircraft were sighted and the flight on to Malta was routine.

We were put up in a hotel with a real bed and the first decent meal we had had in several months.

The next morning, July 31, we were to escort Ike back to Tunis. Well at 7:00 AM planes pre-flighted and refueled, here came a lone open topped car toward the P-38s. It was Ike in the back seat. He got out and headed for the P-38s. We popped to, and he motioned to rest.

He came on up and said he came out to thank us for the good job, but he wouldn't need us, as he was going on to Cairo, Egypt and would have a British Spitfire escort.

He asked each of us where we were from and of course I said Kansas. He replied "Well I'm from Abiline, and I said, "Yes sir, I know."

Each of us had some conversation with him, and he said he must get back to town for breakfast.

We returned to our home airfields west of Tunis without event.

AN OUTSTANDING FIGHTER-THE P-38

by Robert H. Carey-474th FB Group
Col. USAF Retired

My fascination for the P-38 started the day I saw a picture in a new release in January, 1939 of the XP-38 on the parking ramp at March Field, CA. I was determined at that moment that somehow, some day I would fly that big "bird". That revelation came true when I was sent to Santa Ana Army Air Base in California a month and a half after I graduated from flying school at Williams Field, AZ. Thus began my career with the fabulous P-38. Among hundreds of tremendously exciting and even exhilarating experiences, probably the most memorable occurred on July 17, 1944. My P-38 Group, the 474th FB Group commanded by Col. Clinton C. Wasem, but lead by his vice Lt. Col. Henry B. Darling, was on an Armed Recon. Mission over Northern France just west of the Siene River. The invasion armies were driving the Germans out of the area and our mission was to cut rails, bomb bridges, or strafe any rolling stock we sighted.

We had launched the entire Group of 48 aircraft from our base at Warmwell in southwest England. We had just made land fall in the vicinity of Evreux, France at 10,000 feet, when the clandestine "Y" service operating in occupied France advised a combat controller in England that 77 FW-190s and ME-109s were scrambling to intercept us. I was flying Tail End Charlie in the high squadron and was first to call out this tremendous gaggle of enemy fighters as they swarmed like bees into the attack. In a matter of seconds the most spectacular and deadly melee of some 125 fighters filled the sky. I noted the bounce by the Germans began at exactly 10:00AM. The fight was over at 10:32, when Col. Darling called for a check in and regroup and headed for home base. The sky was now clear of all enemy fighters. The air battle was truly an amazing spectacle. At no time could anyone relax for a second, for everywhere high and low one could see P-38s on the tail of FW-190s or 109s and vice versa. When the battle was over and we tallied the score, the 474th had downed 27 enemy aircraft, damaged 19 and claimed nine probables. The 474th lost three. Fortunately, one of the 430th Fighter Squadron pilots, who went down on the initial bounce survived, evaded and eventually returned to the squadron. He stated that he kept count of the fighters that smashed into the ground all around him. His count essentially confirmed the kill claims of the group. My own efforts resulted in one severely damaged FW -190 and possibly one FW 190 destroyed. The latter buried itself in the ground, as I raced after him. My camera film confirmed the crash, though one of my squadron mates claimed his gun fire sent the FW-190 into its high speed near vertical dive. There was no smoke so it's possible that the pilot was dead.

What a truly outstanding fighter - The P-38!

FLYING ON INSTINCT

by Merle E. Blanchard
338th FS, 55th FG, 8th AF

Besides four single-engine returns from deep penetrations over Europe, I found a special reason to love the P-38. On a particularly hairy mission over Holland in early 1944, I had already been fired on by eight different FW-190s, managed to shoot one down and was racking it around with the ninth in a tight 360 degree climbing left turn. This was before Aileron boost and dive flaps were added and that FW-190 was inching inside of me and began firing. Not enough deflection yet but close enough to set my mind racing on how to extricate myself from this hazardous position. From this tight turn, I shoved the column to the wall simultaneously kicking full bottom rudder and went through an inside snap roll. When I rolled out, there were no Jerries in sight. I doubt that any single-engined A/C could follow a 38 though this maneuver even though I had never done it before or since. But it remained in my bag of tricks and gave me a lot of confidence on future missions.

I GOT ME A TRAIN

by Gene W. Sidwell

Most productive mission happened on the way back from a recon mission to Germany. The squadron was over the Austrian alps when someone said over the radio, "Hey there's a train down there." I was flying "Tail End Charlie" to guard the rear and at that point in time the squadron had already passed the train which was just about to come around a mountain. I could see the smoke from the engine and although I couldn't see the train yet I knew by swinging an arc - to the right and then left - I would line up right on it. The P-38 you could give a few degrees of dive flaps whereby you could fly level but the nose would be pointing down slightly. Located in the nose were four - 50 cal. machine guns and a 20 mm cannon. I would guess I flew down to that train about 100 ft above it with guns blazing from engine to caboose. When I pulled up and looked back it was quite a sight. It must have been an ammunition train as the whole thing had completely blown. There were other missions a whole lot tougher but none more satisfying. That was the day I got me a train.

Capt. Hersey on a test hop in #75. Early 1944. (Courtesy of Gene Hersey)

Plane and pilots of the 8th Photo Squadron, Leyte, 1944. First squadron of P-38s in the South Pacific 1942. (Courtesy of W. Kennedy)

The 347 FG, 1944 Pacific, Middleburg Island off N.E. New Guinea. L to R: 68 FS C.O. Leonard Shapiro, 67 FS C.O. Don Lee, Group OPS Shelby England (KIA), Group C.O. Leo Dusard, Dep. Grp. C.O. Robert "Westy" Westbrook, (KIA), 339FS C.O. Zachary Endress (KIA). (Courtesy of Leo Dusard/ Cefaratt)

P-38 ACES

(Five or More Confirmed Aerial Victories While Flying P-38s)

Name and Rank	Unit (s)	Victories
Major Richard I. Bong	9th Sq., 49th F.G., 5th A.F. (21)	
	Hdqts.-5th Fighter command-5 (19)*	40
Major Thomas B. McGuire, Jr.	431-475-5	38
Colonel Charles H. MacDonald	Hdqts.-475-5	27
Lt. Col. Gerald R. Johnson	9-49-5 (11)	
	Hdqts.-49-5 (11)	22
Major Jay T. Robbins	80-8-5 (21)	
	Hdqts.-8-5-(1)	22
Lt. Col. Thomas J. Lynch	39-35-5 (13)	
	Hdqts.-5th F.C.-5 (4)**	17
Lt. Col. Bill Harris	339-347-13 (14.5)	
	Hdqts.-18-13 (1)	15.5
Major Edward Cragg	80-8-5	15
Major Cyril F. Homer	80-8-5	15
Lt. Col. Robert B. Westbrook	44-18-13 (8)	
	Hdqts.-347-13 (5)	13
1st Lt. Michael Brezas	48-14-15	12
Captain Kenneth G. Ladd	80-8- (10)	
	36-8-5 (2)	12
Captain Daniel T. Roberts, Jr.	80-8-5 (2)	
	432-475-5 (3)	
	433-475-5 (7)	12
1st Lt. William J. Sloan	96-82-12	12
Captain Francis J. Lent	431-475-5	11
Lt. Col. William L. Leverette	37-14-12/15	11
Lt. Col. John S. Loisel	432-475-5 (10)	
	Hdqts.-475-5 (1)	11
Captain Murray J. Shubin	339-347-13	11
Captain Cornelius M. Smith, Jr.	80-8-5	11
1st Lt. Kenneth C. Sparks	39-35-5	11
Captain James A. Watkins	9-49-5 (10)	
	Hdqts.-49-5 (1)	11
Captain Walter F. Duke	459-80-10	10
Major Paul M. Stanch	39-35-5	10
Major Elliot Summer	432-475-5	10
Major Hampton E. Boggs	459-80-10	9
Captain Fredric F. Champlin	431-475-5	9
Captain Perry J. Dahl	432-475-5	9
Captain Grover E. Fanning	9-49-5	9
Captain Joseph M. Forster	432-475-5	9
Captain William K. Giroux	36-8-5	9
Captain Allen E. Hill	80-8-5 (8)	
	36-8-5 (1)	9
Flight Officer Frank D. Hurlbut	96-82-12	9
Lt. Col. Meryl M. Smith	Hdqts.-475-5	9
Captain George S. Welch	80-8-5 (5)	
	Hdqts.-8-5 (4)	9
1st lt. Louis E. Curdes	95-82-12	8
1st Lt. Fernley H. Damstrom	7-49-5	8
Captain Frederick A. Harris	432-475-5	8
1st Lt. Kenneth F. Hart	431-475-5	8
Captain John L. Jones	80-8-5	8
Captain Thomas E. Maloney	27-1-15	8
1st Lt. John G. O'Neill	9-49-5	8
1st Lt. Philip E. Tovrea, Jr.	27-1-15	8
Captain Richard L. West	35-8-5	8
Major Maxwell H. Glenn	459-80-10	7.5
Major Jack M. Ilfrey	94-1-12 (5.5)	
	79-20-8 (2)	7.5
Major Robert W. Aschenbrenner	8-49-5	7
1st Lt. Walter J. Carroll, Jr.	96-82-15	7
1st Lt. Zach W. Dean	432-475-5	7
1st Lt. John S. Dunaway	36-8-5	7
Captain Vincent T. Elliott	431-475-5	7
Captain Jack A. Fisk	433-475-5	7
1st Lt. Lee O. Gregg	71-1-12 (1)	
	449-23-14 (5)	7
1st Lt. Claude R. Kinsey, Jr.	96-82-12	7
Captain Ward A. Kuentzel	96-821-12	7
Major Warren R. Lewis	431-475-5 (2)	
	433-475-5 (5)	7

Name and Rank	Unit(s)	Victories
2nd Lt. Lawrence P. Liebers	96-82-12	7
Captain John E. Purdy	433-475-5	7
Major Herbert E. Ross	48-14-12	7
Captain Meldrum L. Sears	71-1-12	7
1st Lt. Richard E. Smith	39-35-5	7
Major Harley C. Vaughn	96-82-12	7
1st Lt. Edward T. Waters	96-82-12	7
Captain Calvin C. Wire	433-475-5	7
Captain James M. Morris	77-20-8	6.33
Captain Burnell W. Adams	80-8-5	6
1st Lt. Charles E. Adams, Jr.	95-82-15	6
1st Lt. David W. Allen	431-475-5	6
Captain Laurence E. Blumer	393-367-9	6
1st Lt. Richard A. Campbell	37-14-12	6
2nd Lt. Ray Crawford	97-82-12	6
1st Lt. Edward J. Czarnecki	431-475-5	6
2nd Lt. Edwin L. DeGraffenreid	80-8-5	6
Major William C. Drier	8-49-5	6
1st Lt. Hoyt A. Eason	39-35-5	6
Major Charles S. Gallup	39-35-5	6
1st Lt. Billy M. Gresham	432-475-5	6
1st lt. James D. Holloway	95-82-15	6
Captain James C. Ince	80-8-5 (2)	
	432-475-5 (4)	6
Captain Verl E. Jett	431-475-5	6
Captain Donald D. Kienholz	94-1-12/15	6
1st Lt. John H. Lane	39-35-5	6
Captain Paul W. Lucas	432-475-5	6
Captain Henry Meigs II	6th NFS-13 (3)	
	339-347-13 (3)	6
Captain Armour C. Miller	27-1-15	6
Captain Paul C. Murphey, Jr.	80-8-5	6
Captain Joel B. Paris III	7-49-5	6.3
1st Lt. John Pietz, Jr.	431-475-5	6
1st Lt. Horace B. Reeves	431-475-5	6
1st lt. William J. Schildt	95-82-12	6
1st Lt. John C. Smith	433-475-5	6
1st Lt. Virgil H. Smith	48-14-12	6
1st Lt. Thomas H. Walker	339-347-13	6
Major Thomas A. White	97-82-12	6
1st Lt. Charles J. Zubarik	96-82-12	6
1st Lt. Burdett C. Goodrich	459-80-10	5.5
Captain Lindol F. Graham	79-20-8	5.5
1st Lt. Robert H. Adams	80-8-5	5
2nd Lt. Ernest J. Ambort	9-49-5	5
1st Lt. Leslie E. Andersen	96-82-12/15	5
1st Lt. Truman S. Barnes	339-347-13	5
2nd Lt. Aaron L. Bearden	459-80-10	5
1st Lt. Louis Benne	49-14-15	5
Captain Gerald Brown	35-55-8	5
Captain Harry W. Brown	9-49-5 (1)	
	431-475-5 (4)	5
Captain Robert L. Buttke	343-55-8	5
2nd Lt. Nial K. Castle	8-49-5	5
Captain George T. Chandler	339-347-13	5.5
Captain Vivian A. Cloud	432-475-5	5
2nd Lt. Paul R. Cochran	49-14-12 (1)	
	96-82-12 (4)	5
Captain Henry L. Condon II	432-475-5	5
1st Lt. Warren D. Curton	9-49-5	5
Captain Rodney W. Fisher	71-1-12	5
Major Claud E. Ford	95-82-12 (1)	
	97-82-15 (4)	5
Major Warner F. Gardner	95-82-15	
1st Lt. Robert C. Griffith	97-82-15	5
1st Lt. Cheatham W. Gupton	9-49-5	5
2nd Lt. Harry T. Hanna	37-14-12	5
2nd Lt. Herbert B. Hatch, Jr.	71-1-15	5
Captain Cotesworth B. Head, Jr.	44-18-13	5.9
1st Lt. Besby F. Holmes	339-347-13	5
Captain Clarence O. Johnson	96-82-12 (4)	
	436-479-8 (1)5	5

Aces (Continued)

Name and Rank	Unit(s)	Victories
1st Lt. Warren L. Jones	49-14-15	5
Major Wallace R. Jordan	9-49-5 (1)	
	Hdqts.-49-5 (4)	5
Captain Daniel Kennedy	27-1-12	5
Major Charles W. King	39-35-5	5
Captain Marion F. Kirby	431-475-5	5
1st Lt. Lenton F. Kirkland, Jr.	429-474-9	5
1st Lt. Carroll S. Knot	49-14-12	5
Major Thomas G. Lanphier, Jr.	70-18-13	5
2nd Lt. Franklin C. Lathrope	94-1-15	5
1st Lt. Richard J. Lee	94-1-12/15	5
Captain Marlow J. Leikness	49-14-12	5
1st Lt. Virgil W. Lusk	49-14-12	5
1st Lt. Lowell C. Lutton	431-475-5	5
Captain John A. MacKay	27-1-12	5
Captain Keith Mahon	449-51-14	5
1st Lt. Jack C. Mankin	9-49-5 (1)	
	431-475-5 (4)	5
2nd Lt. Milden E. Mathre	7-49-5	5
Captain T.H. McArthur	95-82-12	5
Captain John W. McGuyrt, Jr.	48-14-15	5
Captain Joseph E. Miller, Jr.	48-14-12 (4)	
	429-474-9 (1)	5
1st Lt. Robert C. Milliken	429-474-9	5
Major John W. Mitchell	339-347-13	5
Captain Franklin H. Monk	431-475-5	5
Captain Paul V. Morriss	431-475-5	5
1st Lt. Jennings L. Myers	80-8-5	5
Captain Robin Olds	434-479-8	5
Major Ernest K. Osher	95-82-12	5
Major Joel A. Owens, Jr.	27-1-12 (3)	
	Hdqts.-14-12 (2)	5
1st Lt. Charles B. Ray	80-8-5	5
Captain Newell O. Roberts	94-1-12	5
1st Lt. Gerald L. Rounds	97-82-12	5
Captain Louis Schribert	80-8-5	5
Captain Robert B. Schultz	449-51-14	5
1st Lt. Robert K. Seidman	49—14-15	5
Colonel Oliver B. Taylor	Hdqts.-14-15	5
Captain John A. Tilley	431-475-5	5
Captain Herman W. Visscher	97-82-12	5
Captain Ralph H. Wandrey	9-49-5	5
Captain Sidney W. Weatherford	48-14-12	5
Major Willard J. Webb	459-80-10	5
Captain Darrell G. Welch	27-1-12	5
Captain Arthur E. Wenige	431-475-5	5
2nd Lt. Paul H. Wilkins	37-14-12/15	5
Major Ralph L. Wire	9-49-5 (3)	
	449-51-14 (2)	5
Captain Lee V. Wiseman	71-1-12	5
Captain John L. Wolford	27-1-12	5
Captain Max J. Wright	48-14-15	5
Total P-38 Aces		170

* *Victories on freelance missions and with various 5th Fighter Command squadrons.* * *Victories on freelance missions.*

L to R: Ed "Beacon" Campbell, Bob Connolly, Bill Starke, Bob Mitchell. 44 FS, Aug. 1944. Morati.

P-38 squadrons which produced five or more aces - pilots who scored five or more confirmed aerial victories while assigned to the squadron in question:

431st Sq., 475th F.G. , 5th A.F.	16 aces
80th Sq., 8th F.G., 5th A.F	15 aces
96th Sq., 82nd F.G., 12th/15th A.F.	10 aces
432nd Sq., 475th F.G., 5th A.F.	10 aces
9th Sq., 49th F.G., 5th A.F.	9 aces
39th Sq., 35th F.G., 5th A.F.	9 aces
27th Sq., 1st F.G., 12th/15th A.F.	7 aces
49th Sq., 14th F.G., 12th/15th A.F.	7 aces
95th Sq., 82nd F.G., 12th/15th A.F.	7 aces
339th Sq., 347th F.G., 13th A.F.	7 aces
48th Sq., 14th F.G., 12th/15th A.F.	6 aces
433rd Sq., 475th F.G., 5th A.F.	6 aces
459th Sq., 80th F.G., 10th A.F.	6 aces
94th Sq., 1st F.G., 12th/15th A.F.	5 aces
97th Sq., 82nd F.G., 12th/15th A.F.	5 aces

USAAF fighter squadrons which scored 100 or more confirmed aerial victories with the P-38:

431st Sq., 475th F.G., 5th A.F.	221 victories
9th Sq., 49th F.G., 5th A.F.	215 victories
80th Sq., 8th F.G., 5th A.F.	206 victories
95th Sq., 82nd F.G., 12th/15th A.F.	197 victories
96th Sq., 82nd F.G., 12th/15th A.F.	195 victories
27th Sq., 1st F.G., 12th/15th A.F.	187.5 victories
432nd Sq., 475th F.G., 5th A.F.	167 victories
48th Sq., 14th F.G., 12th/15th A.F.	159 victories
339th Sq., 347th F.G., 13th A.F.	150.5 victories
97th Sq., 82nd F.G., 12th/15th A.F.	147 victories
49th Sq., 14th F.G., 12th/15th A.F.	145 victories
94th Sq., 1st F.G., 12th/15th A.F.	136.67 victories
39th., 35th F.G., 5th A.F.	121 victories
433rd Sq., 475th F.G., 5th A.F.	121 victories
71st Sq., 1st F.G., 12th/15th A.F.	115 victories
37th Sq., 14th F.G., 12th/15th A.F.	105 victories

Major Dick Bong, top American Ace of WW II had 40 victories.

P-38 WWII Combat Units

(97 Squadrons, 37 Groups)
Steve Blake P-38 Association Historian provided this listing

EUROPEAN THEATER OF OPERATIONS

8th Air Force
13th PRS, 7th PRG
14th PRS, 7th PRG
22nd PRS, 7th PRG
27th FS, 1st FG
27th PRS, 7th PRG
38th FS. 55th FG
48th FS, 14th FG
49th FS, 14th FG
55th FS, 20th FG
71st FS, 1st FG
77th FS, 20th FG
79th FS, 20th FG
94th FS, 1st FG
338th FS,55th FG
343rd FS, 55th FG
383rd FS, 364th FG
34th FS, 364th FG
384th FS, 364th FG
434 th FS, 479th FG
435th FS, 479th FG
436th FS, 479th FG
8th Air Force Totals: 21 squadrons (17 fighter, 4 photo recon), 7 groups (6 figher, 1 photo recon)

9th Air Force
30th PRS, 10th PRG and 67th PRG
31st PRS, 10th PRG
33rd PRS, 67th PRG, 10th PRG and 363rd TRG
34th PRS, 10th PRG, Prov RG and 69th TRG
39th PRS, 10th PRG and Prov. RG
392nd FS, 367th FG
393rd FS, 367th FG
394th FS, 367th FG
401st FS, 370th FG
402nd FS, 370th FG
428th FS, 474th FG
429th FS, 474th FG
430th, 474th, FG
485th FS, 370th, FG
9th Air Force totals: 14 squadrons (9 fighter, 5 photo recon), 8 groups (3 fighter and 5 recon)

ICELAND
27th FS, 1st FG
50th FS, 342nd, CG
Iceland Totals: 2 squadrons (both fight er), 2 groups (1 fighter, 1 composite)
European Theater Totals: 36 squadrons(27 fighter, 9 photo recon), 16 groups (9 fighter 6 recon, 1 composite)

PACIFIC THEATER OF OPERATIONS

5th Air Force, Southwest Pacific
6th NFS
7th FS, 49th FG
8th FS, 49th FG
8th PRS, 6th PRG
9th FS, 49th FG
25th PRS, 6th PRG
26th PRS, 6th PRG
35th FS, 8th FG
36th FS. 8th FG
36th PRS, 6th PRG
39th FS, 35th FG
80th FS, 8th FG
418th NFS
431st FS, 475th FG
433rd FS, 475th FG
547th NFS
5th Air Force Totals: 17 squadrons (10 fighter, 3 night fighter, 4 photo recon), 5 groups (4 fighter, 1 photo recon)
7th Air Force, Central Pacific
19th FS, 318th FG
28th PRS
73rd FS, 318th FG
333rd FS, 318th FG
7th Air Force Totals: 4 squadrons (3 fighter, 1 photo recon), 1 group (fighter)
11th Air Force, Alaska
11th FS, 343rd FG
18th FS, 343rd FG
54th FS, 28th CG and 343rd FG
344th FS, 343rd FG
11th Air Force Totals: 4 squadrons (all fighter), 2 groups, (1 fighter, 1 composite)
13th Air Force, South Pacific
6th NFS, attached to 18th FG
12th FS, 18th FG
17th PRS, 4th PRG
38th PRS, 4th PRG
44th FS, 18th FG
67th FS, 347th FG
68th FS, 347th FG
70th FS, 347th FG and 18th FG
339th FS, 347th FG
419th NFS, attached to 18th FG
13th Air Force Totals: 10 squadrons (6 fighter, 2 night fighter, 2 photo recon), 3 groups (2 fighter, 1 photo recon)
Pacific Theater Totals: 34 squadrons (23 fighter, 4 night fighter, 7 photo recon), 11 groups (8 fighter, 2 photo recon, 1 composite)

CHINA/BURMA/INDIA THEATER OF OPERATIONS

10th Air Force, India/Burma
9th PRS, 8th PRG
40th PRS, 8th PRG
58th FS, 33rd FG
59th FS, 33rd FG
60th FS, 33rd FG
459th FS, 80th FG and 33rd FG
10th Air Force Totals: 6 squadrons (4 fighter, 2 photo recon), 3 groups (2 fighter, 1 photo recon)

14th Air Force, China
9th PRS, 8th PRG (detached)
21st PRS
35th PRS
449th FS, 23rd FG and 51st FG
14th Air Force Totals: 4 squadrons (1 fighter, 3 photo recon), 3 groups (2 fighter, 1 photo recon)
China/Burma/India Theater Totals: 9 squadrons (5 fighter, 4 photo recon), 5 groups (4 fighter, 1 photo recon)

MEDITERRANEAN THEATER OF OPERATIONS:

12th Air Force
5th PRS, 3rd PRG
12th PRS, 3rd PRG
15th PRS, 3rd PRG and 5th PRG
16th PRS
23rd PRS, 5th PRG and 3rd PRG
27th FS, 1st FG
32nd PRS, 5th PRG
37th FS, 14th FG
48th FS, 14th FG
49th FS, 14th FG
71st FS, 1st FG
91st FS, 81st FG
92nd FS, 81st FG
93rd FS, 81st FG
94th FS, 1st FG
95th FS, 82nd FG
96th FS, 82nd FG
97th FS, 82nd FG
111th RS
154th RS, 68th OG
345th FS, 350th FG
346th FS, 350th FG
347th FS, 350th FG
12th Air Force Totals: 23 squadrons (15 fighter, 8 recon), 8 groups (5 fighter, 3 recon)
15th Air Force
5th PRS, 5th PRG
15th PRS, 5th PRG
16th PRS
27th FS, 1st FG
32nd PRS, 5th PRG
37th FS, 14th FG
37th PRS, 5th PRG
48th FS, 14th FG
49th FS, 14th FG
71st FS, 1st FG
94th FS, 1st FG
95th FS, 82nd FG
96th FS, 82nd FG
97th FS, 82nd FG
145th T/WRS
15th Air Force Totals: 15 squadrons (9 fighter, 6 recon), 4 groups (3 fighter, 1 recon)

French Air Force
II/33 G de R (attached to 3rd PRG USAAF
French Air Force Total: 1 squadron (recon)
Mediterranean Theater Totals: 25 squadrons (15 fighter, 10 recon) 8 groups (5 fighter, 10 recon), 8 groups (5 fighter, 3 recon)

GUIDE TO ABBREVIATIONS

PRS/PRG - Photo Recon Squadron/Group
FS/FG - Fighter Squadron/Group
TRG - Tactical Recon Group
Prov RG - Provisional Recon Group
CG -Composite Group
NFS - Night Fighter Squadron
OG -Observation Group
T/WRS - Tactical Weather Recon. Squadron
G de R -Groupe de Reconnaissance

P-38 FIRSTS

First P-38 Pilot to shoot down enemy aircraft:
-2nd Lt Elza E. Shahan, 27FS, 1st FG

First Squadron to receive P-38s in 13AF Southwest Pacific (SWPA)
-339FS 347FG (Nov. 13, 1942)

First Group to receive P-38s in 5AF Southwest Pacific
-49FG

First Tactical Squadron to receive P-38:
-27FS 1FG (Selfridge Field)

First Pilot to ferry first P-38 to Tactical Squadron:
-B/Gen Robert S. Israel, USAF ret.

First Ace in 8AF:
-Jack Ilfrey, 1FG

First P-38 Pilot to be rescued by another P-38:
-Col Dick Willsie, USAF, Ret. 96FS 82FG

First Pilot to have a prop ripped off in ocean, recover and return:
-John Cappo, 95FS 82FG

First American airplane in Philippines:
-P-38 5AF

First P-38 Pilot in Philippines:
-Capt Bill Gardner,5AF

First 435FS Pilot to be shot down by flak:
-Cal Butler, 435FS 479FG

First 431FS to get 3 victories:
-Harry Brown, 431FS 475FG

Believe to be the First F-4 Squadron to fly B-25s on recon when their F4 losses were high:
-9PRS 8PRG - By Philip O. Robertson 9PRS 8PRG

First victories by P-38s:
Shortly after noon on August 4, 1942, 2nd Lts. Kenneth W. Ambrose and Stanley A. Long of the 54th Fighter Squadron, 28th Composite Group, each shot down a Kawanishi 97 ("Mavis") flying boat near Atka Island in the Aleutians.

Last victories by P-38s:
On August 14, 1945, the day before the Japanese surrender, five Lightnings of the 35th Fighter Squadron, 8th Fighter Group, 5th Air Force were providing cover for two rescue planes over the Inland Sea, Japan, when six "Frank" fighters attacked them shortly after noon. Five of the Franks were shot down and the other damaged, as follows: 1st Lt. Dwight E. Hollister - one destroyed, Capt. Raymond F. Meyer, Jr. - two destroyed and .5 damaged, Capt. Billy G. Moore. - one destroyed, and 2nd Lt. George I. Stevens, Jr. - one destroyed and .5 damaged. Lt. Duane L. Keiffer was shot down and killed in this action, his being the last P-38 lost in aerial combat.

Highest Number of Missions

Robert W. Wood-9 FS, 49 FG-188
Lt. Gen. Earl C. Hedlund-474 FG-103
Steve Sobak-8 FG-83
Frank E. Birtciel-343 FS,55 FG-77
Warren G. Campbell-94 FS, 1 FG-68

First Military Pilots to Fly the YP-38

First Pursuit Group-1 PG

Col. Oliver G. Cellini (maybe the 17th)
Col. Dale D. Brannon (ret.)
Lt. Col. Byron A. Dobbs (ret.)
Col. Phillip Owen Robertson (maybe the 5th)

Lightning High Hours

Jimmy Mattern (Lockheed)-3000
Vernon Baird (Fairchild Aerial)-3000
Verne M. Yahne-95 FS, 82 FG, 15 AF, MTO-2000
Cecil H. Rigsby-8 PRS, 6 PRG, 5 AF, SPWA-1100
John D. Pace-1 FG-1000+
Robert E. Lacy-384 FS, 364 FG, 8 AF, ETO-700

Prisoners of War

Lt. Gen. Earl C. Hedlund (USAF-ret) 474 FG, 9 AF, ETO; Lt. Col. Clair E. Buskirk (USAF-ret) 55 FG, 8 AF, ETO; Frank Mullinax, 49FS, 14 FG, 8 AF, ETO; Cal Butler, 435 FS, 479 FG, 8 AF, ETO; Hal Rigney, 94 FS, 1 FG; Evan Mecham, 7 PRG; Richard Spellerberg, 7 PRG (at the time was with the 326 BS, 92 BG); Clark W. Smith, 48 FS, 14 FG.

Served in Two Theaters

Lt. Gen. Earl C. Hedlund (USAF-ret) 54 FS
F, 28th CG and 343rd FG, 11 AF, Alaska
CO, 428 FS, 474 FS, Glendale, CA and 9 AF ETO
CO 474 FG, 9 AAF, ETO

Brig. Gen. C. M. Isaacson
8 FS, 49 FG, 5 AF SWPA
96 FS, 82 FG, 15 AF, MTO

Col. Philip O. Robertson (USAF-ret)
9 PRS, 8 PRG, 10 AF, CBI
Photo Recon., ETO

Col. H.C. Vaughn
CBI
ETO

Edwin R. Weaver
N. Africa
431 FS, 475 FG, 5 AF SWPA

W.E. Winans
48 FS, 14 FG, 15 AF, N. Africa
449 FS, 23 FG, and 51 FG, 14 AF, CBI

Ross Winks
48 FS, 14 FG, 15 AF, N. Africa
449 FS, 23 FG, and 51 FG, 14 AF, CBI

Tom Lanphier
ETO
339 FS, 327 FG, 13 AF, SWPA

NOTES ON VICTORIES

Compiling lists such as this is an inexact science, at best. The modern researcher is forced to rely on the original records kept/compiled by the units in question forty-plus years ago. Many of those records are, unfortunately, incomplete, inaccurate, vague and/or confusing. Due to the overwhelming mass of material/records involved, the researcher is also subject to making errors of his own. To date two massive and detailed studies have been done of confirmed (?) aerial victories by USAAF fighter units in WWII. The first was the USAF's Historical Study #85, published in 1978. A tremendous research aid at the time, Study 85 has nevertheless been proven to contain many errors. During the past few years computer scientist/aviation historian Dr. Frank J. Olynyk has compiled and published a series of Victory Lists, including one for each USAAF theater of operations - Pacific, CBI, European and Mediterranean. Frank has consulted all pertinent and available records, mostly obtained from the Air Force on microfilm: squadron and group diaries/histories, mission reports, victory credits boards findings, awards citations, etc. While much more detailed and accurate than study, it also contains errors and is thus not 100% so. Due to the above reasons, no such lists really can be, in all probability.

Those historians who have studied and written about one particular group or squadron (at a time, at least) are usually able to compile the most accurate list of that unit's victories. A case in point is the many hours of work I have spent compiling such a list for the 82nd Fighter Group in the Mediterranean, of which John Stanaway and I are preparing a detailed history. I strongly feel that my list of 82nd FG victories is the most complete and accurate that is humanly possible, based on the available records/sources. Thus, I have used my figures for the 82nd on the enclosed list. For all the other units, I have used Frank Olynky's.

As an example, I would like to refer to the scores of the top two P-38 groups in WWII- the 82nd FG, MTO, and the 475th FG in the Pacific. Study 85 credits the 82nd with 548 victories - 7 for Group Hdqts., 199 for the 95th FS, 196 for the 96th FS and 146 for the 97th FS. Frank Olynyk's total for the 82nd is 553 - for Hdqts., 198 for the 95th, 201 for the 96th and 149 for the 97th. My total comes to 547 - 9 for Hdqts., 196 for the 95th, 195 for the 96th and 147 for the 97th. The group actually claimed 554 victories at the end of the war.

Study 85 credits the 475th with 547 victories - 43 for Hdqts., 217 for the 431st FS, 167 for the 432nd FS and 120 for the 433rd FS. Frank's total is 552-43 for Hdqts., 221 for the 431st, 167 for the 432nd and 121 for the 433rd. In his excellent book FIFTH AIR FORCE STUDY, historian Kenn Rust credits the 475th with 545 victories, including 225 for the 431st Sq.

One area of confusion lies in the crediting of victories for the various group Hdqts. pilots. Technically, these victories should be/have been credited to the Hdqts. if the pilot was assigned thereto (Group C.O., Deputy C.O., Operations Officer). The records, however, can be confusing and some of the researchers/compilers/ record keepers have evidently been ambivalent, resulting in some of these victories being credited to the squadron with which the pilot was actually flying when the victory was scored. I personally try to keep a record of the squadron the Hdqts. pilot was flying with when he scored his victory, while actually crediting the victory to Hdqts. On a large 82nd FG victory board during the war, the victories were listed by squadron only. Thus if, for example, the Group C.O. scored a kill while leading the 96th Sq., this victory appeared on the 96th Sq. listing.

Thus, it must be kept in mind that this unit victory list reflects the unit to which the pilot was actually assigned, not necessarily the unit with which he was flying when the victory was scored. For example, in late 1942 several pilots of the 9th FS were sent to the 39th FS in New Guinea to gain combat experience in the P-38 before the 9th Sq. traded in its P-40s for Lightnings of its own. They scored a number of victories while flying with the 39th (including Dick Bong's first five), but these are credited to the 9th Sq. Bong later scored many of his victories while assigned as an operations officer to Hdqts., 5th Fighter Command, usually while "tagging along" with various 5th AF squadrons. Some of these squadrons added these victories to their own score (unofficially), thus creating some confusion later. A good example of this is the 80th FS ("Headhunters"), which is listed with 206 victories in P-38s. Twelve other victories were scored by 8th FG and 5th FC Hdqts. pilots while flying with the 80th, including three by Bong. Thus, veterans of the 80th Sq. might well consider their squadron's score to be 225 (including seven victories scored early in the war with P-39's).

It should also be remembered that most of the P-38 units listed also operated - and scored victories with - other aircraft types during the course of the war. The only units which operated P-38s exclusively were the 475th FG, 5th AF, in the Pacific; the 54th FS, 11th AF, in Alaska; the 459th FS, 10th AF, in Burma/India; the 449th FS, 14th AF, in China; the 474th FG, 9th AF, in Europe; and the 1st, 14th and 82nd FG's in the MTO (12th/15th AF).

On the list, the totals for the 1st, 14th and 82nd Groups are broken down by the air forces they served with - the 12th and 15th (the 27th FS also scored a half victory while based briefly in Iceland). I have given the 82nd FG totals earlier; those for the 1st and 14th are as follows:

1st FG - 440.5 (1.33 Hdqts., 187.5 27th FS, 115 71st and 136.67 94th FS)

14th FG - 426 (17 Hdqts., 105 37th FS, 159 48th FS and 145 49th FS)

Flying over Mines Field, CA 1942 (Courtesy of Robert E. Kirtley)

ENEMY AIRCRAFT DESTRUCTION CLAIMS

PACIFIC THEATER

5th Air Force, Southwest Pacific:	
Hdqts., 5th Fighter Command	26
8th Fighter Group	309
Hdqts.	13
35th F.S.	36
36th F.S.	54
80th F.S.	206
39th F.S., 35th F.G.	12
49th Fighter Group	353
Hdqts.	19
7th F.S.	65
8th F.S.	54
9th F.S.	215
418th Night Fighter Squadron	2
475th Fighter Group	552
Hdqts.	43
431st F.S.	221
432nd F.S.	167
433rd F.S.	121
547th Night Fighter Squadron	1
5th Air Force Total:	1,364
13th Air Force, Southwest Pacific:	
Hdqts., 13th Fighter Command	2
18th Fighter Group	120
Hdqts.	6.5
6th N.F.S.	7
12th F.S.	30.5
44th F.S.	44.5
70th F.S.	24.5
347th Fighter Group	179
Hdqts.	11.5
67th F.S.	7
68th F.S.	9
70th F.S.	1
339th F.S.	150.5
419th Night Fighter Squadron	1
13th Air Force Total:	302
Southwest Pacific Theater Total:	1,666
7th Air Force, Central Pacific:	
318th Fighter Group	22.33
Hdqts.	2
19th F.S.	7
73rd F.S.	4.33
333rd F.S.	9
7th Air Force/Central Pacific Total:	22.33
11th Air Force, Alaska:	
Hdqts., 343rd F.G.	1
54th F.S., 28th Composite Group	2
54th F.S., 343rd F.G.	17
11th Air Force/Alaska Total:	20

Total Pacific Theater P-38 Claims:	1,708.33

CHINA-BURMA-INDIA THEATER

10th Air Force, India/Burma:	
459th F.S., 80th F.G.	82
58th F.S., 33rd F.G.	3
10th Air Force Total:	85
14th Air Force, China:	
449th F.S., 23rd F.G.	8
449th F.S., 51st F.G.	63.75
14th Air Force Total:	71.75
TOTAL P-38 CLAIMS AGAINST THE JAPANESE:	1865.08

EUROPEAN THEATER OF OPERATIONS

8th Air Force:	
20th Fighter Group	89.5
Hdqts.	5.83
55th F.S.	217.5
77th F.S.	28.17
79th F.S.	28
55th Fighter Group	102
Hdqts.	1
38th F.S.	40
338th F.S.	38
343rd F.S.	23
364th Fighter Group	37
Hdqts.	1
383rd F.S.	12.5
384th F.S.	15.5
385th F.S.	8
479th Fighter Group	52
434th F.S.	35
435th F.S.	2
436th F.S.	15
Unidentified P-38 units	1.5
8th Air Force Total:	282
9th Air Force:	
367th Fighter Group	77
392nd F.S.	37.5
393rd F.S.	17.5
394th F.S.	22
370th Fighter Group	41
Hdqts.	2
401st F.S.	16
402nd F.S.	15
485th F.S.	8
474th Fighter Group	96
Hdqts.	5.5
428th F.S.	26.67
429th F.S.	33
430th F.S.	30.83
9th Air Force Total:	214
Iceland:	
27th F.S., 1 st F.G.	.5
50th F.S., 342nd Composite Group	2
Iceland Total:	2.5

MEDITERRANEAN THEATER OF OPERATIONS

12th Air Force	
1st Fighter Group	222
Hdqts.	.33
27th F.S.	77
71st F.S.	72
94th F.S.	72.67
14th Fighter Group	198
Hdqts.	6
37th F.S.	56
48th F.S.	69
49th F.S.	67
93rd F.S., 81st F.G.	1
82nd Fighter Group	338
Hdqts.	4
95th F.S.	115
96th F.S.	132
97th F.S.	87
350th Fighter Group	10
346th F.S.	4
347th F.S.	6
12th Air Force Total:	769
15th Air Force:	
1st Figther Group	218
Hdqts.	1
27th F.S.	110
71st F.S.	43
94th F.S.	64
14th Fighter Group	228
Hdqts.	11
37th F.S.	49
48th F.S.	90
49th F.S.	78
82nd Fighter Group	209
Hdqts.	5
95th F.S.	81
96th F.S.	63
97th F.S.	60
154th T.R.S.	1
15th Air Force Total:	656

Standing in front of Miss-V, personal mount of Burma ace, Walter Duke, left to right: Lts. William Lyon, William Baumeister, Joseph Smith, William Behrns, Oscar Garland, unknown, William Pauley, Burdette Goodrich, unknown, and Capt. Walter Duke. 459th FS, 10th AF, CBI Chittagong, India-Burma border. 1944. (Courtesy of Len Boyd)

94th (Hat-in-the-Ring) Ftr. Grp. with the Famous Polish 303. Ground Level L to R: Earl Hille, Robert Williams, George Sutchiffe, B. Baran, Glen Hubbard, John Zumbach, J. Palusinski, Zygmunt Wodecki, James Harman (KIA). First Step: Donald Starbuck, Jack Kille, Robt. B. Wilson, Zarebski, Robert Neale, Clifford Malzahn, Newell Roberts, Clarence Rimke, J. Ostrowski. Second Step L to R: A. Wojda, Everett Humphrey, L. Majewski, Powierza?, Wesley Pringle, M. Szelestowski, Norman Widen, S. Socha, M. Kolubinski, Richard McWherter, M. Lipinski, Sidney Pennington, ? Jennings, Jack Ilfrey. Back Row: Lewis Murdock, B. Drobinski. (Courtesy of Jack Ilfrey)

P-38 Biographies

Thanksgiving Dinner-November 1944. 401st FS, Charleron, Belgium. Standing left to right: Burwell, Brown, Heppler, Collier, Morey, Ford, Roberts, Greenwood, Coffin, Burgess, Frank, Nelson, Newmann, Murany, Cilli, Anderson, Doyle, Everett, Croxton, Hoffpauir, Morton, Johnston, Paladino. Lloyd, Noon, Parry, Croddock, Greenley, Ham. Seated left to right: Kuhn, Harrison, Barrett, Lagier, Osborne, Kruger, Thompson, Powers, Christman, Berg, Petersen, Crooks, Bayle, Randel, Hammerich, Caldwel, Oberstebrink, Kilcrease, Owens. (Courtesy of Wm. J. Noon)

LAWRENCE ABEL, was born at Hickory Withe, TN, way out in the country, into a family of five boys and two girls. He joined the Air Corps Nov. 11, 1940 at Memphis, TN; assigned to McDill Field, FL for basic training. No runways there, only a few buildings and lots of tents. Later sent to Jefferson Barracks, MO on a troop train to await entrance to Chanute Field, Rantoul, IL. A lot of K.P., tent guard, wood and coal pile duty at old J.B

Graduated from Mechanic School at Chanute Field in September 1941. He found a new home at Portland Air Base working on P-43 Fighters. At last, after war broke out, formed a P-38 unit at Paris Field, WA. From there on to Elmendorf Field, AK and on down the Aleutian Chain to Cold Bay, Umnak, Adak Amchitka and finding residence at Shunya to finish his stay in the Islands. He returned to States March 1944 and on to Miami, FL for R&R. Then to Lincoln, NE for reassignment; from there to Peterson Field, CO on a P-40 assignment as a section chief in the hangar.

He bided some time on P-63s after this at Bruning, NE, Dyersburg, TN and Pyote, TX. Discharged with accumulated points October 1945 at Maxwell AFB, AL.

Tired of civilian life, bored or otherwise, he re-entered the Air Force at Memphis in August 1950. Had a tour there on C-46s, dropping troops and cargo around the country. Now getting his assignment to Japan and eventually Korea in 1951 till March 1953 and back to the States. He was a Crew Chief on F-84s through the far east assignment.

Stateside assignment was to Perrin AFB, TX to Dock Chief on T-33s. Several years later had his assignment to RAF station Bentwaters, England. He took his family with him and resided three years, working on T-33 Dock again. F-101s were the basic primary operation.

Back to the States January 1964 to be assigned to Myrtle Beach, AFB on F-100s making M/Sgt. there and being assigned Aircraft Chief Insp. (A.C.I.) on F-100 Docks.

Now time for the old boy to hang it up, October 1966. In the ensuing years worked on various aircraft mod programs ending a stint here with Lockheed A/C, GA. Retired later from the State of Georgia in the maintenance field.

He has a wife named Joyce, five children and 13-1/2 grandchildren. He and family have traveled all over the States and overseas before and after military duty.

All told, he has military experience in a lot of aircraft, but for its day and time he'll take the old P-38. It really got the job done though there were many hours spent out in bitter cold keeping them flying in a kind of war that had to be won and everyone joined in with no thought of backing up or losing.

ALES R. ABNA, JR., born Dec. 29, 1922, East Point, GA. Graduated Russell High School, East Point, GA. After discharge attended University of Georgia.

Enlisted: Army Air Corp, 15th Air Force Dec. 14, 1941. Discharged Sept. 10, 1945, Rank of Sergeant.

Training: Keesler Field, MS, Long Beach, CA 97th Fighter Sdn., 82nd Fighter Gp..

Stationed: Ireland, North Africa, Sicily, Italy,

After discharge employed as Club Manager, American Legion Post 51; Atlanta General Depot, Radio Repair. Began service with City of East Point Fire Department in 1951. Retired in 1977, with rank of Captain after 26 years of service. Owned and operated a General Merchandise Business for nine years.

Married Katheryn R. Bradley (Army Brat) on Feb. 10, 1949.

Children: Gary C. Abna, born September 1950, Leone Abna Grizzard, born October 1952, Ales R. Abna, III, born July 1956.

Helped raise, loved and guided two nephews: Michael B. Whaley, born 1951, Patrick H. Whaley, born 1953.

Grandchildren: G. Roger Grizzard, born 1974, Christi Grizzard, born 1976, G. Cleve Abna Jr., born 1979.

Attended reunion in Denver, CO, in 1984, and talked nonstop for three glorious days.

Died June 10, 1986, at age 63.

JOHN W. ACKLEY, Captain, AFUS, Retired, was born on May 8, 1922 in Athens, OH. He was employed with Glen L. Martin Company in Baltimore when he joined the Army Air Corps Aviation Cadets in May 1942. Cadet training bases included: Santa Ana, Ryan Field-Hemet, Minter Field-Bakersfield, and Williams Field where he was commissioned in October, 1943 Class of 43-J. He flew P-39s and P-38s at Concord and Santa Ana then to Europe in March, 1944. He flew 63 combat missions with the 430th Sq. 474th Fighter Gp., from fields at: Warmwell, England; Site A-11 in Normandy, Site A-43 at St. Marceau, Site A-72 at Mons en Chaussee, in France; and at Site A-78 Florennes, Belgium. He shot down two ME-109s, damaged another and shared a FW-190 destroyed.

Awards included the DFC and Air Medal with 12 clusters. After separation from the service he studied Agricultural Engineering at Oklahoma A&M and Ohio State University, and was employed 33 years with Deere & Company, Moline, IL before retiring in November, 1984.

CHARLES E. ADAMS JR., was born in New Orleans, LA on Sept. 22, 1919.

He joined the Army in 1942 and transferred to the Air Corps at Santa Ana, CA. After Primary flight school at Blythe, CA, Basic at Lemoore, CA and Twin Engine Advance at Williams Field, AZ he graduated in the class of 43-G. He flew P-38s at Muroc, CA and at Lomita Flight Strip, Lomita, CA prior to a 21 day troopship ride to North Africa. After a short stay in the Fighter Training Command at Constantine, Algeria he was transferred to the 82nd Fighter Gp., 95th Fighter Sdn. at Foggia, Italy in January 1944. During his tour he was temporarily assigned to the Royal Air Force Gunnery School at Ballah, Egypt flying Spitfires and Hurricanes. Upon return to Foggia he flew bomber escort, strafing, dive-bombing, and fighter sweeps to complete 50 missions. He was credited with aerial destruction of six enemy aircraft and joined the ranks of the fighter aces. He received the Distinguished Flying Cross and the Air Medal with 12 clusters. His group was awarded three Presidential Unit Citations. Returning to the States he toured with a Bond Drive group before serving as basic instrument and flight instructor at Minter Field, CA.

After discharge from the Service in August 1945 he completed studies and graduated as a Petroleum Refining Engineer at the Colorado School of Mines in May 1948. He joined Chevron Corp. and after 35 years of service in Colorado, Louisiana, Australia, and Texas, he and his wife Dorothy, retired to Perry Park, CO.

LEWISE COLEMAN ADIE, was born in Billings, MT, but was raised in Oakland, CA when her family moved there in 1925.

She attended the University of California and after taking the C.P.T. course at Mills College and the Oakland Airport, she decided to pursue a career in aviation. After obtaining employment, she continued flying and working toward an instructor's rating.

After Pearl Harbor, it was necessary to move to Reno, NV to continue private flying and it was from there that she was accepted in the second class of the Army Air Forces Flying Training Detachment in Houston, TX.

After graduating in May 1943, she was ordered to the Ferrying Div. of A.T.C. at Long Beach, CA where except for temporary duty at Pursuit School in Palm Springs, CA, she remained until deactivation in December, 1944. Though she was checked out and qualified to fly 16 different types of aircraft, most of her deliveries were pursuit class, and of these, in her opinion, the P-38 was the absolute ultimate.

JOHN C. AIKEN, Lt. Aiken, 97th Sdn., 82nd Fighter Gp. 15th Air Force, first flew in the College Pilot Training program (C.P.T) from 1939-1940. He joined the Air Corps and graduated with Class 43-D, Craig Field, Selma, AL.

In 1948 John joined United Airlines, retiring in 1978. The last ten years as Flight Manager or Check Pilot.

MARTIN ALGER, (Pete), was born in Ft. Dodge, IA. He graduated from Ft. Dodge Junior College and learned to fly as a member of the first CPT class held by the college in the summer of 1940 and received his wings at Stockton Field, CA January 1942. His first assignment was flying security patrol in P-40s at Wheeler Field, HI until October 1942. He then joined the 49th Fighter Gp., 9th Sdn. in Australia flying P-40s until the Squadron changed to P-38s in New Guinea.

He was awarded the Silver Star for gallantry while leading a flight of P-38s during a daylight attack by 45 Japanese bombers escorted by about 60 Zeros over Port Moresby. After shooting down one bomber he noted five Zeros attacking his comrades. He attacked breaking up the formation but ran into enemy fire that riddled his lightning with holes. With one engine dead he dove about 17,000 feet at more than 400 miles per hour managing to land safely. His total combat score was four Japanese planes and two probables.

After 23 years in the Air Force he retired as a Lieutenant Colonel in Tulare, CA where he lived until his death Jan. 4, 1988.

JOHN ARTHUR ALLEN, was born June 12, 1920 in Toledo, OH. The son of an industrial engineer, he attended about 20 different schools prior to graduation from Washington Gardner High in Albion, MI in 1939.

John enlisted in the Army Air Corps on October 10, 1939. His first duty station was Selfridge Field, MI. His first assignment was with the Radio Section of the 17th Pursuit Sdn. of the First Pursuit Gp. He attended Air Mechanic Schools at Scott Field and Chanute Field in 1940. Upon returning to Selfridge Field, he was assigned to the 71st Pursuit Sdn., as the 17th Pursuit Sdn. had been shipped out to the Philippines in his absence.

Prior to maneuvers in 1941, John was married to Ann Burton from Albion, MI. At the close of maneuvers, he was on leave when Pearl Harbor was attacked. His first duty station, after war was declared, was the Naval Air Station at North Island, CA. The First Pursuit Gp. was assigned a few Lockheed 322s for training purposes. The British had rejected the 322s because of poor performance. The 322 did not have turbochargers.

After moving to Grand Central Air Terminal in Glendale, CA in January of 1942, he was assigned an AT-9 to crew and later a P-38. He attended training classes on the P-38 at the Lockheed Aircraft Company.

About mid-1942 the First Pursuit Gp. was "split" to organize a new group. Much of this reorganization was done at Muroc, CA. John was assigned to the 96th Fighter Sdn. of the newly formed 82nd Fighter Gp. The First Fighter Gp. was shipped overseas a short time later.

John made Tech Sergeant prior to being shipped overseas with the 82nd Fighter Gp. He was on the Queen Mary when she cut a British Corvett in half, with a loss of two or three hundred lives. During the tour in North Ireland, John crewed for Col. Loe. From Oran, North Africa to Foggia, Italy John was crewing one P-38 or another. One of his planes, "Sweetness," was considered for a trip home for a Bond Tour because of its record. Toward the end of the war in Europe, he spent time as Flight Chief and Squadron Inspector. Besides the medals John received with the 82nd Fighter Gp., he was also awarded the Bronze Star.

Upon returning home in June of 1945, he spent some time working with his father building houses and starting a family, three daughters and one son. He started working for U.S. Air Force Procurement in the early 1950s. He was a Quality Assurance Rep at a number of suppliers of aircraft parts and support equipment. For a number of years he had the responsibility for the review and validation of T.O.s and Engineering Data procured from contractors in the Grand Rapids area.

John retired from the Department of Defense in October of 1977. Since that time, he has worked as an inspector and consultant for Intertek Corporation, a manpower service. John is also a member of the Experimental Aircraft Association and attends the annual convention when possible.

FRANK AMAYA, was born in Stockton, CA, on Sept. 11, 1923. He applied for Aviation Cadet Training and entered active duty on February 1943.

During his aviation cadet training he flew Stearmans, BT-13s, and AT-6s at Thunderbird Field #1, Marana Air Base and Williams Field, AZ, respectively; graduating with Class 44-F in June 1944.

After graduation, Frank flew P-39s during transition training at Victorville Air Base, Victorville, CA.

He was later assigned to the 4th Air Force and took P-38 training with the 434th AAF Base Unit at Santa Rosa Air Base, Santa Rosa, CA. Frank continued flying P-38s at Paine Field, WA until honorably discharged in November 1945.

Frank attended the University of Southern California and since then has followed a career as an Aerospace Engineer. He is presently with the Northrop Corporation and lives in Hawthorne, CA. He has two sons, Frank and Ronald, and a daughter, Sharon.

Frank is a proud charter member of the National P-38 Association.

WAYNE D. ANDERSON, was born in Manila, IA, Oct. 25, 1919. Attended grade school there, then moved to Harlan, IA, and finished high school. Attended the University of Iowa. Enlisted in the Air Corp at Omaha, NE the last of February 1942. Was sent to Santa Ana, CA in April of 1942 for pre-flight, Dos Palos, CA for primary flight, secondary flight training at Lemoore, CA, finishing at Marana, AZ. Advanced at Williams Field, AZ, and graduated class of 43D. Muroc, CA for tac training in P-38s. First overseas base was in New Caledonia and joined the 339th Fighter Sdn. Then to Guadalcanal and assigned to 13th AF 18th Fighter Gp. and 12th Fighter Sdn. Operated through the Solomans, New Guinea and Philippines campaign accumulating 131 missions and shooting down some enemy planes. Discharged in 1945, stayed in the Reserves and retired in 1980 as Lieutenant Colonel. Married in Australia 1945. Lived in California and raised three boys and one daughter. Presently is the owner and president of Southern California Machinery Co. in Irwindale, CA.

LLOYD M. ANDES, Lt. Col., enlisted in the Air Force Feb. 8, 1941 at Hamilton Field, CA. After basic training he was sent to Air Mechanic School, then reassigned to 54th Pursuit Gp., Paine Field, WA. He volunteered for a fighter team formed for duty in Java. The carrier with the aircraft aboard was sunk and the teams were all reassigned in the Southwest Pacific. Lloyd was assigned to the 70th Sdn., Fiji Island. Pilots from this squadron flying P-38s were credited with shooting down Admiral Yamamoto with his Betty aircraft in 1943. After serving 14 months in the Pacific, he was returned to the ZI for pilot training. He was in the graduating class of 44-B at William Field as a P-322 pilot. After R.T.U. training in P-39, P-63 and P-38s he was assigned 367 Gp. 394 Sdn. 9th Air Force.

After tours in P-38 and P-47 and the end of the European Conflict, he was discharged in 1946 as a Captain.

He was recalled during the Korean War and on September 8, 1950 was assigned to Mather AFB, CA as operations training officer flying B-25s and T-29s. Four years later he was assigned to MATS as operations officer in C-124 Sdn. His next assignment was with the Air Force Academy as flying training officer. Other assignments were Vietnam with 2nd Air Division as plans officer, Kirkland AFB as Chief Stan Eval and retired at Sandia Base, NM in January 1970 with 9000 hours flying time.

Decorations include; Silver Star, Distinguished Flying Cross, Purple Heart and Air Medal with ten clusters.

BILL ARMSTRONG, was born in Raleigh, NC and raised primarily in Kansas City, MO. He joined the AAF in 1942 and entered active duty in 1943. Becoming an Aviation Cadet at Santa Ana, CA, he soloed at King City, CA. He was selected for an accelerated pilot class with AT-6s for basic at Marana, AZ and then P-38s in advanced at Williams Field in Chandler, AZ. Upon graduation his assignment was Victorville, CA and Moses Lake, WA for P-39 transition moving to Ontario, Canada for final P-38 training.

He flew 28 combat missions with the 49th Sdn., 14th Fighter Gp. in Italy. They included dive and skip bombing, chaff drops, fighter sweeps, armed reconnaissance, photo recce, supply drop and bomber escort. He received 11 decorations.

He received a B.S. in Aeronautical Engineering in 1948. He held various engineering and management positions with McDonnell Aircraft, LTV and Martin-Marietta. He established American Business Advisors in 1967 and is currently President of that consulting firm in Maitland, FL.

His wife is Mabel Bernard, of Kansas City, MO and they have a son Dr. Steve Armstrong of Jacksonville State University, Jacksonville, AL and daughter Nancy Ilgenfritz of Marietta, GA. There are four grandchildren.

FREDRIC ARNOLD, see (Kohn) pg. 102

ROBERT W. ASCHENBRENER, was born in Fifield, WI in Nov. 22, 1920, not far from Poplar where America's ace of aces Dick Bong was raised. Aschenbrener attended college in Iowa from 1939 to 1941, enlisting in the Army Air Force before Pearl Harbor. He was commissioned a second lieutenant upon completion of flight training in September 1942.

Assigned to the 49th Fighter Gp.'s 8th Sdn., 1/Lt. Aschenbrener began flying combat in the P-40 at New Guinea in 1943. His first two victories were scored on November 15, followed by a single on Feb. 15, 1944. Promoted to captain, Aschenbrener became the 8th FS operations officer and experienced his most intense period of combat during the Philippine campaign of late 1944.

By then all three squadrons of the 49th Gp. were flying P-38s, and Aschenbrener took to the big twin-engine fighter. He became an ace with four kills on November 24, adding his eighth victory four days later.

When the 7th Fighter Sdn. moved to Lingayen Aschenbrener promoted to Major, became CO. He claimed his ninth and tenth victims on December 11 and 21. Flying at 50 feet altitude, Aschenbrener's Lightning took 20 mm hits and he crashed into a rice paddy only about two miles from the field. He was picked up by Huk guerillas who hid the injured pilot at the foot Mt. Aryat for two weeks since he was unable to stand. However, his back injury improved and the Huks were joined by the crew of Navy torpedo plane and Hellcat ace Alexander Vraciu, who had been shot down by flak on December 20. Captain Aschenbrener returned to his unit after 27 days and then assumed command of a pilot replacement center at Nadzab, New Guinea.

Completing his education at the University of Missouri in 1946-47, the double ace gained a bachelor's degree in journalism. With his wife Ann he raised a family of four sons and daughters, and retired in California.

Enlisted September 1941 - Primary Flight Training Gallinger , TX - Basic, Goodfellow Field, San Angelo, TX - Advance training Moore Field Mission, TX. Also flew P-51s, P-47s and the last of the P-43s (McDill Tampa, FL)

345 combat missions in two tours.

Tally record: Ten kills, all single-engine fighters.

Decorations: Distinguished Service Cross, four DFCs, ten Air Medals, Purple Heart.

ROBERT BALDWIN, was born and raised in Los Angeles, CA. He was commissioned at Kelly Field, TX, June 1940 and immediately assigned as a basic flight instructor at Randolph Field, TX. After 18 months and numerous requests for transfers saw a career change to engineering and flight test duties including AFGD #3, Naples, Italy. Late in WWII he was transferred to the First Fighter Gp. as CO of the 71st Fighter Sdn., Foggia, Italy which was flying escort and air to ground missions throughout southern Europe in P-38L aircraft. As CO of the 51st Fighter Gp., Suwan, Korea, he flew 85 combat missions in the F-86E and F and was credited with the destruction of five Mig 15 aircraft.

Subsequent to the Korean Campaign he served as group and wing commander of the Air Defense Weapons Training Center, Yuma, AZ. After an assignment on the Air Staff at Headquarters USAF as chief of the Air Defense Division of Operations he was assigned to NATO as Chief of Air Defense for the Mediterranean Area. He retired from the USAF June 1966.

GEORGE BANCROFT, was born in South Windsor, CT and is currently living on the family farm where he was brought up. When WWII was declared, he left Cornell University and went to Texas and graduated from Ellington A.F.B. in the Class of 42-H. From there he was sent by boat to North Africa where he joined the 1st Fighter Gp. 27th Fighter Sdn. and checked out in P-38s . He flew 50 missions and shot down two enemy aircraft. He received the D.F.C. and eight air medals. During the Korean War, Bancroft was recalled to active duty at Elgin A.F.B. for two years as a project officer. After completing college, Bancroft worked in an Aerospace Company for 31 years where he was involved with the development of the Space Suit Assembly currently being used on the Space Shuttle.

He, his wife, Beverly, and two sons are active pilots and their daughter, a prospective flyer. They fly Cessnas' out of their private farm flight strip, next door to Bradley International Airport. All P-38 pilots are welcome visitors.

EDWIN C. BAQUET, was born in Unionville, MO and was raised in South Dakota and Oregon. He joined the AAC Cadet Program in February 1943 and received his pilot wings at Luke Field, AZ in December (43-K). After training in the P-40, P-39 and P-38 he was assigned as a P-38 pilot with the 14th Fighter Gp., 49th Fighter Sdn. near Foggia, Italy. On his first mission he spun out of a tight turn at altitude and learned much about the P-38. Baquet damaged a couple FW-190's during his combat tour but could not confirm victories. He returned to the states and began flying the B-59 (Bell Aircomet) in preparation for P-80s and a tour in the Pacific - the war ended and the P-59 unit was disbanded.

He continued to fly single/twin engine aircraft and completed a combat tour in Korea flying the B-26. Approximately five years was spent in a Ferrying Group where he flew most of the single engine conventional/jet fighters through the F-100. He led the first group of 16

new F-86s over the North Atlantic High Flight route to Europe. Baquet retired from active duty in December 1964 in the grade of Lieutenant Colonel.

However he was not through with the USAF and entered the Civil Service system specializing in accident prevention work. He completed 20 years and retired as a Safety Engineer in June 1984. He spends much of his time now enjoying reunions and working with the Order of Daedalians.

Married 40 years to Oresa Adams and have three children: Carole, Kelly, and Kevin. Currently live in Riverside, CA.

WILLIAM D. BARONE, Flight Surgeon, Major, USAF. Enlisted as commissioned office October 1942. Graduate of OTS, FL. He is a graduate of the School of Aviation Medicine, TX. Flight Surgeon for the following: 2nd Air Force, USA; 8th Air Force, England; 9th Fighter Command (Africa-England); 9th Air Force (XIXth Tactical Air Command) England, France, Germany, Austria, Luxemburg, Czechoslovakia. Planes flown in: L-4s and 5s, B-25s, B-24s, B-17s, A-20s, AT-6s, Fairchilds, P-38 (Piggy back).

William supported Gen. George Patton's 3rd Army from Normandy to Germany. Commanding Officer of 36 Bed Mobile Air Force hospital with 9th Air Force, England and France. Landed on Omaha Beach D-22, Normandy, France. 12th occupation Air Force Nuremburg, Germany, December 1945. Attached to the United States Navy, May 1944 pre-invasion training on English Channel. He did Public Relations for 9th Air Force in France and Germany.

He was Assistant to Chief of Medicine, Amarillo Air Force Hospital, Amarillo, TX, 1943. Assistant GP Surgeon of 19th Bomb Gp. (B-17s), Pyote, TX. Flight Surgeon, Airbases at Dalhart, TX, Gulfport, MS, Randolph and Kelly Fields, San Antonio, TX. Flight Surgeon, Airfield of U.S. Military Academy, Newburg, NY, Bedford Airbase at Bedford, MA.

Commanding Officer of 188th Air Force Mobile Hospital. Attached to the 358 FTR Bomb Gp. Cretteville, Normandy, France, under the XIXth TAC was assigned to Gen. George Patton's 3rd Army, July to October 1944.

Reassigned to the 12th Tactical Recon. Sdn. of the 10th Photo Recon Gp. as Flight Surgeon, still under Patton's Army. At Furth German Airbase, we became part of the Occupation 12th Air Force where I remained until December 1945.

Promoted to Major in 1945. Discharged active military service April 1946. Air Force Reserve 1946-1954. Served as Flight Surgeon on active duty. West Point Air Force Base of the Military Academy at Newburg, NY. Bedford Airbase, Bedford, MA.

William received all Battle Stars of WWII in Europe.

Commendations include: Gen. George Patton Jr. (3rd Army), Gen. O.P. Weyland XIXth TAC -March 1944 and 1945, Maj. Foy, Troop Ship Commander, January 1944, Presidential Citation, 12th Tactical RCN Sdn., 358 FTR Bomb Gp. January 1944, Normandy, July 1944.

MONTE G. BASBAS, Wayland, MA, born Manchester, NH. Married to Audrey Vagiates, three children. Dartmouth College, AB; B.U. Law School, JD.

Former Mayor, City of Newton. Presiding Justice, Newton District Court. Member Massachusetts Bar, New Hampshire Bar and U.S. Supreme Court; Past Pres. Massachusetts City Clerk's Assoc.; Past Pres. Mass. Mayors' Assoc.

Enlisted May 29, 1942; Oxnard, CA, Flew STEARMAN; Marana, AZ, Flew "VULTEE VIBRATOR;" Williams Field, AZ, SEF School, CLASS OF 44E Flew AT-6, AT-11, RP-322 (castrated P-38). Reconnaissance Training, Coffeyville, KS; South Pacific Nov. 8, 1944; 25th Photo Recon. Sdn.; 300 combat hours/62 missions.

NEW GUINEA, Biak; PHILIPPINES, Tacloban, Mindoro, Clark, Lingayen, Dulag; OKINAWA, Kadena, Yontan; JAPAN, Chofu, Tachikawa. Awards received include the Distinguished Flying Cross; Air Medal/clusters; Eight Battle Stars. Reserves 1945-1959.

Left engine crippled over Kyushu, Japan. Returned safely to Lingayen, Luzon, P.I. with even greater love and respect for P-38. Participated in bomb assessment reconnaissance of Hiroshima and Nagasaki bombings.

Racked up F5 and took exceptional shots of Hiroshima "mushroom"-mailed them home from Japan-Never arrived.

CLIFFORD J. BAUN, was born in Portland, OR on Aug. 17, 1919. He grew up and attended elementary and high school in Mtn. Home, ID.

He was not a flyer. He volunteered for military service in January, 1942, and served as airplane maintenance technician in the Army Air Corps in the South Pacific (New Guinea, Australia, etc.). There was no time for much, if any, formal training and soon after enlistment, he was on the Queen Mary, headed for the war zone. Battles and Campaigns: Bismarck-Archipelago Papua GO 33 WD 45 - New Guinea GO 40 WD 45. He received an honorable discharge in September, 1945, with the rank of Staff Sergeant.

After returning to Mtn. Home, he worked several years in a dry cleaning business and later worked for the U.S. Postal Service for 20 years, until his retirement in 1979.

He was married to Evelyn Nixon on Mar. 21, 1949. A son, Larry, and a daughter, Brenda, were born to this union and there are now three grandchildren.

After Cliff's retirement he suffered from emphysema several years, until his death on Sept. 7, 1986.

Cliff loved the out of doors, camping and fishing. He was an avid rockhound and enjoyed the lapidary arts, exhibiting in many area rock and gem shows.

BEN BEACH, was born in Seattle, WA. Being schooled in the Puget Sound area he graduated from Snohomish High in 1940. Three months later his National Guard unit, the 248th Coast Art., was federalized and located at Ft. Worden, WA where he observed P-38s carrying out mock runs on the coastal gun emplacements. Mid 1942 brought him acceptance into the Air Corps Pilot Training Program, carrying him through Nashville, Santa Ana AB, Ryan Field, Minter Field and Luke Field graduating as a single pilot, class 43-H. Transition training in P-39s flew him through Tonapah, NE, Bucannon Field and Hamilton Field in California.

Ten hours of transition in the P-38 at Orange Co., AB brought him orders to CBI. Assigned to the 459th Ftr. Sdn. for an 18 month tour, he flew 84 combat missions throughout Burma. He scored one confirmed and was awarded the DFC with one cluster and the AM with two clusters, leaving the theatre as Captain.

Dismissed from the service in 1945, he passed through the University of Washington, San Jose State and San Francisco State in California to earn an MA and several teaching credentials. He is married with four daughters and three grandchildren. While teaching school he participated in the AF Reserves until retirement. Today he is semi-retired, operating a small business and talks P-38s to anyone who will listen.

ROBERT B. BERGANDER, (Bergie), Honey Creek, WI, joined the USAAF at Chanute Field, 1940. Completing the Aircraft Mechanics Course, he taught aircraft hydraulics until Pearl Harbor. Bob joined the 27th FS, 1st FG, arriving in England June 1942. The North African Invasion found Bob and his squadron wading ashore near Oran, Nov. 8, 1942. For three years he worked in Flight Engineering as the 1st FG fought the war across Africa, Sardinia and Italy. As Flight Engineering Chief, T/Sgt. Bergander was awarded the Bronze Star Medal by Gen. Struthers (15th AF) for innovative aircraft repair methods under combat field conditions. He returned to the States, May 1945.

Discharged in 1945, Bob completed college, followed a technical training career with USAF

Training Schools (aircraft), The Martin Company (ICBMs) and retired from the Raytheon Company (Submarine Sonar) as Division Training Manager in 1982.

JOHN A. BERRY, born in Cedar Valley, UT Jan. 4, 1916. Educated at Lehi, UT and Santa Monica, CA. While in college and working at Douglas Aircraft, Pearl Harbor changed his plans for the future.

Although exempt from the draft, he made several attempts to join the Air Force and finally succeeded in February 1943; Graduating from Luke Field, Class 43K in single engine fighters. He was assigned to a unit with P-39s and P-51s heading for England. He was among ten pilots who were selected to get 20 hrs of P-38 time before leaving and found himself being shipped to Port Moresby, New Guinea and assigned to the 475th Fighter Gp., 433rd Fighter Sdn. where he learned real fast in handling a twin engine fighter and staying alive.

He remained with the"Satans Angels" Gp. through Nadzab, Hollandia, Biak, Leyte, Mindoro, Clark Field and Lingayen Gulf. Made 133 missions. Shot down seven E/A and got official credit for two.

Berry was flying wingman for Maj. Richard Bong when he got his 40th E/A on Dec. 17, 1944, north of San Jose, Mindoro. Transferred to the U.S. November 1945, was assigned to the Air Force Exhibition Unit at Wright Patterson AFB where he flew at exhibitions and toured with AF exhibits all over the U.S. While stationed there he met and married his wife Anita and raised three children.

Berry separated from the service January 1950 as a Major and returned to the Air force as a civilian aviation engineer and flight control technician at overseas bases for the next 20 years.

He retired in 1972 and is presently living in Mesa, AZ.

WILLIAM L. BISHOP, born Dec. 26, 1921, in Salem, OR, joined Aviation Cadet Program June 1942. Graduated Class 43-I, Williams Field, Chandler, AZ, TE fighter pilot. After completing two fighter training assignments at Salinas, CA and Ellensburg, WA, entire squadron sent to photo-reconnaissance training at Will Rogers Field, Oklahoma City. Sent to Fifth Air Force, 91st Recon. Wing, 6th Photo Gp. and assigned to 26th Light Recon. Sdn. based at Nadzab, New Guinea. First mission Apr. 19, 1944; squadron moved successively to Hollandia, Dutch New Guinea, Biak, DNG, then attached to Maj. Gen Walter Kreuger's 6th Army as photo recon support unit based at Lingayen strip following 6th Army's Lingayen Gulf assault landing on northern Luzon, PI.

From Squadron Operations, Officer 26th Photo, moved to Asstistant Group Operations Officer 6th Photo Gp. based Clark Field; then assigned Squadron CO of sister squadron, 25th Photo, based at McGuire Field, Mindoro Island, PI. The 25th immediately became Group's forward unit, moving to Yontan Strip, Okinawa, until war's end; then the 25th occupied Cho-Fu Strip, Japan, a suburb of Tokyo. From Okinawa I flew what should have been (but wasn't) the last official combat mission of WWII, reconnaissance photo coverage of the second atomic bomb dropped on Nagasaki. Combat hours: 327, Missions: 76. Received Air Medal w/three OLC, eleven Battle Stars, Two Unit Citations. Attained rank of Major.

Am now a professional magician and operate a magic shop at 3242 S. Harlem Ave., Riverside, IL 60546.

ROBERT C. BLACK, was born in Lexington, MS Dec. 11, 1923, graduated from Drew, Miss High School in 1940 and entered Army Air Corps in May, 1943. He had basic training at Sheppard Field, TX and Pre-Flight at Santa Ana Army Air Base, CA in October, November 1943, beginning flight training with Class 44-F at Hemet, CA December, 1943. He proceeded with flight training at Merced and Luke Field, gunnery training at Gila Bend and P-39 transition at Victorville, CA. He received P-38 training at Santa Rosa, CA December, 1944 to July, 1945. He bailed out of a burning P-38 near Clear Lake, CA without injury. He was assigned to the 343rd Fighter Gp., 54th and 344th Fighter Sdns., August, 1945 to June, 1946 at Attu, Anchorage and Shemya.

He married Jane Taylor of San Jacinto, CA and is presently practicing as a Certified Public Accountant in Laguna Beach, CA.

MERLE E. BLANCHARD, was born in Campbell, CA on Nov. 16, 1919 and worked as a riveter for Lockheed at Burbank, CA in 1939-40. He worked on the first flush-riveted P-38 (XP322) and fell in love with it.

He went into the Air Corps in January 1942 and began aviation cadet training June 1942, graduating March 1943. He was assigned to the 55th Fighter Gp. McCord Field, WA to check out in the P-38 and then to Port Angeles, WA with the 338th Fighter Sdn. before embarking for England, September 1943.

He flew 77 missions over Europe, completing his tour August 1944. He commanded the 495th Training Sdn., Acham, England until his return to the United States, December 1944.

He is credited with two enemy aircraft destroyed and two damaged. His decorations include two Distinguished Flying Crosses and five Air Medals. He retired in 1965 as a Lieutenant Colonel.

WAYNE L. BLUE, was born and raised near Tecumseh, NE. Attended University of Nebraska 1938 to December 1941 when he signed for pilot training with an Aviation Cadet Team who visited the campus. Primary training-Tulare, CA; Basic-Gardner Field, CA; Advanced-Williams Field, AZ. Commissioned Aug. 27, 1942. His first assignment was 2nd Photographic Gp., 6th Photo Recon. Sdn., Peterson Field, Colorado Springs. Initial introduction to the P-38 there. Upon completion of training April 1943 he received orders to report to Memphis-Morrison Field, Miami, FL to pick up a P-38 for delivery to Marratcha-Casablanca. May 1943 assigned to 14th Fighter Gp., 37th Fighter Sdn., Telergma, Africa.

As a pilot of Gen. Doolittle's Northwest African Strategic Air Force he completed 50 missions in the North African invasion, Tunisian campaign, bombing of Pantelleria, Sicilian campaign and escort missions over Italy. Was credited with one confirmed victory - a JU-88. His decorations include Air Medal, ten Oak Leaf Clusters and the Distinguished Flying Cross.

Retired in 1962 from the USAF at Selfridge AFB, MI. He has since resided in Lincoln, NE with wife Jan. Is now retired after 21 years with the University of Nebraska. He has two daughters, one son and five grandchildren. He was a member of the 1941 Rose Bowl Football Team from Nebraska.

BRADLEY J. BOARD, born in Richfield, MN. Enlisted as an Aviation Cadet in October 1942.

Preflight at Santa Ana, Basic at Pecos, TX, Williams Field for Advanced. Graduated October 43-I.

To Salinas, CA for 1st Phase Fighter in P-38s, Burbank momentarily, then transferred to Photo-Recon, Will Rogers Field, in January 1944.

Overseas in April to the 10th Photo Gp. 31st Photo Sdn. Flew 14 missions before and after D-Day, was awarded Air Medal and DFC. The Squadron was awarded the Presidential Citation for Invasion filming.

Discharged in November 1945. Went directly into the Ski Resort business for 22 years, then to Tucker Sno-Cat in Oregon. Field Representative for ten years.

Retired to restoring Old and Classic cars. Have three daughters with my Salinas wife, Bette.

We have visited all seven remaining original Squadron Pilots at their homes or at Reunions.

Attended 'Gathering of the Eagles' and P-38 Convention, Burbank.

JOHN F. BOEYE, was born Nov. 21, 1922, at Webster City, IA. While he was working for North American aviation he decided to enlist and joined the Air Force in Los Angeles, CA. In November of 1942, he was sent to Cadet Center in San Antonio, TX; then to Parks Air College in East St. Louis for Primary. His Basic Training was at Garden City, KS, where he had a forced landing in a wheat field. He graduated with the Class of 43I on Oct. 1, 1943, while at Eagle Pass, TX. This training included ten hours in a P-40. From Texas he went to Tallahassee, FL, on Feb. 28, 1944, to Thomasville, GA, for RTU, and Hawaii 7th Air Force (P-39s). On Apr. 17, 1944, he was transferred to Ondonga, New Georgia, where he joined the 70th Fighter Sdn. called the White Knights. There he began flying P-38Gs. After that he went to Munda, New Georgia, Sansapor, New Guinea, Morotai, Luzon, Lingayen Gulf, Mindora, Zamboanga, Laoag, Luzon and Manila in the Philippines. He came home on Dec. 9, 1945, and was discharged at Camp Grant, IL, with a rank of Captain.

He took part in the longest flight by a P-38 on Sept. 18, 1943 (seven and one-half hours, ran it at 2200 r.p.m. with 15 inches manifold pressure). He was involved in a crash of a P-38J on Jan. 15, 1945, at Lingayen Gulf. He was credited with one Hamp Kill on Nov. 25, 1944, over Negros in the Philippines.

After his military career he attended the University of Iowa and obtained his B.S.C. and L.L.B and has been a practicing attorney in Red Oak, IA, since 1950.

WILLIS DANIEL BOONE, (DAN), was born in Flora, MS and raised in Jackson, MS. He lived in Chicago, IL until enlisting in the Army Enlisted Reserve Corps in December, 1942. He received Basic Training at Shephard Field, TX

and later attended the University of Denver (CTD) and received pre-flight training at Santa Ana, CA.

He received Primary, Basic and Advanced Flight Training at Thunderbird I Field, AZ, Marana Army Airfield, AZ and William Army Airfield, AZ, respectfully. He graduated in the class of 44-D. He was given Gunnery Training at AJO, AZ, P-39 transition at Victorville, CA and P-38 (RTU) Training at Santa Maria, CA.

He was then assigned to the 344th FTR Sdn. located at Shemya in the Aleutian Islands until the war ended in September, 1945. He remained in the Aleutians flying P38s until they were "grounded" in the summer of 1946 following a crash when the "Booms" separated during a low level pullout from a loop.

His group transitioned to the P-51 and later, the P-80. he returned to the States in 1947 and remained on active-duty until retiring at Langley AFB, VA in October 1971.

RICHARD R. BOOTH, born Mar. 30, 1923. Attended public schools in New York, New Jersey and Florida. Graduated from McDonogh School (semi-military), MD, 1941; attended Cornell University 1941-1942. School of Mechanical Engineering; JD degree, University of Miami, School of Law, Cum Laude, February, 1949.

Aviation Cadet, Class of 43-K; received pilot's wings and commission as a 2nd Lieutenant, Luke Field (now Luke AFB, Phoenix, AZ), Dec. 6, 1943. Flying instructor, basic, (aerobatics, formation flying, cross-country and night flying, and Fighter Pilot. Forty-three missions with the 9th Air Force (Tactical) Europe, flying P-38 Lightning and P-51 Mustang aircraft. Resigned from AF Reserve, rank of Captain.

Enjoy playing tennis, golf, sailing, calligraphy (Olde English illuminated manuscripting); and public speaking.

He is a member of the Florida Bar. Dade County Bar Association, South Florida Chapter of Federal Bar Association, the VFW, and the Military Order of World Wars.

JAMES D. BORGE, the only offspring of Dora and Joseph Borge, was born in Yerington, NV Jan. 2, 1921.

While attending the University of Nevada the buzzing of the Campus by an ATO Fraternity Brother in a P-38 prompted him to volunteer for the USAAF in June of 1942.

He was called into cadet training in February 1943. Later that year he married his college sweetheart, Amy (Turano) at the home of Commanding Officer Col. Newton Crumley in Tucson, AZ.

His tactical instructors on P-38s at Santa Maria, CA, Maj. William P. Moore and Capt William A. Robertson, chose Jim to join them when they returned for their second tour of combat in the 449th Fighter Sdn., 51st Fighter Gp. (only P-38 Squadron in China) under Gen Claire Chennault.

After flying 230 P-38 combat hours strafing and bombing Japanese transportation in French Indo-China, the war in the Pacific and China was won. Jim returned to the U.S. to resume his studies at the University of Nevada where he graduated Phi Kappa with a B.S. in Chemistry June 1948.

In 1952 he graduated from New York State University Medical School Cum Laude. As Board Certified Surgeon and sub-specialty of Cancer Surgery, he joined Dr. Dwight H Murray Sr. (then President of the American Medical Association) in Napa, CA until retirement 1985.

He invented the Borge Surgical Clamp in 1977. He is the father of three children Greg Lorna, and Tab. Christmas Eve he and Amy celebrate their 44th wedding anniversary.

TEMPLE BOWLING, was born Dec. 24, 192[illegible] in Hogansville, GA. He is a graduate of Auburn Univeristy, where he began flying and obtained a civilian license in the spring of 1941.

He entered Aviation Cadet Training at Corsicana, TX on Sept. 26, 1941 and went through Primary Training and on to Basic Training at Perrin Field, Sherman, TX in December of 1941. He was sent to Moore Field, McAllen, TX for Advanced Training and got his wings there in Apr. 30, 1942, as a member of the class of 42-D.

His first assignment was to the 50th Fighter Sdn., 14th Fighter Gp., Hamilton Field, CA. Six weeks later he flew a P-38 from California across the country to Presque Isle, ME, and from there on to Iceland. Supposed to be going to North Africa and supposed only to spend the night in Iceland, he and his squadron remained in Iceland for 19 months.

He returned to the States in April, 1944, and went back overseas to Foggia, Italy in October 1944, where he was assigned to the 49th Fighter Sdn., 14th Fighter Gp.

He flew all types of missions over Germany, Austria, Czechoslovakia, and Northern Italy. He flew a total of 50 missions in Italy and Iceland and was awarded the Distinguished Flying Cross and the Air Medal with five clusters.

He retired from the Air Force after 30 years' service on May 31, 1971 as a Full Colonel and lives in Montgomery, AL.

He married Joan Douglas of Corsicana, TX on Sept. 26, 1944. They have four children and four grandchildren.

JOSEPH B. BOYD, was born in Rockwood, TN, Feb. 20, 1920. His father was killed in an accident when he was four and young Boyd grew up in an orphanage. Shortly after graduation from high school, he enlisted in the Army Air Corps and was a Sergeant at the time of Pearl Harbor.

He applied for aviation training, was accepted and received his wings in Class 42-G as an enlisted pilot. Two months later, he was in the Mediterranean Theatre as a C-47 pilot. After completing a tour in C-47s, he asked for P-38s and was assigned to 71st Sdn., 1st Fighter Gp. in Sicily and Italy.

In total, he had over 1200 hours of overseas flight time and about half of which was in P-38s, with 72 missions and holds the Distinguished Flying Cross, Purple Heart, Air Medal with 13 Oak Leaf Clusters, and six unit citations.

It was at that time he received a battlefield commission as a 2nd Lieutenant and 18 days later was promoted to 1st Lieutenant. He returned stateside in June, 1944, and was assigned as Air-Corps test pilot in California.

In 1946, he went to Okinawa for a year as operations officer for the 418th Night Fighter Sdn. and was discharged in 1947 as a captain. His night fighter experience got him recalled to active duty during the Korean War, to train pilots for the all weather program.

He was discharged again in 1952 and he and his wife, Frances, are retired and live in Rockwood, TN.

LEONARD BOYD, born Dec. 12, 1921, Edmonton, Canada, oldest son of Harvey Boyd and Jean Gillies Boyd. Reared, educated at San Francisco and Oakdale, CA. Attempted to join RAF late 1941. Studying for entry into AF at time of Pearl Harbor attack. At age 20, enlisted as Aviation Cadet May 30, 1942.

Preflight: Santa Ana, CA; Primary: Pacific Air School, Ft. Stockton, TX; Basic: Pecos, TX; Advanced: Luke Field, Phoenix, AZ. Graduated Class 43-G, July 28, 1943. Graduated Flight Instructor School Randolph Field, San Antonio, TX. Graduated Instrument Flight Instructor School, Bryan AFB, College Station, TX. Flight Instructor at Chico AFB, CA. OTU: P-39,

Buchanan Field, Concord, CA; P-63, Portland AFB, Oregon; P-38, Metropolitan Airport, Van Nuys, CA, 1944.

Entered combat during British invasion of Ramree Islands in January 1945 with the 459th Twin Dragon FS, hottest P-38 outfit in the CBI with most Aces and victories. Squadron Presidential Citation. He flew 26 combat missions out of Chittagong and Rhumkapalong, India, strafing and dive-bombing enemy targets from Mandalay to Rangoon, Burma. Awarded Air Medal and two Battle Stars. When atomic bomb fell, he was at Dhudkundi AFB, India, flying P-47 Thunderbolts with the 89th FS, 80th FG, training for the coming east China coast invasion.

Engineering Officer, Provost Marshall. Ajo Gunnery Base, AZ, 1946. Left active duty, Williams Field, AZ, January, 1947.

A charter officer to organize northern California Air National Guard. Provost Marshall, Base Operations Officer, Commanding Officer of the Fighter Detachment, 244th ABG, Hayward, CA. Awarded Commendation Medal. With the end of the Korean War Captain Boyd left the military service, 1953.

Father of Michael, Daniel, Kevin. Grandfather of Shannon, Joey, Shawn, Casey. Married to the former Maryland McCarthy of New Orleans, LA.

Auto Dealer over 20 years. Past State President of the Independent Auto Dealers Association of California. Home in San Jose, CA.

EDWARD I. BRADDOCK, was born on Feb. 10, 1919 in Collinston, LA.

He enlisted in the Army Air Corps on Mar. 21, 1941 at Barksdale Air Force Base, LA. He received an appointment to Air Corps Flying School in August 1941. Graduated from Kelly Air Force Base in class 42-C, the first class of Flying S/Sergeants.

Was a member of the 82nd Fighter Gp. when it was activated at Muroc in June 1942. Had no twin engine time when given a cockpit check, and later soloed from Muroc, the P-38.

He flew submarine patrol from Mines Field, now Los Angeles International Airport. He was commissioned 2nd Lieutenant in September 1942.

The 82nd went overseas in September 1942. The P-38s were shipped via surface vessels and the personnel traveled to England aboard the Queen Mary. About 2 P.M. the day the ship was to dock, she struck a light cruiser and limped into port with only damage to ship and no personnel injury.

Upon receiving the P-38s the 82nd proceeded to North Africa. After 50 combat missions, Braddock returned to the United States. He remained in the U.S. Air Force and retired as Colonel on Oct. 1, 1970. He flew 82 different military aircraft during his Air Force career.

Braddock married his childhood sweetheart in April 1942 and has lived happily ever since.

CEDRIC T. BRISTOW, (LTC), (Duke) USAF (RET), born in Winnipeg, Canada in 1960 moved to San Francisco, CA at the age of three. His long military involvement (19 years Naval Reserve and National Guard and 21 years active service in the U.S.Army Air Corps and USAF) began at age 14 in the Junior ROTC as a high school student. He then enlisted in the Naval Reserve for four years, later transferring to the California National Guard, Coast Artillery. In 1927, Duke received his commission as a 2nd Lieutenant. He was a member of the Horse Drawn Field Artillery, California National Guard from 1932 until January 1942, when he transferred to the Army Air Corps 14th Pursuit Gp. (P-38s). He served with the Group in 11 campaigns; England, North Africa and Italy from August 1942 'till May 1945. Returning to the U.S, Duke was assigned as Assistant Chief of Staff to Commanding General, Army Air Force Training Center, Orlando, FL until April 1946 at which time he reverted to "M" Day Assignee, USAF, AOC. The beginning of the Korean Conflict brought the return to active duty with the USAF. Following a variety of assignments, in 1960 Duke was ordered to MAAG as advisor to the Chinese Air Force in Taiwan, ROC. His final period of active duty found him at the Pentagon, as Chief, M&O, Div., OSI, Inspector General's Office until retirement in September 1966.

During recurring periods as a civilian his occupations included: banking, accounting, and owner of a General Insurance Agency.

He assisted in the formation of the Peninsula Chapter (San Francisco and San Mateo County) Reserve Officer Association, and was elected its President in 1948 and again in 1949. Duke is a member of the American Legion, life member of TROA, ROA and MOWW. His hobby is philately and is a life member of the American Philatelist Society.

WILLIAM G. BROADFOOT, JR., was born Aug. 17, 1918 in Selma, NC. Graduated Phillips Exeter Academy 1936; University of North Carolina, Chapel Hill, 1941. Joined U.S. Army Air Force December 1941. Primary training Dorr Field, FL; basic, Gunter Field, AL; advanced, Napier Field, AL. Commissioned Aug. 5, 1942. Assigned to P-47s, Farmingdale, Long Island, NY. Group shifted to P-40s and sent to India with 80th Fighter Gp., arriving June 30,

1943. He had 21 combat mission in P-40s. November 30, 1943 transferred to P-38s, 459th Fighter Sdn., as flight leader. He had no twin engine time.

With two flights "around the field" in P-38s, totalling 2 hours, 20 minutes, he led his flight into combat against Zeros, "over the hump." February 28, 1945 he returned Stateside after 113 combat missions totalling 408.8 combat hours.

These included three (3) single engine returns, the longest Rangoon to Chittagong. His decorations include Air Medal, two Oak Leaf Clusters; DFC; Silver Star; British DFC. He served as Executive Officer of 459th and attained the rank of Major.

WILLIAM G. BROOME, was born Nov. 19, 1917 in Marked Tree, AR. After attending the University of Tennessee and several years as a Plantation Assistant Manager, he enlisted as an Aviation Cadet in January 1942 at Pine Bluff (AR) Army Air Base. He received flying training at King City, Chico and Victorville; graduating in Class 42-H. He was assigned to the 78th Fighter Gp. (P-38s) at Hamilton Field, which went to Goxhill RAF Station, North England in November 1942.

In February 1943 he was transferred to the 48th Fighter Sdn., 14th Fighter Gp. which was regrouping at Mediouna, French Morocco. The group moved up to Telergma, Algeria and began combat missions in May. Progressive moves were made to Tunisia and Sicily. He completed his 50th mission (215 combat hours) on September 9, 1943, D-Day of Italy. During this overseas tour he claimed several victories and earned the Distinguished Flying Cross, and Air Medal with nine Oak Leaf Clusters. Lt. Col. Broome retired in December 1971.

GEORGE PAT BROWN, was born in Houston, TX in 1921, but now calls Wortham, TX his home. Pat joined the USAAF at Randolph Field in August 1940 as an aircraft mechanic, and graduated from pilot training at Williams Field, AZ with Class 43-D. He was a member of the 82nd Fighter Gp., 96th Fighter Sdn. in North Africa, Sicily, and Italy. During this time he had two confirmed, two probable, and one damaged in aerial combat; plus three destroyed on strafing missions. He received the DFC, Air Medal with 11 Oak Leaf Clusters, and the Presidential Unit Citation. After WW-II, he served in Aircraft Maintenance, and later as a Missile System Analyst with the Thor Missile Programs retiring in 1960.

Pat and his wife Lavada were married in 1942 and have three children - Bobby, Melinda, and Mark. He graduated from East Texas State University with the bachelor and master degrees. And also from Asbury Theological Seminary with a master's degree. For the past 20 years, Pat has taught at Azusa Pacific University in Azusa, CA, and Texas State Technical Institute in Waco, TX. During this time he has also served as an FAA Designated Pilot Examiner for all pilot and flight instructor certificates, and also as a FAA Designated Written Test Examiner. Pat retired in November 1987.

GERALD BROWN, was born in Globe, AZ and raised in Phoenix from where he entered the Air Corps in February, 1941. Upon receiving his wings in March, 1943, he joined the 38th Fighter Sdn., 55th Fighter Gp. at Paine Field, WA. The Group went overseas to England and went operational Oct. 15, 1943. The mission was bomber escort which proved to be difficult because of weather and the then formidable Luftwaffe. Brown was hit several times by flack, shot up twice by fighters but was able to become an Ace by shooting down two 109s, two 190s and Hellk while flying the P-38. Finishing his first tour in April, 1944, and after a brief stint as Operations Officer at a RTU Base he joined the 4th Fighter Gp. for another tour in P-51s.

Nothing remarkable happened on this tour with no more victories but he did have to bail out of a burning 51 and became an evadee on Sept. 5, 1944. Brown went to Japan in 1948 and joined the 35th Fighter Gp. He was the Commander of the 39th Fighter Sdn. when it went operational July 3, 1950 in the Korean War. He was shot down by ground fire on November 30 and remained a POW until released on Sept. 7, 1950. After several assignments in the States and Europe he retired Nov. 1, 1967 as Wing Commander of Luke AFB, AZ.

HARRY W. BROWN, was born in Amarillo, TX and attended public schools. He joined the Texas National Guard when he was 13 years of age and was a gunner corporal on a 75 mm. cannon when most kids his age were Boy Scouts. After completing prerequisite college training, he "acknowledged" his "correct age" in 1940 and was honorably discharged so that he could become an Aviation Cadet. He underwent Primary flight training at Cal-Aero Academy in Ontario, CA; thence to Goodfellow Field in San Angelo, TX for Basic training and, finally graduated in the class of 41-F at Kelly Field in August 1941.

He was promptly transferred to Wheeler Field, T.H., and assigned to the 47th Pursuit Sdn., 15th Gp., flying P-36s and P-40s. He shot down two Japanese aircraft flying a P-36 during the attack on Pearl Harbor and became the first Texan decorated in WWII. He subsequently was transferred to the 45th Sdn. and then to the Fifth Air Force, flying P-40s. On arrival in Australia, he was assigned to the 9th Sdn., 49th Gp., and, after a while, transitioned to P-38s. He secured his third kill during the Battle of the Bismarck Sea.

When the first all P-38 Group, the 475th, was formed, he was transferred to the 431st Fighter Sdn. as flight leader and became that fabled group's first Ace, scoring three kills over Tsili Tsili in August 1943. He scored his seventh kill while leading the squadron over Rabaul.

He was the Commanding Officer of McGuire Air Force Base when he was released from active duty in 1948 and returned to school receiving an M.B.A. at the University of Denver. Completing prerequisites for the Ph.D. at the University of Oregon. He eventually managed the Personnel function of a major west coast corporation until retirement.

As to anecdotes, he has maintained that one of the most intersting specifications of that wonderful airplane is that the capacity of the ammo cans was exactly four cases of Australian gin and the most unpopular man in the squadron was the guy who landed wheels-up on a fuel stop at Iron Range while transporting that precious cargo!

CLAIR E. BUSKIRK, was born Jan. 8, 1923 in Parkersburg, WV. After graduation from Parkersburg High School I enlisted in the Army Air Corps in August 1941 as a Private.

After graduation with the first Army Air Corps airplane mechanics class from Sheppard Field, Wichita Falls, TX in February 1942, 25 of us new mechincs were sent to Lockheed Aircraft Corp. at Burbank, CA during March 1942 for a special course on the P-38 Lightning Fighter Plane.

In April 1942 my first assignment as an airplane mechanic was with the 55th Fighter Gp., 37th Fighter Sdn. based at Turmwater, WA. Maj. Jack Jenkins was the Squadron Commander. They had just transitioned from P-43s to P-38s.

In a few months my advancement was to Corporal then on to Sergeant as a P-38 Crew Chief.

During August 1942 my appointment as an Aviation Student Pilot sent me to SAAAB, C

(Santa Ana Army Air Base) for pilot preflight training with Pilot Class 43E. (I met my wife, Doris Creighton there).

Primary flight training was performed at Ryan Field, Hemet, CA in Ryan PT-22s.

Next came basic flight training at Gardner Field, Taft, CA flying the "Vultee Vibrators" BT-13s.

Advanced flight training was completed at Williams Field, Chandler, AZ flying AT-17s, AT-9s, AT-6s and RP-322s (P-38s).

I graduated with Class 43-E as a Flight Officer and received my Silver USAAF Pilot Wings May 20, 1943.

Doris and I were married on July 3, 1943.

P-38 Fighter transitional flight training was flown at Glendale Grand Central Airport and Muro Air Base (now Edwards Air Force Base) during June through September 1943.

Assigned to the 55th Ftr. Grp., 338th Ftr. Sqd. based at Nuthampstead.

Within a few weeks of flying combat in P-38 Lightnings over Europe and into Germany escorting B-17 and B-24 Bombers and strafing ground targets, I had shot down one German Me-109 fighter plane, flew home to England in my P-38 on one engine after being hit by German Flak over Kiel, Germany, and made an emergency bailout of my P-38 over England.

During the past months I had promotions to 2nd and then 1st Lieutenant.

My last P-38 combat mission (my 79th) in July 1944 was flying with a crew member, Bombardier Lt. Stroud in our P-38 glass nosed "Droop Snoot," CL-X. We dropped some bombs on a target in France and that mission completed my first combat tour of 300 hours.

I returned by boat to England and once again to my 55th Fighter Gp., 338th Fighter Sdn., promoted to Captain. We were now based at Wormingford, west of Colchester, England. I was to fly my second combat tour of 200 hours in P-51 Mustang Fighters.

On Jan. 6, 1945 on my 106th mission, our 338th Fighter Sdn. was flying home to England from a bomber escort mission. As we flew over the German Airdrome Gebelstadt at about 5000 feet, my Squadron Commander instructed me to take my flight of four P-51 Mustangs and strafe the airfield. On this strafing run my P51 was hit by ground anti-aircraft fire. My wingman Richard Hebst flew on up to 5000 feet with me, then continued on up to make a radio contact for a fix home. My engine started smoking and loosing power. After a few minutes I was at just a few hundred feet altitude. I bailed out once again with my trusty old back pack parachute. That completed my 425 hours of combat.

The Germans captured me right away and I finally ended up in Stalag Luft 1, Barth, Germany as a POW (Kreigesgafaganon) until May 13, 1945. The Russian Army liberated us.

Our B-17 bombers arrived at Barth, Germany. They flew us into France and we were rehabilitated at Camp Lucky Strike near La Havre.

After a few weeks we took a troop ship cruise to a port in Virginia, Good Old USA. The journey on home to my wife Doris in California continued by troop train.

After processing in Santa Monica, CA, I was discharged from active military duty in December 1945 and transferred to the USAAF Reserves for over 20 years plus two years with the California Air National Guard based at Van Nuys.

I retired from the USAF Reserve in 1969 as a Lieutenant Colonel.

After 30 years service I retired from the Los Angeles Fire Department in 1980.

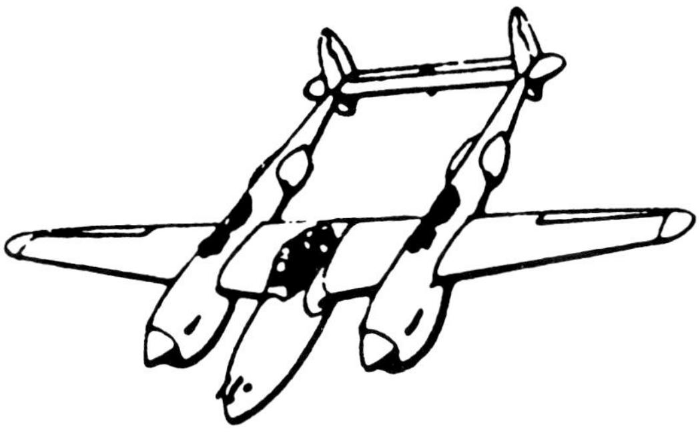

Doris and I have two daughters and three granddaughters.

Doris and and I have enjoyed living in Irvine, CA since 1973.

COLIN J. CAMPBELL, learned to fly at Aberdeen, SD in 1933. He is a member of the OX-5 Club (Aviation's Pioneers), barnstormed the upper mid-west during the 30s; with Hanford Airlines, a feeder line airmail contractor using Lockheed Vega's, dusted in northern California, with Maj. Mosley and the CPT program at Alhambra, CA, and joined Lockheed Aircraft Corp. in 1939 as an inspector.

Assigned first to the BAM Hudson Bomber and then to the YP-38 and P-38 contract until the end of the war. Transferred into Engineering on the P-80, T-33, F-94, A,B,C program and finally the F-104. Left Lockheed in 1958 and spent the next 25 years in management with Aerojet General Corp.

ROBERT E. CAMPBELL (Bob), Lt., from Ada, OK, had the good fortune to fly both the P-38J Lightning and P-51D Mustang while serving with the 79th Fighter Sdn., 20th Fighter Gp., 8th Air Force at Kings Cliffe, England, from May 25 through Oct. 3, 1944. His P-51D "Wanda Honey" was named for his high school sweetheart Wanda West. They married Oct. 3, 1943, shortly after Class 43-I graduation at Luke Field, AZ, and they still are.

Bob completed his European Theater tour of 61 missions at 275 hours of combat flying time which was almost equally split between these two illustrious fighters: 143 hours in P-38Js and 132 hours in P-51Ds. Bob's "Form 5" shows his longest P-38J mission of 5:25 was escorting bombers to Munich on July 12, 1944. While escort duty was primary, he flew P-38Js in English Channel patrol during the D-Day invasion, dive-bombing, strafing and even a 'Droop Snoop' bombing mission.

The 79th Fighter Sdn. was the last of the 20th Fighter Gp. to change over to P-51Ds. Bob was lucky and was credited with destroying two Messerschmidt ME-109s (air) and damaging a third (air) while flying his P-51D in August 1944. He was awarded the Air Medal with five oak leaf clusters and the Distinguished Flying Cross. He returned to the U.S. and instructed instruments and transition in AT-6s at Perrin Field, TX. He was discharged from the U.S. Army Air Corps in September 1945. Bob is an Engineer and retired from project management in 1984.

There is no doubt in Bob's mind that the P-51D was the best fighter of WWII, particularly in the European Theater; however the P-38Js had to be a close second. His hat is off to those who flew P-38s from the beginning of Fighter operations until P-51's were made available in mid 1944.

NICHOLAS CAMPESE, (Nick), (of the 475th Fighter Gp.) was born in Italy.

He enlisted in the USAF in 1942. In March, 1943, he participated with a group to form the 475th Fighter Gp. at Amberley Australia. He served in New Guinea, Biak, Leyte Invasion, IE Shima Okinawa.

He was a S/Sergeant in communications in the 431st Fighter Sdn., of the 475th.

He was discharged from the USAF in November 1945.

He received a BEE and MEE degree from Rensselaer Poly, in Troy, NY.

As an engineer for General Electric in Schenectady, NY, he particapted in the Development of the Vulcan (Gatling) Machine Gun, and other Aircraft Systems.

Since 1958 he has participated in the development of the Polaris, Poseiden, and Trident Missile Guidance Systems as a Computer Design Engineer.

He retired from GE in 1986 and has since been involved in Financial Planning and Tax Consulting.

He lives with his wife and three of his six children in Latham, NY.

JAY VICTOR CANNON, born in Jacksboro, TN; finished high school there and went to Wayne University, Detroit, MI. Joined the USAAF December 1942, graduated from Airplane Mechanic School at Sheppard Field, TX, then entered Aviation Cadets 1943.

Jay received his wings at LaJunt, CO, December 1944 and was assigned to P-38 training. He received Photo Recon training in F-5s at Coffeyville, KS. Overseas assignment was to 8th Photo Recon. Sdn., Okinawa. From Motobu Airfield combat missions were flown to Tokyo to program all bombing missions and invasions. Next operation was from Chofu Japan in the Occupation Forces to map the entire Japan.

During the Korean War Jay flew B-29s from Guam. Returned to the States to join SAC to fly Helicopters for rescue and C-123J from the Greenland Ice Cap. After commanding the only Helicopter Sdn. at Reno, NV, he became the manager of the world wide F4-C (McDonnell Phantom) training program for three years. His final assignment was Commander 38th Rescue Sdn., DaNang Air Base, Vietnam.

As Commander he flew 278 missions in HH-3 Sikorsky (Jolly Green Giants) and assisted in saving 609 aircrewmen. Awards were: Bronze Star, Flying Cross with three Clusters, Air Medal with four Clusters and Unit Citation.

During his 25 year career he flew 30 different aircraft. He rates the P-38 first, B-25, Sikorsky H-5 and HH3E Helicopters, Lockheed T-33 Trainer and McDonnell F4C in that order.

Jay retired from the USAF July 1967, lives in Amarillo, TX, is married and has five children. Four sons include a city fireman, state trooper, U.S. government meat inspector, and patent engineer and agent.

Maj. Cannon is presently a Certified Flight Instructor and Instrument, Airplane and Powerplant Mechanic, Instructor at Texas State Institute, Counselor and Historian for Experimental Aircraft Association. He has over 17,000 hours flying and was 70 years old July 17, 1988.

WILLIAM L. CAPRON JR., (Bill), was born in Cedar Rapids, IA Feb. 12, 1921. Raised in Cedar Rapids and attended Coe College. He enlisted in the service in February 1942, was commissioned 2nd Lieutenant Infantry at Ft. Benning, GA in October 1942, and assigned to the 96th Inf. Div. He transferred to the Air Corp and received his wings with class of 44-B at Williams Field, AZ. He was assigned to duty with the 428th Ftr. Sdn., 474th Ftr. Gp., Europe.

He was shot down on first mission August 25, 1944, "Black Friday." He evaded capture and lived with a French family northwest of Beauvais, France until recovered by British Troops. He returned to duty and finished the war with 36 missions in P-38s. He was separated from the service in February of 1946 and made a career in the farm machinery business in California until retirement to Mesa, AZ in February 1986. He now lives just three miles due north of the end of the runway at "Willie" Field.

JOHN A. CAPUTO, born in New York City Mar. 17, 1919, John A. Caputo began his military career Dec. 15, 1941 as a Flying Cadet at Maxwell Field, AL.

An experimental group in Class 42-G, he graduated from both single and twin engine flying schools. Unknown at the time, the group was destined for assignment to P-38s, heretofore reserved only for single engine pilots.

As a fighter pilot and flight leader in North Africa and Italy he flew 50 missions in P-38s earning the DFC and the Air Medal with nine Oak Leaf Clusters. His combat score-all ME-109s—two confirmed victories, one probably and one damaged.

John's first aerial combat and first victory took place over Gabes, North Africa, Feb. 8, 1943. Eight victories were claimed that day, however one P-38 and four bombers (two B-25s and two B-26s were lost to enemy fighters and flak over the target.

During his varied career, John served as a P-38 fighter instructor after returning from overseas—both at NAS North Island and at Santa Rosa, CA. After transferring to the Military Air Transport Service in 1945, he qualified as C-54 aircraft commander at the ripe old age of 26. He went on to fly 176 trips to Berlin carrying coal, flour and food stuffs to the beleaguered German population during the Berlin Airlift.

Close to retirement his last two years of service were as operations officer for the Airborne Warning and Control Wing (AWACS) at Sacramento, CA. He also managed to squeeze in a few combat missions in Vietnam in C-121 airborne radar aircraft.

Following retirement in October 1969 as a Lieutenant Colonel, John joined the California State Department of Business Taxes as a tax representative, then under the administration of Gov. Ronald Reagan.

Retiring a second time after ten years of civil service in May 1981, John settled in Sacramento where he can indulge in his favorite outdoor activity—skiing.

JOSEPH A. CARDIMONA, was born in Wilkes-Barre, PA, Dec. 25, 1917. He graduated from Coughlin High School in 1935. He enlisted in the Army Air Corps on Jan. 2, 1942, applied for Cadet Training and was accepted in September of 1942. He went through Santa Ana preflight, Visalia, CA, primary, Gardner Field, Bakersfield, CA, basic and got his wings at Williams Field, Chandler, AZ with Class of 43G.

Cardimona flew 56 missions with the 82nd Fighter Gp. out of Foggia, Italy during which he was credited with two victories. He was rotated back to the states in August of 1944 and was stationed at Bend, OR as a P-38 instructor pilot. In the spring of 1945 he was sent to Boise, ID, where he checked out in the P-63 and flew as a target for bomber crew gun cameras out of Mountain Home AFB. From there he went to Ontario, CA, where he was an instructor pilot. While there he was awarded the DFC for his Foggia, Italy, activities.

On Sept. 16, 1944, he married Aline Trent of Hoopeston, IL, whom he met while he was stationed at Chanute AFB, Rantoul. He was

blessed with four sons, one of them has a doctorate in Lazer-Physics, the youngest is a doctoral candidate in Geophysics, one has a masters degree in psychology and one is a musician.

He was released from the Air Force in December 1945. He re-entered the Air Force in 1948 and was stationed at Bangor, ME, where he flew P47s and F-84 jets. He was released again in 1950, entered college and earned a degree in Mechanical Engineering. He was employed by General Motors in Milwaukee, WI.

He is presently enjoying his retirement, since 1983, in Greendale, WI, a garden community on the outskirts of metropolitan Milwaukee.

ROBERT CAREY, born in Spokane, WA, May 11, 1916 and educated at Opportunity Grade School in the Spokane Valley, North Central High in Spokane and graduated from Washington State University with an Engineering degree and a Reserve commission in the Corp of Engineers in 1938. Was trained in Industry at Caterpillar Tractor Co., Peoria, IL.

Was called to active duty in August 1941. Spent one year in Alaska and the Aleutians before entering flight training. Graduated with wings from Williams Field, Chandler, AZ, December 1943. After OTU Training in P-39s at Santa Rosa, CA and P-38s in Santa Ana, CA, I went overseas in March 1944, and joined the 430th Fighter Sdn. and the 474th Fighter Gp. After the war in Europe ended, returned to the States. Received a regular commission in July 1946. Spent 11 years in R&D at WPAF, Ohio and Baltimore, MD. Was assigned duty with HDQ TAC, Alaska Air Command, 5th AF in Vietnam and HDQ ADC in Colorado Springs. Retired as Colonel in July 1969.

I flew 54 different USAF aircraft in my career.

ROBERT N. CARLTON, was born in Morgan Hill, CA on Oct. 20, 1917. He attended San Jose State College, CA, and graduated with a Degree in Business. After working for Lockheed Aircraft Corp. he was accepted for pilot training in April, 1941. He graduated from Luke Field, AZ on Dec. 12, 1941.

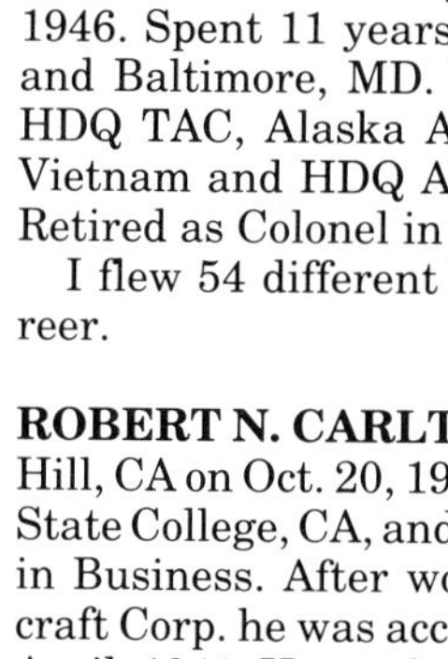

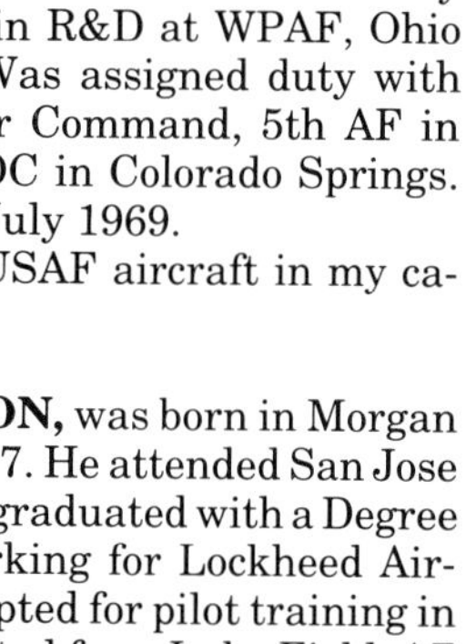

His first assignment was to March Field, CA where he became a member of the 49th Fighter Sdn., 14th Fighter Gp. Initially he flew P-40s and P-66s. Early in 1942 the squadron moved to Mills Field (now San Francisco International Airport), where he checked out in P-

38s. After training at San Francisco and Hamilton Field, CA, the 14th Fighter Gp. departed California for England. This was the second Group to fly P-38s to England.

After training in England for three months, his Group departed England and participated in the invasion of North Africa. Lt. "Legs" Carlton flew 42 combat missions in North Africa in P-38s and was credited with two aircraft shot down and one aircraft destroyed on the ground. He was awarded the DFC with one cluster and the Air Medal with three clusters. He was one of the first P-38 pilots to successfully bail out of a P-38 during a high speed dive.

He stayed in the Air Force after the war and retired as a Lieutenant Colonel in 1966 after 26 years of service. Since his retirement he has worked for the City of Burbank and for aerospace industries as a Logistics Engineer.

BRUCE CARNACHAN, when I graduated from High School in June 1942 I worked at Lockheed on P-38 construction. (Also Fornock was leadman—met him again in 1956-57 when he joined our sports car club, but transferred to Sunnyvale. Met a Mr. Scrermmer last Sunday at a car show for Heart Association. He was foreman of that department where I worked in 1942—small world.)

I went into the Air Force in December 1942, went to India in August 1943, then to China with a service squadron as an Armorer. Traveled about China with the 54th Service Sdn. Flew some missions as Gunner with B-25 Sdn. on loan, then in 1944 transferred to the 449th Fighter Sdn. as Armorer. This was only P-38 Sdn. in China. (There was another, 448th, in India.) Was with 449th till war's end, and even got a ride in one (radio equipment removed) after secessation of hostilities. Our base at this time was an old A.V.G. strip on Indo-China border at Mengtze. The Japanese military commander was flown into our field in a Betty Bomber (painted white with Swiss markings) to surrender.

As you can see, I look back on the P-38 with fond memories—also have respect for P-40s, as I worked on many in China and have a twinge whenever I see a B-25.

BOB CARPENTER, is now Chairman of the Board of a large plumbing and heating supply company with seven branches in Michigan.

Bob is a former 38 pilot with the 95th Fighter Sdn. of the 82nd Fighter Gp. The following citation hangs in his office.

Distinguished Flying Cross

"For extraordinary achievement in aerial flight as pilot of a P-38 type aircraft. On May 1944, Lieutenant Carpenter participated as escort to heavy bombers on a mission against enemy installations in Austria. After the bomber formation had completed a highly successful attack on the target, Lieutenant Carpenter's squadron strafed targets of opportunity enroute to base. Observing several trains enroute, Lieutenant Carpenter immediately attacked and personally destroyed four (4) locomotives and a number of rolling stock. Despite intense and accurate enemy anti-aircraft fire, Lieutenant Carpenter strafed an enemy airdrome and accounted for four (4) enemy aircraft destroyed on the ground. His aircraft was severely damaged by enemy flak, however, Lieutenant Carpenter courageously continued his attacks until his ammunition was expended and brought his crippled aircraft to base. By his outstanding professional skill, courage and devotion to duty, as evidenced throughout over thirty-eight (38) successful missions against the enemy, Lieutenant Carpenter has reflected great credit upon himself and the Armed Forces of the United States of America."

The citation left out the fact that he flew top cover over a buddy, Jack Jolly, on this mission returning to base. ME-109's hovered overhead. If they only knew that Jolly was so badly shot up that he couldn't maneuver, and that Carpenter was out of ammunition, there might be a different story to tell today.

DICK CARPER, born and raised in north east Oregon, Dick flew B-25s in advanced pilot training. A call for Photo-recon pilots put the entire class in P-38s. He was stationed in Florence, Italy, 12th Air Force, 5th Photo-recon Sdn. and spent most of his missions photographing the Brenner Pass. In late March 1945, returning from a mission, he decided to buzz the swimming pool where he knew his buddies would be hanging out. He was in a nose high position looking for the pool and rolled into a split "S," therefore coming in much lower and faster than planned. As a matter of fact, trim-tab pulled him out. The guys at the pool heard a big boom. Everyone in the pool jumped out, and everyone out jumped in. There happened to be one General at the pool who is still looking for Dick. He was moving too fast and low to be identified.

Dick stayed in the service one year after the war assigned to the "General Staff and Command School" in Kansas during which time he flew many types of aircraft. To him the P-38 was in a class all of its own.

JANE EMERSON CARTER, born in San Antonio, TX, but grew up in Kansas City, KS. Graduated from Stephens College, Columbia, MO. Earned a commercial, instructor rating in January 1940. Taught three classes of C.P.T., one of which was pre-glider training.

Joined WASP in December 1942 and graduated May 1943. Stationed at Love Field, Dallas with the 5th Ferrying Gp. until deactivation December 1944. Checked out in P-38 on Oct. 16, 1944 at Love Field. Picked up P-38 at Long Beach, CA and flew to Dallas Dec. 20, 1944. Watched parade honoring and deactivating WASPs from the air in a P-38. It seemed appropriate. After deactivation, worked as Airway and Tower Controller.

WILLIAM H. CAUGHLIN, was born January 19, 1922 in Calexico, CA. Aviation Cadet November 1942, graduated s/engine pilot November 3, 1943 Luke Field, Phoenix, AZ. Flew P-40s, P-39s, P-38s. Joined P-38 27th F Sdn., 1st F/Gp. 15th A.F., Foggia, Italy March 1944. Flew 50 missions, bomber escorts, f/sweep strafing d/bombing.

Destroyed three German fighters, and several on ground. Awarded DFC, Air Medal four/clusters, unit three Pres./citations. Roughest mission June 10, 1944 1st F.G. & 82 F.G. dive bombing Ploesti, Rumania. Home September 1944. Instructor pilot P-38s Santa Rosa, CA.

Overseas 1946-47 Straubing Germany flying P47s. Home May 1947 W/Patterson, Dayton, OH, F/Test Div. Flew jets, twin and four/engine bombers. H/discharge November 1948. Spent civilian life in cotton and textile business. Married 43 years to Blanche Batzer, have three sons, three daughters, eight grandchildren. Home Spring Valley, CA. Member 1st F. Gp. Assoc. 27th F. Sdn. H/Assoc. and National P-38 Assoc., alternate yearly reunions with great times shared by pilots, crew members and wives.

GIL CEFARATT, was born in Los Angeles, raised in Burbank and the San Fernando Valley. He attended Remsen Street (now Glenoaks) grammar school, San Fernando Jr. High and Verdugo Hills High in Sunland, all in California. He worked at Vega-Lockheed on P-38 wings in the summer of 1943. After a short stay at Walter Lantz Cartoon Studio and before graduating he enlisted in the U.S. Navy serving in the Radar section on various Ships and flying with a PB4Y-2 and PBM squadrons out of Sangley Point, Philippines. He earned his high school diploma after his enlistment. Three years after Naval service he went back to Lockheed working as an Electronic Technician in Engineering Flight Test, flying with some of the

greats like Tony LeVier, Herman "Fish" Salmon, Sammy Mason, Stan Beltz and Billy Robinson.

Gil served in Field Service as a Tech Rep. on P-2V Neptunes for the U.S. Navy, Tech. Rep on the F-104 Starfighter for USAF, Italian Air Force and the Turkish Air Force After being on the early Flight Test portion of the SR-71 Blackbird he served as a Tech. Rep on the P-3 Orion for the U.S. Navy. After being assigned to the C-5A Galaxy proposal he went into the Service Engineering group working as a Technical Writer on the P-2V Neptune, P-3 Orion, U-2 and SR-71. He helped to develop the Non Destructive Inspection (NDI) manuals for USAF on the T-33, F-104 and Constellation airplanes. He is currently in the Skunk Works Div. at Lockheed.

Gil is active in Air Shows in the Southern California area. He was active on the P-38 First Flight Golden Anniversary at Santa Maria and Burbank in January 1989. He conceived and was President of the P-38 National Convention in May 1987. He was active in the P-38 Fortieth Anniversary Symposium in September 1977. He was asked to be on the committee for the 1979 Los Angeles reunion of the P-47 Thunderbolt Pilots Association.

He is an Honorary member of the Pearl Harbor Survivors Association Chapter 12.

Member of the Van Nuys based AT-6 Condor Sdn. for 20 years.

He is a member of the Western Regional P-47 Pilots Association and an Honorary member of the 474FGA, 20FG and 55FG Associations.

OLIVER CELLINI, I graduated from Kelly Field (at that time it was called "West Point of the Air") Feb. 1, 1939 as a Pilot, 2nd Lieutenant Army Air Corps Reserve. I was assigned to the 17th Pursuit Sdn., First Pursuit Gp., Selfridge Field, MI. In 1939, 1940 we flew mostly P-35s, P-36s, P-39s, P-40s and other miscellaneous aircraft. I got my Regular Army Air Corps Commission in August 1939.

Apparently Wright Field, Patterson Field did not have enough test pilots to do all the work that was necessary especially after the Air Corps started expanding. Some of the key people down there at that time were Mark Bradley, Eston Schuster, George Price, Maj. Gilkey I'll never forget him, Ernie Warburton used to come in and others. Periodically they use to select some of us from Selfridge (Hot Pilots) to go to Wright Pat and perform what they use to call Service Test Flying.

I first checked out in the YP-38 on Apr. 25, 1941. I was told that I was the 17th military pilot to ever check out. As you can see from the record we all flew whatever had to be flown both at home base (Selfridge) and at Wright Pat.

In May of 1941 the 58th FG was formed from the 1st FG. I was made CO of the Headquarters Squadron. We were assigned to other units of the First for flying.

The 27th PS, 1st PG, Selfridge got the first YP 38 in June of 1941. Although I was assigned to the 58th FG (we had no airplanes) I flew with the First. I flew P 38s until I was transferred to Baton Rouge, LA in September, 1941. After that I rarely ever flew a 38 except when I was able to get over to Eglin Field. I had many friends over there so I got to fly whatever models they had for testing.

KAY GOTT CHAFFEY, born in Nampa, ID, and took Civilian Pilot Training (CPT) at the College of Idaho. She joined Civil Air Patrol and had Commercial Pilot's license before joining the Women Airforce Service Pilot (WASP) program December 1942, in the second group to be trained at Houston, TX. Class 43-W-2 helped move the school, flying the AT-6's to Sweetwater, TX for graduation. Then assigned to Air Transport Command, Ferrying Div., Love Field, Dallas, TX. Moving PT's, BTs, AT-6's, UC 78's and twin engine Beechcraft to where they were needed. After transition to heavy aircraft at Love Field Instrument School in B-18 and C-47, went to Pursuit School February 1944 at Palm Springs, CA. Began in BC-1, the P-40, P-47, P-39 and P-51. Then five women and 35 men scrambled upon graduation for their first assignment was to move to P-51's from Long Beach to the East Coast!

Later came P-63 check-out at the Bell factory and many trips to East Base, Great Falls, MT. Next, transition on B-25 and finally the P-38. Love Field had a Lockheed modification plant, and the WASP barracks rattled every night with P-38 guns firing next door. Sent to Orlando, FL for Officer's Candidate School but the WASP remained on the Civil Service Status and were deactivated Dec. 20, 1944. Kay returned to college, completed her degrees and taught until retirement.

GEORGE THROCKMORTON CHANDLER, born in Wichita, KS Feb. 1, 1921, joined the Army Air Corps in November of 1941. Graduated 42G from Luke Field and went immediately to Hawaii to fly P-40s. Flew P-40s in New Guinea with 5th Air Force, then to Guadalcanal to fly P-38s with 339th Sdn., 347th Gp. of the 13th Air Force. Fought through Solomon Islands to Rabaul. Credited with five Zero fighters shot down, awarded the DFC and Air Medal with 11 clusters. Returned to the United States to the 4th Air Force P-38 Training Base at Chico, CA where he stayed until the end of the war. Discharged as a Major in January of 1946, and returned to Kansas and entered the banking business.

From 1948 to present date, has moved from Executive Vice President to President to Chairman of the Board of the First National Bank in Pratt, KS, where he still resides with his wife, Barbara Ann — whose name "BARBARA-ANN " was painted on the nose of each of the P-38s he flew in the Solomons.

George and Barbara have three sons, one daughter and eight grandsons. One son is currently President of the First National Bank in Pratt.

FRANCIS H. CHAPMAN, was born at Auburn, CA on June, 23, 1924. At the age of three moved to and was raised in Fallon, NV. While attending the University of Nevada at Reno applied for Aviation Cadet Training on Dec. 9, 1942. He graduated from Williams Field, AZ in the Class of 44-C.

He flew P-39s first at Victorville Army Airforce Base, CA and at Moses Lake AAF Base, WA and finally to the P-38s at Ontario A. AF Base, CA. In September of 1944 was sent overseas to the ETO, in November was assigned to the 82nd Fighter Gp., 96th Fighter Sdn. in Foggia, Italy. Received the Air Medal with two Oak Leaf Clusters and finishing out the war with them. Returned to the U.S. in September of 1945.

In civilian life flew with the Nevada Air National Guard, being recalled to active duty for the Korean Conflict. Flew P-51s, and F-86s well into the 60s. Due to being transferred while working for McKesson Corporation (from 1947 till the present) finished his military career in California at Hamilton A.F. Base and has attained the rank of Lieutenant Colonel.

He and his wife of 43 years live in Monterey, CA and will retire there after next year.

GENE H. CHATFIELD, born Apr. 27, 1919, graduated from Single Engine Fighter School at Spence Field, GA, Class 43-D. A typical Army SNAFU sent him to join the 97th Sdn. of the 82nd Fighter Gp. in Lecce, Italy, to fly P-38s. Never having been a twin-engined plane, and never having seen a P-38 until then made Chatfield's first mission a memorable occasion. It was to be a "milk-run" escorting B-17s to Athens Greece. The Group was jumped, and in the ensuing dog fight, Chatfield found himself head on with a FW-190. Fortunately, the superior turning ability of the P-38 quickly put Chatfield on the tail of the FW-190. Repeated bursts of the 50 Calibers and 20 MM Canon did not set the enemy on fire or blow him up, as seen in the training films, so Chatfield kept on shooting until the Focke-Wulf crashed and burned in a field N.W. of Athens, and that was just the first day.

Chatfield attained the rank of Captain and was Operations Officer of the 97th Fighter Sdn. He flew 50 missions doing high altitude escort, straffing, fighter sweeps and dive bombing, over Italy, Austria, Germany and the Balkan States. He had three confirmed aerial victories and three probable, plus three locomotives and numerous planes on the ground. He was awarded the Air Medal with nine clusters, the D.F.C. and the Silver Star. He was separated from the Service in September 1945 while serving as Asst. Operations Officer at Van Nuys Air Base.

DONALD CHESSMAN

JAMES CHILDS INCE, was born at Paragould, AR on Mar. 28, 1919. Grew up in Boulder, CO. Left Colorado University to join AAF. Graduated from Luke Field, AZ, Class of 42-A. Sent immediately to Hawaii and joined the 6th Pursuit Sdn. at Wheeler Field. Flew P-26s, P-36s and P-40s on patrols searching for Japanese invasion fleet.

Ordered to New Guinea in July 1942, and joined the 80th Fighter Sdn. "Headhunters". Flew P-39s and earned the nickname "Impossible" by getting out of a number of tight situations with Zeroes.

Early 1943 the 80th received the first P-38s shipped to the Pacific. Scored 1st victory over Salamua in P-38s first fight. Joined 475th Fighter Gp. in late Summer 1943. Promoted to Captain and became Flight Leader in 432nd Sdn. based at Dobodura. Began long escort flights to Rabaul, Wewak and Hollandia.

In November 1943, while flying cover for Finchaven landing, Ince's Sdn. intercepted 20 Betty bombers escorted by 40 Zeroes. Bettys heading directly for shipping. Took P-38s straight thru Zeroes down to deck after bombers. Ince shot down one Betty, but had left engine shot out. As he pulled up was attacked by Zero. Ince made head-on attack and set the Zero on fire, barely missed being rammed. Feathered prop and returned home.

Ince returned to U.S. in March 1944. Was in Korea, Japan and Taiwan from 1951 thru 1955.

Retired as Lt. Colonel in 1965.

Tally: six confirmed four probables. Shared one troop transport at Bismark Sea.

Decorations: Silver Star, DFC w/one OLC, Air Medal w/four OLC.

LEO CHILDS, worked as project engineer for Lockheed Corporation, specifically in the modification of the P-38 into F-5s for photo reconnaissance missions. After the war, as a member of Gen. George Goddard's Air Force, Childs continued to pursue an extensive career in the development of aerial and space borne intelligence systems utilizing the latest development in remote sensing technology. He was a member of the design teams for the Northrup RB-49 flying wing, the RB-36 and RB-37 planes. Childs spent the last 22 years of his career with the NASA Johnson Space Center, Houston, TX, and played an active role in the development of the *LANDSAT* satellite systems for conducting earth resource surveys from space.

Childs retired from NASA in 1985 and devoted full-time to managing family owned timber land and other interests.

ELMER CHRISTENSEN (Ed), born Dec. 20, 1920 in Stoughton, WI. Graduate of South Texas Law School in 1959, and a practicing attorney in Houston, TX. Enlisted in the U.S. Air Force in 1942. Pre-flight in Santa Ana, CA, primary at Rankin Areo Academy, basic at Lancaster, CA graduated from Williams Fields in Decmber 1943. Joined the 367th Fighter Gp. in Europe in February, 1944 assigned to the 393rd Fighter Sdn., returned Feb. 25, 1945. Flew 70 missions in P38s mostly low level straffing or dive bombing. Awarded the air medal with ten Oak Leaf Clusters, Unit Presidential Citation with five medal bronze battle stars. Instructor pilot in P38s at Santa Maria Air Force Base and then at Barstow Air Force Base till discharge in October, 1945. Re-enlisted in May 1946 as a T./Sergeant and spent three years as a supply sergeant at Westover Air Force Base. Remained in the Air Force Reserve and flew 119s for the 446th Troop Carrier Gp. until discharge in 1976 as a Lieutenant Colonel.

Current Address is 14923 Bramblewood, Houston, TX 77079.

KENNETH B. CLARK, born in Kansas, MO Nov. 18, 1923. Entered Aviation Cadets from Seattle, WA, graduating at Luke Field, AZ, class 43-I. Completed Central Instructors' School, Randolph Field, and instructed at Marana and Williams Fields, AZ before entering P-38 transition, completed at Santa Maria AAF, CA in the Summer of '44. Joined the 9th Fighter Sdn., 49th Fighter Gp., 5th AF at Biak Island. Flew 101 combat missions in P-38s in the Philippines and Okinawa and was in MacArthurs "Air Honor Guard" of eight Lightnings-first fighters to land in Japan at war's end. Remained in service until retirement in the grade of Colonel, June 1974. Most of that time was spent in Air University and Military Airlift Commands. Flew C-124 "Old Shakey's" out of Tachikawa, Japan, (6th Troop Carrier Sdn.) from '58 to '61; was MAC Chief of Safety from '68 to '71 and later Wing/Base Commander at Scott and Ramey AF Bases.

Settled on a few acres near Macon, GA, and worked part-time for the Georgia Chamber of Commerce for ten years before becoming self-employed as an association management consultant. He and his wife of 43 years, Arlene, a native of Oregon, share interest in horses (Paso Finos), birds, wildlife in general and country living.

THOMAS J. CLELAND, was born in Los Angeles, CA, second oldest of four children, Jan. 27, 1920.

He learned to fly during Civilian Pilot Training (CPT) at Grand Central Airport, Glendale, CA in 1940. Primary training was accomplished in a Porterfield and secondary in a WACO-10. He joined Lockheed in 1940 and and was assigned to the P-38 Program. After two years at the factory he was asked to be a Technical Representative with the 8th and 9th Air Forces in the European Theater (ETO). P-38 bases he was assigned included: Gox Hill, Kirton in Lindsey, Burtonwood and High Ercall in England, and Langford Lodge and Altergrove in Northern Ireland.

Tom often flew piggy-back in the P-38. He returned to the States in 1945 and entered the U.S. Navy. After Navy service he continued with Lockheed and retired after 45 years. Tom is a widower and has one daughter, Kimberly.

HERB COCHRAN, Decatur, IL. A month after Pearl Harbor took tests from a Traveling Cadet Board. Went to flight school August 1942 in the southwest. Graduated from Williams Field, AZ in Class of 43-D.

Took P-38 training at Muroc (Edwards) and Santa Ana, CA, with classmates. He joined the newly formed 475th Gp. 433rd Sdn. in Australia. Took part in the Wewak, Hollandia and Rabaul Strikes.

Shot down two planes at Rabaul and another at Alexishafen, New Guinea. Became Captain and Flight Leader.

Experienced engine failure over the Wewak target. Flew home and landed on one engine with no problems. Also made single engine landing due to propeller trouble. On a mission take-off from Biak Island, fully loaded, had double engine failure from gas line problem; crash landed in ocean.

Completed 125 missions and rotated home in February 1945.

Retired after 38 years as foreman in a grain processing company.

OSCAR THEODORE COFFELT, Maj. (Ted), from Jackson, MI. Attended Jackson Jr. College. Flight schools Muskogee, Randolph, Kelly, Class of 41-G. In 441st AAFBU at Metropolitan Airport, Van Nuys, CA.

Married Catherine. Children: Candy, Caroline, Theodore. Died in Portland, OR, of natural causes Dec. 23, 1966.

GEORGE L. COFFEY, born Idabel, OK, reared in Norman, OK. Attended University of

Oklahoma for two years 1939 -1942. Took Civilian Pilot Training (CPT) and was licensed as pilot before enlisting in Army Air Corps as aviation cadet in May 1942. Called to active duty in November 1942 and graduated from Luke Field, AZ and commissioned 2nd Lieutenant Oct. 1, 1943 and rated Pilot Single Engine. This class of 43-I was utilized to provide pilots for the 367th Fighter Gp. at Santa Rosa, CA. Received 150 hours of P-39 operational training. Then transferred to North Island, San Diego to P-38 transitional training and shipped to England May 1942 for further training and reassignment as fighter pilot. Was utilized as a replacement pilot for the 33rd Photo Recon. Sdn. located at Charlgrove, England, flying combat missions from that location June and July. Moved to Normandy in July and to an air field near Pari at time of the liberation of Paris. The squadron moved to Charleroi, Belgium and then again to Y-10 near Brussels. Then to Venlo Germany and became the first USAAF unit to operate from German soil. Completed my 75th mission at this location and became eligible for return to ZI. The exploits of this squadron were published in 1987 by J.B. Woodson a member of the 33rd Photo Recon. Sdn. Association under the title of "Straight and Level."

Decorations included: the Air Medal with 11 OLC; Presidential Unit Citation; the Belgian Fourragere; and ETO medal with Battle Stars.

MILLARD F. COFFMAN, 1st Vice President P-38 National Association "Coffee" began his Air Force career at the Aviation Cadet Classification Center, Nashville, TN in February 1943. Classified "pilot" the next stop was preflight school pilot on Kelly Hill, San Antonio, TX, Class of 44-A. Primary flight training in PT-19, instructed by Mr. Hindrex at El Reno, OK. Basic flight training BT-13 instructed by Lt. Fritz at Strother Field, Winfield, KS. Became ill and was held over to finish with Class 44-B. Advance flight training in AT-6 and P-40 at Foster Field, Victoria, TX. Graduated 2nd Lieutenant.

P-38 combat training was at Santa Maria AAFB. Left for combat duty Jan. 1, 1945 just six days after he and Janet were married. He flew 51 combat missions in the P-38 as a member of the 12th (Dirty Dozen) Fighter Sdn., 18th Fighter Gp., 13th Air Force. Based on various islands of the Philippines. Missions were both local and to Formosa, Indo China, Borneo and Naval Convoy support.

One memorable flight was the ten-hour round trip from Palawan to Borneo May 5, 1945 of eight P-38s, each carrying two 165-gallon belly tanks. Total fuel available was 756, which would be sufficient for eight to eight and a half hours of normal combat missions. Strafing opportunities were excellent and the flight stayed too long on target, so the return was beset by low fuel anxieties.

Even reducing speed to 155 mph, RPM to 1500 and manifold pressure to between 25 to 30 inches, plus jettisoning everything possible, "Coffee" barely made it to Puerta Princesa, the home base. Once he reported going down to ditch when the left engine failed and the remaining power would not support flight, but at the last minute he was able to restart the engine and so made it to land instead of water

He retired from Northrup-Aircraft after 33 years and lives with his wife, Janet, in Burbank, CA.

Can you top this? "Coffee" claims on the longest, if not the longest, P-38 flight. His was ten hours long and was fighter sweeper with take-off from Westbrook Field, Puerto Princesa, at 0745 hours on May 4, 1945, P.I., to Kughing, Borneo, and return at 1745 hours. "Coffee" flew with the 12th Fighter Sdn., 18th Fighter Gp., 13 AF. This report was taken from a copy of the original Flight Report, Forms No. 1 and 1A listing aircraft time 466:04, Crew Chief, T/Sergeant Hauss, airplane P-38L-1-LO #44-24154.

KENT L. COPENHAVER, was born in Sayre, OK moved to Brownsville, TX in 1935. He entered the Air Force as a cadet in January 1943, sent to Jonesboro, AR prior to entering SAACC in San Antonio, TX. He took primary at Sikeston, MO, basic at Winfield, KY, advanced at Eagle Pass, TX graduating with class of 44E. Subsequently he had P-38 training at Chico, CA, P-39 training at Victorville then to 14th AF 449FS flying P-38s in Kunming, China via India.

He remained active in the Air Force Res. taking duty at Grand Prairie, TX flying the AT-6, T-28, and F-80, then to Kelly Field at San Antonio, TX flying C-119, retiring from AF Res. as Lieutenant Colonel in 1983.

He presently resides in Brownsville, TX.

ARCHIE B. CORD, JR., was born June 15, 1918 in Des Moines, IA. He enlisted in the Army Air Force in December of 1941. He trained in Oxnard, CA, and Roswell, NM

In September of 1942, Capt. Cord was assigned to Peterson Field, Colorado Springs. There he flew P-38s in the 24th Photo Recon. In July of 1943 he was assigned to 1st Mapping Gp. at Bolling Field, and from there was assigned to the Northwest Africa Photo Recon. Wing.

As a member of the 90th Photo Recon. Wing in the Mediterranean Theatre, under the com-

mand of Karl F. Polifka, Capt. Cord flew F-5s and B-25s on 13 reconnaissance missions during the Italian Campaign, receiving the European-African Middle East Campaign Medal with four Bronze Stars. Capt. Cord, in addition was awarded the First Oak Leaf Cluster to the Air Medal. Later, he acted as C.O. when Col. Polifka went on leave.

Capt. Cord was relieved from active duty in August of 1945, and became a member of the Air Force Reserve. He was a member of the 8506 Air Force Recovery Gp. in Lubbock, TX, and as a Lieutenant Colonel was commanding officer of the 9973 Air Reserve Sdn. in Odessa, TX.

At the time of his death, in December of 1972, he was commanding officer of the reserve squadron in Longview, TX.

JAMES H. COSTLEY, was born July 15, 1920 in Eveleth, MN.

Entered the service March 1943 at Santa Ana, CA. Class of 44A. Primary training - King City, CA. Basic-Chico, CA; Advanced-Williams Field, AZ; Operational Training, - Ontario, CA.

Sent overseas December of 1944. Assigned to the 7th Sdn. 49th Fighter Gp. Flew first combat mission Jan. 23, 1945. Flew 87 combat missions in all and received the Air Medal with two Oak Leaf Clusters.

Bases stationed overseas were San Jose, Mindoro; Lingayen Gulf, Luzon; Okinawa and Japan.

Departed Japan for home Thanksgiving Day 1945. Returned to railroad service and retired as conductor Apr. 30, 1981.

CHARLES N. COUTURE, was born in Cheboygan, MI. He attended Michigan State University, which he left after 3-1/2 years due to being drafted into the U.S. Army. After spending 1942 with the 10th Armd. Div. at Ft. Benning, GA, he transferred to the U.S.A.A.F. in 1943. Upon receiving his Wings and Commission at La Junta, CO, he transferred from B-25s to P-38s. Following three months in Photo Reconnaissance training, he served two years in combat in the South Pacific with the 26th Photo Recon. Sdn.

The F-5 P-38 Lightning, with all armament and excess weight removed, was the fastest plane in World War II. You had to be a little crazy to be a "Photo-Joe," because you flew mostly alone, without benefit of guns, your only defense being speed, which is why he rates the

P-38 Lightning the tops of World War II aircraft. He survived a nose gear collapse on takeoff, a single engine landing with the other engine on fire, plus 68 missions against the enemy.

His most memorable mission was catching the Japanese Navy in Hong Kong Harbor and dropping empty P-38 drop tanks on their ships, which they thought were napalm or fire bombs. He delighted watching the Japanese sailors set world records dispersing. He saw action out of Nadzab, New Guinea; Biak, Netherlands East Indies; Leyte, Lingayen Gulf, and Clark Field at Manila, Philippines; Okinawa, and Kimpo, Seoul, Korea for which he received 19 decoration. He say she will be forever grateful to his Crew Chief and Mechanics.

After leaving the Air Force, he married Marjorie Langmaid and had two sons and two daughters. They lived in New York City and later Michigan, where he worked for 35 years as a manager for American Tobacco Co. until retirement in 1985.

JOHN E. COX, was born at Bloomfield, MO on Nov. 4, 1918, son of Clarence and Minnie Cox.

He entered service June 10, 1941 at Jefferson Barracks, MO and was stationed at Portland, OR and Olympia, WA. He shipped overseas on Feb. 8, 1943, serving in the European-African-Middle Eastern Theater and was separated at Jefferson Barracks, MO on Aug. 31, 1945. He remained in St. Louis working until October 1956 when he moved to Bernie, MO where he operated a Zepher service station from 1960 to 1976.

John was with the 37th Fighter Sdn., 14th Fighter Gp., Transportation, Special Vehicle Opr. 932 in Tunisia, Sicily, Naples Foggia, air offensive, Europe, Southern France, Rome, Air Combat Balkans, Northern France, Rhineland, North Apennines, Po Valley. He was burned in 1943 and spent 93 days in a hospital in Oran, Italy.

He was united in marriage to Grace Frazier on Sept. 4, 1954. In January 1977 he and his wife retired and moved to Isabella, MO on beautiful Bull Shoals Lake and are now enjoying retired life.

HARRY C. CRIM JR., born 1919, raised in Miami, FL. Entered the Army in June 1941, assigned infantry, served at Ft. Bragg, Camp Wheeler, and Ft. Benning. He became an Aviation Cadet and was commissioned with Class 42D, Foster Field, Single Engine Pursuit P-35. Victoria, TX.

Assignment the 55th Pursuit Gp. Olympia, WA, with the 37th Sdn. P-43s and P-38s. The P-38 were first Ds, then Es, then Modified Es. Harry made one of the first successful bailouts when an engine burned in July 1942. The 37th Sdn. was transferred intact to the 14th Fighter Gp. at La Senia Airdrome, Oran, Africa. The 37th was equipped with F Model P-38. Escort flights for B-17s, B-26s and B-25s as well as dive bombing and close support missions were flown from a series of makeshift fields like Tafarou, Masion La Blanc, Telergma, Mejez-El-Bab and Zaghouan until the end in Africa. Then escort missions were flown to Sicily, Sardina and Italy. An airfield strafing mission was flown from Africa to Mostar, Yugoslovia. THREE FOR ONE missions were flown to Pantelleria, being first a B-17 escort then a B-26 escort and then dive bombing-strafing, all in one flight.

He flew the invasion of Sicily and Italy doing battlefield interdiction, close support and air superiority missions, moving to Sicily when it fell. He participated in the fighter sweeps of the Foggia air field complex. The 14th Gp. was shifted to Gambut #2 in the Lybian Desert and flew cover for a Naval evacuation of personnel from the Grecian Islands. He started the African Campaign as a Flight Leader and progressed to Squadron Commander finishing 50 Missions in November 1943.

1944 - P-38 RTU Ontario, CA, September 1944 to the 21st Gp., 72nd Sdn. as Operations Officer in Hawaii. In January the 21st Gp. converted to P-51s and was shipped by carrier to Guam. It moved to Iwo Jima by way of Tinian and began long range escort of B-29s to Japan. After a Bonzi attack killed and wounded a number of the 21st pilots, he was assigned Squadron Commander of the 531st Sdn., which he led until the war ended.

His decorations include the Silver Star, the DFC with three clusters, and the Air Medal with 19 clusters. Official victories were six Airborne and eight Ground.

After WW II he served in the Reserve while attending the University of Florida. Graduating in Mechanical Engineering in 1947. He was Commissioned in the Regular Air Force in 1947 and assigned as Commander of Maintenance Sdn. of the 20th Fighter Bomb Wing. The 20th converted to F-84 during this period. He was designated to participate in the second series of Atomic Weapons testings in the Pacific at "Sandstone" and directed a radiation detection and quantifying operation there.

From 1952 - 1956 he was assigned to the Pentagon in Research and Development on guns, ammunition, rockets and fire control systems. From 1956-1959 he was an Air Force Advisor to the Ambassador to NATO stationed in Paris, France. From 1959-1964 he was in Engineering, F-11 Weapon System Office stationed at Wright-Patterson AFB as Deputy then Chief Engineer. From 1964-1967 as Division Chief in Research and Development, Headquarters USAF, where he retired as a Regular Air Force Colonel.

He went with Lockheed Georgia Company as a scientist from 1967 till 1971 at which time he formed a company of his own Magnolia Enterprises engaged in Consulting, Construction, Research and Development.

Harry lives with his lovely wife of 44 years in Atlanta, GA, has three grown children and three grandchildren, enjoys excellent health, and is active in his business.

CHARLES J. CROCKER, was born in Pawtucket, RI on Oct. 9, 1920. Attended grade and high school in Marshfield, WI. Attended Ripon College two years at Ripon, WI.

Entered flight training in the U.S.A.C. in Texas September 1941; graduating from Randolph Field April 1942. Served in the North Africa, Sicily, Italy Campaign 1943.

Returned to U.S.A. January 1944 and put in for duty in Pacific arena. Departed Ontario Air Base, CA August 1944 for Iwo Jima. Was in mid-air collision during maneuvers at Oahu on Oct. 18, 1944. After 3-1/2 years in military hospitals was medically discharged.

Retired after 25 years serving as President and Managing Director of the Spokane Industrial Park Inc. in Washington. To Whidbey Island, WA.

Married Hyla Yelle in Olympia, WA in January 1943. Died June 16, 1987. Survived by wife, son Charles J. Crocker, Jr., and daughter Barbara Jane Olstad and three grandchildren.

DEWEY CROWE, Lt., from Johnson City, TN. Attended Georgia Tech. Flight Schools Dallas, Brady, Kelly. Class of 41-G.

Killed in action Sept. 14, 1942, Aleutians 54th FS. He collided with Squadron Commander Jackson over Kiska (his photograph is in the Kelly School "The Gig Sheet").

ALFRED L. CUMMINGS, born in Hugo, MN on Dec. 19, 1918 and graduated from White Bear Lake High School (Minnesota) in 1937. Entered Aviation Cadet Training on Apr. 2, 1942 at Lemoore Air Force Station, CA. After completing Primary and Basic Flight Training at Lemoore, he went on to Advanced Flight at Yuma Army Air Base, AZ, receiving his Wings on Apr. 12, 1943. He attended Fixed Gunnery School in Gila Bend, AZ.

May 1943 until February 1945 he was assigned to the 82nd FIW, Las Vegas, NV flying AT-6, P-39, and P-63 aircraft. In March 1945 he began flying the P-38L at Ontario AAF, CA. On Feb. 27, 1946 he was transferred to t he 475th Fighter Gp., 432nd Sdn., Korea flying P38L5 aircraft, with additional duty as Armament Officer. His last military flight was on Nov. 8, 1946, closing out with 1851 hours of pilot time. He returned to the U.S. on Dec. 19, 1946 and was discharged with the rank of 1st Lieutenant on Feb. 26, 1947 at Camp Beale, CA.

Upon his return to civilian life and Minnesota, he went back to work for his previous employer, the Northern Pacific Railway. Failing health forced an early retirement in 1965 when he underwent two open heart surgeries.

In 1951 he married Jeanette Stone in Thief River Falls, MN. They had three children. Paul, the eldest, is an Industrial Engineer and Aircraft Maintenance Officer with the 148th FIG (ANG); Catherine, who is an Occupational Therapist; and Thomas, a recent University of Minnesota graduate. As his health permitted, Al enjoyed hunting, fishing, and gem collecting until his death in 1985.

MITCHELL S. CWIEK, born in Terryville, CT, Sept. 21, 1922.

Schools attended: Midwestern University, and University of Maryland.

Began flying as Aviation Cadet at Thunderbird Field, Glendale, AZ January 1943. Solo in Stearman PT-17 Aviation Cadet Class 43-F.

Entered Aviation Cadet Training at Santa Ana, CA in September 1942. Primary Flight Training; Thunderbird Field, Glendale AZ. PT-17. Basic Flight Training: Marana Army Air Base, Marana, AZ BT-13A. Advanced Flight Training: Williams Field, Chandler, AZ AT-9, AT-6, P-322. Transition Training: La Junta, CO B-25s.

Combat Crew Training: Columbia Army Air Base, Columbia, SC B-25.

Combat Crew Training: Drew Field, Tampa, FL B-17.

Assigned to 15th Air Force...crew flew new B-17G from Homestead, FL to Foggia Italy, via Fortaleza, Brazil and Dakar, Africa. First mission: Ploesti, Romania on May 5, 1944 (the first high altitude bombing raid on this target).

Flew 11 combat missions in B-17s, assigned to 352nd Bomb Sdn., 301st Bomb Gp. While in hospital with food poisoning in May 1944, five of my B-17 crew were killed in a mid-air collision while returning from a bombing mission.

In July 1944 I was successful in getting approval from 15th Air Force for a transfer from the 301st Bomb Gp. to the 14th Fighter Gp., a P-38 Gp. flying out of San Severo, Italy. This effort to get a transfer was prompted by former flying school mates I met in the hospital who encouraged me to try. As soon as I got out of the hospital I started a campaign to get approval for the transfer, which was eventually successful! The former class mates were P-38 pilots with the 14th F.G.

Transferred to the 14th Fighter Gp. and assigned to the 37th Fighter Sdn. in late July 1944.

After very little refresher training in the P-38 (I had been checked out in the P-322 back at Williams Field, AZ in June 1944) the squadron went on TDY to Corsica to participate in the invasion of southern France. I flew five missions in the P-38 out of Corsica.

Returning to our Italy base in San Severo, was scheduled to fly as wingman for the Group Commander on a bomber escort mission to Austria on Aug. 22, 1944. Over Hungary we encountered enemy fighters and I was shot down during the dog fight.

Captured and spent the next eight months as a POW at Stalag Luft III near Sagan, Germany and later at Stalag 7A near Mooseburg, Germany.

Following liberation (Apr. 29, 1945) and a stay in hospital in Paris, France and Wales, I returned to the U.S. Following return, flew with Proving Ground Command and then entered Aerial Photo School. Later served in Japan and Yokota, supporting the bombing and reconnaissance effort against North Korea, assigned to the 6th Photo Technical and the 548th Recon. Technical Sdn.

Remaining Air Force career was in Reconnaissance/Photo Intelligence, being assigned to the Pentagon and later supervising the Air Force Photo Intelligence School in Wichita Falls, TX. Spent 30 months with the British Royal Air Force as an Exchange Officer, teaching in their Joint School of Photo Interpretation. Retired from the Air Force as a Major, while assigned as Intelligence Staff Officer with the 363rd Tactical Recon. Wing, Shaw Air Force Base, SC.

After the completion of a full 20 years in the Air Force I worked for 15 years for the U.S. Government in Washington, DC. My specialty was strategic reconnaissance.

Was married to Norma Jo Kenworthy in Bixby, OK on Dec. 25, 1946. We had four children (three girls and one boy) and now have three grandchildren with another expected in May 1987.

Still married to the same gal and enjoying life in Escondido, CA.

To keep occupied I run a small business. Am a Registered Representative and an insurance agent.

PERRY J. DAHL, Col., is Deputy Chief of Staff, Plans and Programs, Headquarters North American Air Defense Command and Headquarters Aerospace Defense Command. He assumed this position in August 1977 after

serving as Assistant DCS/Plans and Programs.

Born Feb. 18, 1923 in Saskatchewan, Canada, Col. Dahl migrated to Seattle, WA, at an early age. He entered the Army, Infantry Branch, as an enlisted man in 1940. He applied for aviation cadet training and upon graduation in June of 1943 became a second lieutenant, U.S. Army Air Corps. He served in the SW Pacific Theatre where as a P-38 pilot he downed nine enemy aircraft in air to air combat.

Following relief from active duty in 1945, he attended the University of Washington and Southern Colorado State College where he received his Bachelor of Science Degree.

Recalled to active duty in 1950, Col. Dahl served as a flight test pilot at the Air Force Maintenance Depot in Chateauroux, France, flight testing F-84 and T-33 aircraft.

In 1954 he returned to the United States and became editor of Flying Safety Magazine, Directorate of Flight Safety Research, Norton AFB, CA.

He attended Air Command and Staff College in 1957 and upon graduation became flying safety officer with the Washington Air Defense Sector, Air Defense Command, Ft. Lee, VA.

In 1960 he assumed command of the 734th Aircraft Warning and Control Sdn., Morocco, North Africa, as a remote assignment.

Upon reassignment to the U.S., he became fighter interceptor officer, 25th Air Div., McChord AFB, Washington, equipped with F-106 fighter interceptors.

He was assigned to Headquarters Air Defense Command, Ent Air Force Base, CO in August 1963 as Chief, Base Utilization Branch, DCS/Plans.

In 1966 he joined the Air Staff, Headquarters USAF, with the Directorate of Aerospace Program. While in the Pentagon he was chairman of the group that developed the CONUS Base Posture Study for the Air Force Chief of Staff.

In 1969 he entered training as a forward air controller at Eglin Air Force Base and went to SEA as the Director of the Direct Support Center for III and IV Corps. In this role he had operational control of 15 forward operating locations tasked with performing the forward air controller function in support of U.S. Army and ARVN ground forces. He flew 175 combat missions as an armed forward air controller in the OV 10 aircraft.

Following the tour in Southeast Asia, Col. Dahl was assigned to the United States Air Force Academy where he served as Deputy Commandant for the Cadet Wing and Vice Commandant of Cadets until June 1974.

Col. Dahl returned to Southeast Asia in July 1974 as Commander, 56th Special Operations Wing, Nakhon Phanom, Thailand.

A command pilot, Col. Dahl's decorations include the Silver Star, Legion of Merit, Distinguished Flying Cross, Bronze Star Medal, Meritorious Service Medal with one Oak Leaf Cluster, Air Medal with 15 Oak Leaf Clusters, Air Force Commendation Medal, Purple Heart,

Presidential Unit Citation with one Oak Leaf Cluster, Good Conduct Medal and Vietnam Cross of Gallantry with palm.

Col. Dahl is married to the former Barbara Cox of Seattle, WA. They are the parents of Jeffery and Jody Dahl. Retired June 30, 1978.

LLOYD THOMAS DANIELS, (known as Tom or Dan) was born in Ontario, CA Mar. 23, 1921.

At the age of 20 he enlisted in the Aviation Cadet Program of the U.S. Army Air Corps. He graduated from flight school at Williams Field, AZ on Feb. 6, 1943 with the class of 43-B. After completing P-38 training at Peterson Field he was assigned to the 8th Photo Recon. Sdn. at 14 Mile airstrip Port Moresby, New Guinea arriving there in September 1943. He stayed with this squadron for 22 months during which time he flew 75 combat missions with a total of 333 combat hrs. in F-5 or P-38 type A/C.

During this time Daniels attained the rank of captain and was awarded the Distinguished Flying Cross, the Air Medal with two Oak Leaf Clusters, and the Southwest Pacific Medal with nine campaign Battle Stars. After separation in 1946 he instructed civilian flying for two years and was an Alaska bush pilot for four years. He was re-called to active duty in 1952 and flew F-86 fighters for the South Korean Air Force as an advisor. After a tour in Europe he flew electronic counter measures in B-66 type A/C during the Cuban crises.

He retired in 1963 and now resides at 1263 W. "D" St. Ontario, CA 91762 (714/984-1925).

CHARLES C. DAUGHERTY, was born May 17, 1922, in El Nora, IN, and enlisted for aviation cadet training in September 1942. He completed advanced single engine pilot training at Moore Field, Mission, TX, Class of 44-E, and flew P-39s at Victorville, CA, and the P-63 Kingcobra at Ephrata, WA, where he became a P-38 pilot by a strange quirk of events.

In November 1944, all the P-63s at Ephrata were flown to Great Falls, MT, the "Red Star" painted on them, and they were flown by Russian pilots to Russia via Alaska! The single engine pilots at Ephrata, after a brief ground school introduction and a "piggy-back" ride then became P-38 pilots. Following further P-38 time at Barstow, CA, and Redmond, OR, he and a small number of other P-38 pilots were sent to Italy and he was assigned to the 49th "Hangman" Fighter Sdn., 14th Fighter Gp., 15th Air Force. In addition to logging considerable combat rated flying time, he assisted in dismantling the 49th Fighter Sdn. in August 1945, when all of its P-38s were flown from Foggia down to Bari, Italy, to join acres of then unneeded fighter planes. "It was a strange and lonely feeling to taxi that beautiful monster up to a line of its faithful silent brothers, park him wing-tip to wing-tip, set the brakes, turn off the deep-throated engines, watch the props wind down to a halt, and then climb down and walk away — for the last time!"

Daugherty attended Franklin College, and after the war graduated from Butler University and the Indiana University Law School. He served 26 years as Judge of various courts in Indianapolis, IN, retiring in 1984, and is presently engaged in the practice of law, on a limited basis, in that city.

ROBERT EDWIN DAVIS, was born in Chicago, IL on Sept. 21, 1924. He volunteered for the USAAF in June 1943. One year later in June 1944, he graduated as a B-25 Pilot with the Class of 44G, from Mather Air Force Base in Sacramento, CA. He went to Will Rogers Field in Oklahoma City for Advanced Flight Training in P38s in Photo Reconnaissance. From there it was overseas to the 32nd Photo Recon. Sdn., 15th Air Force, stationed in San Severo, Italy where he flew missions in Yugoslavia, Germany, Czechoslovakia and Northern Italy. He returned to the U.S. in May 1945 and was discharged in September of 1945. Bob graduated from the Illinois Institute of Technology in June 1948 with BA in Civil Engineering and went to work in the Chicago Office of the New York Central RR. He has since worked for Mac Donald Engineering Co., Parsons Brinckerhoff Quade & Douglas and for the past 23 years for the Agency for International Development which is under the State Department.

Bob married Blanche Elizabeth Prucha in 1947. They just celebrated their 40th wedding anniversary. They have two children and five grandchildren. They presently reside in Tuscon, AZ.

RICHARD P. DeGREY, born May 16, 1919 in Salt Lake City, UT. From the time he was 12 years old, his goal was to become an airplane designer. After completing his aeronautical engineering training he was immediately hired at Lockheed on Oct. 5, 1938. After working on three pre-war projects he was assigned in 1940 to the P-38 design group and initially made layouts and drawings for the Model 322, which was a variant of the XP38, for the British Government. December 7, 1941 realigned the whole P-38 Program to the U.S. Army Air Corp. During the period 1940 to V.J. Day September 1945, DeGrey "Fought in the Battle of Burbank." His responsibilities included design work and supervision on hydraulic and fuel systems, aileron booster, powerplant installations and fuel drop tanks systems. At wars end DeGrey was a group engineer. The Production rates had grown to approximately 20 P-38s a day. The 10,000th airplane was near the end of the production line on V.J. Day. The entire Lockheed Burbank plant was completely covered by a camouflage net of chicken wire, poles and cables, chicken feathers and tar. From the air, the plant looked like just another residential section of the city, complete with trees and houses, sidewalks and roads. It was an engineering marvel, but when it rained, DeGrey remembers the odor of millions of wet chicken feathers almost made work impossible for a couple of days.

After the war, he joined Kelly Johnson's famed "Skunk Work" and has participated in the design or management of most of the famous aircraft that group have created over the years. He considers his most interesting and challenging project, the YF-2/SR-71. The P-38 was a close second.

GERALD DeJONG, parachuted to safety out of a disabled P-38 on September 1944 over Joplin, MO. He was a photo recon. pilot in final training at Coffeyville, KS when he experienced oxygen failure at 30,000 ft. Realizing he was passing out from anoxia he rolled his P-38 on its back and dove to reach oxygen and life at a lower altitude. Regaining consciousness at approx. 18,000 ft. he found his airspeed indicating 520 MPH and heading almost straight down. At about 10,000 ft. the tail section between the booms was torn off by air compressibility. He exited the now tumbling P-38 at approx. 5,000 ft. and floated to the ground in a farmers pasture.

Born in Lynden, WA he joined the USAAF Cadet Program February 1943 and was trained in the Western Training Command, seeing training in Utah, California, Arizona and Colorado. He graduated in class 44-F at La Junta, CO as a 2nd Lieutenant.

On Thanksgiving day 1944 he joined the 26th Photo Recon. Sdn. on Biak Island in the Southwest Pacific Theater. He flew combat photo missions from such places as Biak Is., Moratai Is., Tacloban on Leyte Is., Mindoro Is., and Lingayen, Luzon Is. in the Philippines. Photographing and mapping the enemy in the Philippines, the China coast, Formosa and the islands between. He was in a flight of the first planes to reach Shanghai, China at wars end and spent several months in Korea as occupation forces mapping South Korea.

De Jong arrived home two days before Christmas in 1945 and married his high school sweetheart.

He has spent the past 28 years owning and operating a funeral home in Colton, CA.

JULIAN G. DEMERS, was born in St. Johnsbury, VT. He volunteered and was accepted to Aviation Cadet Detachment, AAF on Mar. 20, 1942. He did his single engine combat flight training (P-39s) at Page Field, Ft. Myers, FL, and was assigned to combat duty in July of 1943 at the "tender" age of 25! After less than 16 hours transition time to P-38s in Algeria, "Pappy" Demers flew combat missions out of Lecce and Foggia, Italy with the 95th FS, 82nd FG. He flew bomber escort missions to targets at almost all major points of Axis-held Europe, including Ploesti, Budapest, Wiener-Neustadt, and Munich. On one such mission to Sofia, Bulgaria flight leader 1st Lt. Demers called to another flight and dove to intercept ten attacking enemy aircraft.

Forcing them down to low level, he then closed in to point-blank range on one enemy fighter and shot until it crashed into the ground. So successful were his tactics that five members of the two flights destroyed one enemy aircraft each and damaged two others. "Pappy" Demers was awarded the DFC for aggressiveness and superior leadership shown in the encounter. After 50 combat missions, two confirmed aircraft destroyed, the DFC, ETO ribbon with three Bronze Stars and Air Medal with nine Oak Leaf Clusters, he spent the remainder of the war training P-38 pilots in Santa Maria, CA.

After Vermont Air National Guard duty and the call to active duty in '51, "Pappy" ultimately retired with the rank of Major and 20 active duty years of service to his country. Now living in Fallon, NV with his wife of 40 + years; is retired to his woodworking shop.

ELLIOTT E. DENT, JR., was born Aug. 12, 1921, at Saint Winifreds Freehold, Saint Mary's Co., MD. He attended Charlotte Hall Military Academy, Univeristy of Alabama, and Class of 42-K Cadet Training.

He flew with the 7th Fighter Sdn., 49th Fighter Gp., 5th Air Force in New Guinea. They were stationed at Dobodura, Gusap, Hollandia, Biak, and Tacloban, Leyte, Philippine Islands. He flew P-40s most of the time and switched to P-38s in Biak. He was in the group of 32 planes out of the whole Fifth Air Force that made it through the weather during the invasion of the Philippines, and was shot down over Ormoc Bay on the seventh day after arrival at Tacloban. He lived with the Philippine guerrillas for several weeks, then walked through Japanese lines awarded the Distinguished Service Cross, Dis-

tinguished Flying Cross with three Oak Leaf Clusters, Air Medal with four Oak Leaf Clusters, Purple Heart with one Cluster, and Presidential Unit Citation with one Cluster.

Leyte, Philippine Islands, on Nov. 1, 1944 while on patrol over Ormoc, Layto, Capt. Dent led his flight of four P-38 aircraft in an attack upon 15 to 20 enemy fighters protecting a large Japanese convoy. As he was closing in on one of the enemy fighters, four others came out of the clouds and, though he had become separated from the rest of his flight and was alone, he immediately made a head-on pass at them. For 15 to 20 minutes he battled them fiercely, firing short bursts at each fighter in turn. By skillful maneuvering he escaped their determined attacks and sent two of them crashing into the sea and a third into a mangrove swamp. He was pursuing the fourth enemy fighter when antiaircraft fire from a Japanese destroyer shot out the right engine of his airplane and, with smoke billowing into the cockpit, he was forced to parachute into the water. Though he was in sight of Japanese ships, he managed to escape detection and was rescued that night.

Dent sold his company ten years ago and was retired four days before starting his present company, Advertising Specialties. It's a fun business which leaves free time to play golf, etc.

JAMES P. DIBBLE, born Nov. 3, 1921 in Grand Rapids, MI. Enlisted in Army Air Corps on Apr. 6, 1942. Completed Elementary Training in PT-22s at King City, CA, May-July 1942.

Completed Basic Training in BT-13s at Chico Army School, Chico, CA, July-September 1942.

Completed Advance Training in AT-6s and AT-9s at Luke Field, AZ and received Pilots Wings on Dec. 3, 1942.

Assigned to the 337th Fighter Sdn. at Glendale, CA, December 1942 for training in the P-38.

Transferred May 5, 1943 to the 1st Fighter Gp., 94th Fighter Sdn., based at Chateaudun du Romel, Algeria, and later Mateur, Algeria.

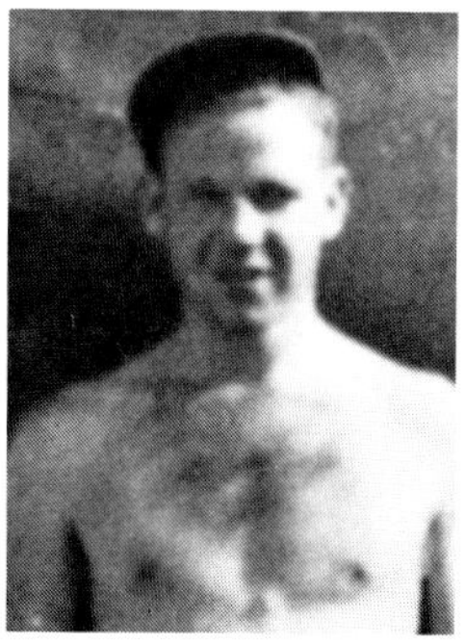

While a member of the 94th Fighter Sdn., Lt. Dibble flew 49 combat missions, totalling 194:40 total combat hours.

He was credited with destroying two (2) ME-109s and probably destroying one other.

He was killed in action, Sept. 9, 1943, while leading his flight in support of the Salerno Invasion. Ordered to strafe an enemy column of over 400 vehicles, Dibble lead his flight behind the mountains down a valley toward a road where the enemy reinforcements were waiting. In spite of their speed of over 300 miles per hour, the flight was hit from all sides by German antiaircraft fire. Dibble was shot down, crashing in a ball of wreckage. His wing man, Lt. Stan Wojcik, was also hit, but was able to pull his burning P-38 up and bale-out the only possible way. Lt. Wojcik stood up in the cockpit, forcing himself out backwards, successfully missing the tail of his plane.

To escape the anti-aircraft fire, the remaining members of the flight had to fly under Lt. Wojcik as he floated toward his captors.

1st Lt. Dibble was awarded the Air Medal with nine oakleaf clusters, and the Presidential Distinguished Unit Citation with an Oak leaf Cluster. He was recommended for the Distinguished Flying Cross just prior to his death, and the family is currently trying to obtain that award.

BYRON DOBBS, (Barney), born Jan. 14, 1920 in Phoebus, VA. Attended technical school at Lowry Field, CO in 1940. Entered training in May 1942. Joined the 1st Pursuit Gp. at Selfridge Field, MI. Student at PT-22 Ryan Field, Hemet, CA. BT-13 Minter Field, Bakersfield, CA and Williams Field, Phoenix, AZ. Advanced training at Williams Field in December 1942, Class 43B. Then assigned to Peterson Field, Colorado Springs, CO.

Assigned to Photo Recon. Training, Peterson, FL. Joined the 5th AF, 8th Photo Recon. Sdn. based in the Southwest Pacific. Fought mosquitos, malaria, jungle rot and Japs. Flew photo and armed recon missions. No kills, but took many nice pictures.

After 13 months and 88 combat missions, in September, 1944 proceeded to Will Rogers Field, Oklahoma City, OK. After that I was sent to the Pilot's Aircraft Maintenance Course at Chanute Field, IL. We returned to Will Rogers Field just in time to close the base and take the birds and Chinese students to Rapid City Army Base, SD. Retired from the Air Force on July 1, 1979 at Kelly, AFB, TX.

For seven years have enjoyed fun, games and travel. Married on Sept. 26, 1942 at Minter Field Chapel to Kathleen Daly. Have four children: Lt. Col. Robert L. Dobbs, USAF, Capt. Mark L. Dobbs, KIA, KC 135 Tanker crash at Castle AFB, CA on Sept. 19, 1970, Sandra Kay and Kathleen Anne. Six grandchildren, three of each: Jimmy, Frank, and James, and Erin, Jennifer and Rachel.

EDWARD W. DOERING, Lt. Col., USAF, Ret., born suburban Detroit, MI, raised in Suburban Philadelphia, PA. Entered Aviation Cadet Service July 1942. Called to Santa Ana, CA, February 1943; graduated Williams Field, Chandler, AZ, Class 44A, Jan. 7, 1944, P-38s. Extended training at Salinas, CA, and Ontario AAB, Ontario, CA, before being assigned to the

20th Fighter Gp., Kingscliffe, England, May 1944. In July 1944, the Group converted to P-51s. On Aug. 27, 1944, was hit by ground fire while strafing in Denmark. Attempted to return to England, but had to bail out over North Sea, and five (5) days later, was rescued by a German minesweeper. Was POW at Stalag Luft I, Barth, Germany. Released by Russians May 1, 1945.

Last P-38 time was in July 1945, at Ft. Worth Army Field, Ft. Worth, TX. Other Aircraft flown: P-51, P-47, RP-322, T-33, F-86F, F5A & B, T-29, C-45, C-46, C-47, C-54, B-25, B-47. B-52D.

Now retired, (30 years Active Duty) he and his wife Jane, (married Jan. 11, 1944) live in Greenbush, MI, near Wurtsmith AFB. They have two daughters and one son, each married, and five grandchildren. Many hobbies include gardening, fishing, travel, and enjoying these grandchildren.

WALTER FRANCIS DUKE, born Aug. 6, 1921 in Leonardtown, MD. Attended Leonard Hall, St. Mary's Academy, and one semester at the University of Maryland.

Enlisted Royal Canadian Air Force July 24, 1942. Preflight at Fingal, Ontario. Elementary at Toronto and St. Catherine, Ontario.

Service Training at Dunnville, Ontario.

Enlisted U.S. Army May 20, 1942. Advanced flight training at Napier Field, AL graduated Aug. 5, 1942.

Assigned first to 56th Fighter Gp. then to 80th Fighter Gp., 89th Sdn., at Mitchell Field, NY. Went overseas with the 80th. Flew P-40N out of Jorhat, Sookerating and Nagaghuli, Assam (India). Transferred to 459th Fighter Sdn. at Kurmitola, Bengal (India) 459th moved to Chittagong, Bengal on Mar. 1, 1944.

Missing in action June 6, 1944. After a fighter sweep to Meiktila and Heho Airfields, Burma, he was unable to locate his wingman, went to look for him. Japanese intelligence information indicates while alone he was jumped by several fighters and shot down three before he was shot down. These have never been credited by the USAF.

Record: Air ten victories, three probable, six damaged. Ground 8.5 destroyed, one damaged.

Decorations: Silver Star, Distinguished Flying Cross w/two OLC, Air Medal w/two OLC, Purple Heart, two DUC.

Married Verja Ann Graham of Washington, DC on Apr. 13,. 1943, no children.

85 Combat missions, totaling 256 hours.

FRANK H. DuRANT, JR., was born in Coconut Grove, FL on Nov. 3, 1923.

He joined the AAF July 19, 1943 as an Aviation Cadet. While stationed at Santa Ana, CA saw his first P-38 and the love affair began. He earned his wings and was commissioned at LaJunta, CO on Dec. 23, 1944. Reported to Coffeyville, KS where he was trained in the P-38 (F-5) as a Photo Reconnaissance Pilot. He was stationed on Okinawa in the 25th Photo Recon. Sdn. when the first Atom bomb was dropped. After seeing the first pictures of the devastation, they knew the war would be over in a matter of days—and it was. He spent another year overseas during the occupation of Japan photographing communication centers in Japan and Korea while assigned to the 8th Photo Recon. Sdn. After coming home, Oct. 1, 1946, he joined the AAF Reserve; stayed in Reserves 21 years and retired as a Major.

As a civilian he married the late Georgia M. Cook from Cambridge, MD on July 26, 1947. They had two children, Douglas F. and Connie Jo. He married Diane A. Scarboro on Sept. 19, 1981. After working for United Ins. Company of America 38 years, retired in 1985. September 11-14, 1987 attended class reunion in Colorado Springs, CO, pilot Class 44-J.

JOE DURHAM, served in the Royal Air Force for some two years during the early years of the war, and I was transferred into the USAF in September of 1942. I was then assigned to the original 82nd Gp., stationed at that time in Northern Ireland. As you probably know, when the North African invasion got underway in December, the group was ordered to fly our aircraft from Lands End in England to an airfield near Oran in Algeria. During this low-level flight unforeseen headwinds caused us to run short of daylight and we were forced to land on Gibraltar.

This part of the trip for my squadron (95th) was generally without incident, but there was one rather amusing aspect. Some days before the departure from England, the thin pad normally serving as a cushion on our parachutes was removed and replaced with a much thicker folded rubber dinghy (liferaft). The P-38 is equipped with a small urinal cup attached to the end of a rubber hose (relief tube). This hose is fastened directly beneath the cockpit seat and was designed originally with the exact length to permit the urinal cup to fit effectively in the crotch of the pilot. During the rather long flight to Gibraltar some of the pilots discovered - to - their alarm -that the extra thickness of the new parachute seats had not been taken in to consideration by installers, and the hose was several inches too short to serve its purpose.

We had a big laugh at the description of some of the pilots of the methods they devised for dealing with this unfortunate circumstance. One pilot had packed a large towel with his meager belongings and relieved himself into it. He told us further that he had to wring the towel out several times onto the floor of cockpit before he was finished. However, I think the most ingenious method was devised by the pilot who had packed a five-cell flashlight. He explained that, by removing the batteries and both ends of the flashlight, he was provided with an extension tube that fitted near-perfectly between his crotch and the shortened urinal cup. Most of the group managed to "hold it" until we landed but it took a while to walk around that puddle that I deposited on that Gibraltar strip as soon as my feet touched the ground.

Since we were among the first groups in the Afro-European theatres, to be equipped with P-38s, the performance of the aircraft's engines had not been fully proven under combat conditions. There was some concern among Lockheed designers that the Allison engines on the first models might suffer considerable damage if driven at full power for extended lengths of time. This ridiculous reasoning prompted them to equip the earlier models with "throttle stops" limiting the engines to some 80% of their total power.

This power restriction of engine power concerned me greatly, since I had learned from my earlier combat experience with British Hurricane and Spitfire squadrons that the wise fighter pilot always engages and/or flees the enemy with all possible power and speed. Fortunately, after discussions with the group commander and squadron leaders, these "stops" were removed from the throttles. The performance of these fine engines during later combat proved conclusively that the earlier concerns were totally unfounded.

Mechanics and ground crews were required to follow us from Northern Ireland to North Africa by ship, in a slow moving convoy, and we had been on active duty for several weeks before they managed to catch up with us. During these weeks our ground crews were enroute, all pilots were required to make daily inspections of their own aircraft. Although, at the time, I had flown hundreds of hours in dozens of different types of planes, my knowledge of aircraft engines was - and remains today - absolute zero. Each day, when I removed the cowlings from the engines, I hadn't the foggiest notion of what I was looking for. If I saw no broken pipes or loose wiring I declared the aircraft fit for full flying duties. Needless to say, the final arrival of our ground crews was celebrated by all flying personnel.

LEO DUSARD, Maj. Gen., was born in Kirkwood, MO on Nov. 2, 1915. After graduating from Washington University, St. Louis, he entered the service as a flying cadet in June, 1938.

He served as a Pursuit Pilot, Pursuit Sdn. Commander and Fighter Gp. Commander

through WW II. These assignments included commanding the P-38 Replacement Training Unit at Glendale, CA in the early years of WW II and modifying the first Operational P-38 to a "piggy-back" to be used in the furtherance of training pilots in that aircraft. In addition, he was CO of the 347th Fighter Gp. in the Pacific, with 76 combat missions. He participated in nine major campaigns and was credited with downing two enemy aircraft. Subsequently, after varied command and staff assignments, he retired from the Air Force in 1970.

RUFUS DYE, JR., was born and raised in Brookport, IL. He joined the Army Air Corps in October 1941. He graduated from pilot training at Luke AFB, AZ in November 1943 and flew two tours in WWII with the 392nd Fighter Bomber Sdn., 367th Fighter Bomber Gp. in the E.T.O. Dye had 3-1/2 confirmed air victories.

Col. Dye remained in the Air Force after WWII and flew P-51s in Korea and F-105s out of Korat, Thailand during the Vietnam War. Dye's F-105 survived a direct hit from an air-to air-missile on a raid into North Vietnam.

After returning to the U.S. Dye retired in 1971 at Bergstrom AFB, TX. He received five DFCs, 28 Air Medals, Purple Heart, Bronze Star, Meritorious Service Medal and the Joint Service Medal.

VERNON H. ELARTH, COL., USAF, (ret.) born Aug. 10, 1917 in Minot, ND. Died July 4, 1987 in Weaverville, CA. Married Ruth Lockridge July 2, 1942. Survived by wife Ruth, sons Vernon and Norman, and two grandsons.

Entered Air Force training in February 1943. Retired October 1969 in Grade of Colonel. Flew P-38 in the Aleutians in 1945 with 18th Ftr. Sdn. 343rd Ftr. Gp. He also served overseas during Korean and Vietnam conflicts. In addition to flying fighter planes, he held the positions of Chief of Maintenance and Director of Material during the later years of service.

JACK A. ERICSSON, was born Apr. 11, 1921 in Chicago, IL. Attended Northwestern University, Evanston, IL. Received Primary training at Bonham, TX, Class 42-H; Basic training at Randolph Field, TX; and Advanced training at Moore Field, Mission, TX. He was assigned to the 36th Fighter Sdn. 8th Fighter Gp. and sent to the theater of operations, SWPA, New Guinea. He flew escort, dive bomb/strafing and fighter sweeps making 172 missions and 470 comat hours. Jack is credited with one Japanese Val Dive Bomber (kill) and one Japanese Oscar (damaged). Retired May 1966 (Reserve) with rank of Colonel.

Civilian occupations include: co-pilot, United Airlines; military sales, McDonnell Douglas; and life insurance agent. On Sept. 26, 1980 he married Mary L. Ericsson. Jack has a son, John D. Ericsson and a daughter, Mrs. John D. Barnes. Seven grandchildren—two boys, five girls.

WALTER A. ERNST, was born in Philadelphia, PA on Jan. 31, 1923.

Enlisted in the Army and assigned to the USAAF in November 1942. Joined the 28th Photo Recon Squadron at Peterson Field, CO in August 1943. Served as Asst. Crew Chief and later as Crew Chief.

As a unit of the Seventh Air Force the 28th PRS covered the Central Pacific, the Marshall, Marianas, Caroline, Bonin, Palau, and Ryukyu Island Groups.

Pre-invasion photographs of Tinian, Guam and Iwo Jima assisted the Army, Navy and Marine Corps in their operations. The 28th also supported them with front line photographs at Saipan, Tinian, Guam, Peleliu, and Okinawa. From Okinawa, China, Formosa, Korea and Japan were covered.

After discharge at Andrews Field, MD, December 1945, he returned to the Hosiery Industry as a mechanic. A career change led to the position as the Manager of Plant Engineering Services for a major office furniture manufacturer.

Married, one daughter and three granddaughters.

Presently retired.

Enlisted - Philadelphia, PA, Nov. 18, 1942. Reception Center - New Cumberland, PA, November 1942.

Air Corps Unassigned. Basic Training - St. Petersburg, FL, November - December 1942. Hospital - MacDill Field - Tampa, FL, December 1942; Technical Training, A&E - Keesler Field, MS, January-May 1943. Laredo Field - Laredo, TX, June 1943; Peterson Field - Colorado Springs, CO, July 1943.

28th Photo Reconnaissance Sdn. Third Air Force: Peterson Field - Colorado Springs, CO, August - October 1943; Will Rogers Field - Oklahoma City, OK, November - December 1943.

28th Photo Recon. Sdn. Seventh Air Force: Kipapa Field - Oahu, HI, January 1944, Hickam Field - Oahu, HI, February 1944; Kipapa Field - Oahu, HI, March - May 1944; Aslito Field, - Saipan, Marianas Islands, June 1944; Kwajalein Field - Kwajalein, Marshall Islands, July - August 1944, Peleliu Field - Peleliu, Palau Islands, September - November 1944; East Field - Saipan, Marianas Islands, December - March 1945; Yontan Field - Okinawa, Ryukyu Islands, April - July 1945.

28th Photo Recon. Sdn. Eighth Air Force: Yontan Field - Okinawa, Ryukyu Islands, July - November 1945 Andrews Field - MD, Discharged - Dec. 18, 1945.

EDWARD O. ESSER, born 1915 in Toledo, OH. He applied for Cadet Training one year before Pearl Harbor and graduated from Luke Field, AZ in the class of 42K. He patrolled the coast of California flying out of Los Angeles for one month and then went to Africa being assigned to the First Fighter Gp. 94th Sdn.

On June 20, 1943, while flying over Sicily his plane was shot down and he parachuted into the Mediterranean Sea. He spent seven days in a dinghy before washing into Sicily and becoming a prisoner of war in Germany for two years.

Edward O. Esser presently is a beer distributor for Anheuser Busch and Stroh's in Toledo, OH.

WILLIAM C. FLORENTINE, was born to William C. and Frances (Willis) Florentine Nov. 24, 1916, Downers Grove, IL. He had one younger brother. Obtained pilots license 1941 through CPT Program. Joined Army Air Corps April 1942. Preflight Santa Ana, CA, Primary T-bird II Pheonix, Basic Minter/Bakersfield, Advanced Williams Phoenix, graduated Class 43-C. Assigned 55th FG McChord Tacoma, WA. Attended Chandelle College Portland checking out in P-38. Member 338th Ftr. Sdn. stationed Port Angeles, WA. Reassigned with the 338th to McChord, Gp. HQ. Shipped to England September 1943, Station 131 at Nuthampstead, Herts. The 55th went operational October 15, 1943.

On his 9th mission, Nov. 13, 1943 (the 2nd to Bremen) he was shot down along with six other

pilots (one of three survivors) and taken prisoner. Freed 18 months later he returned home. Retired as Major. Married, father of three children.

THOMAS K. FOLLIS, was born in Palo Alto, CA. He attended schools in Glendale, CA. Enlisted as Pvt. U.S.A.A.F., December 1942. Reported for basic training Lincoln, NE, March 1943. Sent to Montana State College, Bozeman for CTD. On to Santa Ana A.A.B. for classification as Pilot class 44E. Primary flight school at Visalia, CA flying Ryan PT-22. Due to appendectomy, dropped back to 44G at Lemoore A.A.B. (Basic BT15 and AT17). Next to La Junta, CO advanced pilot school in B-25s.

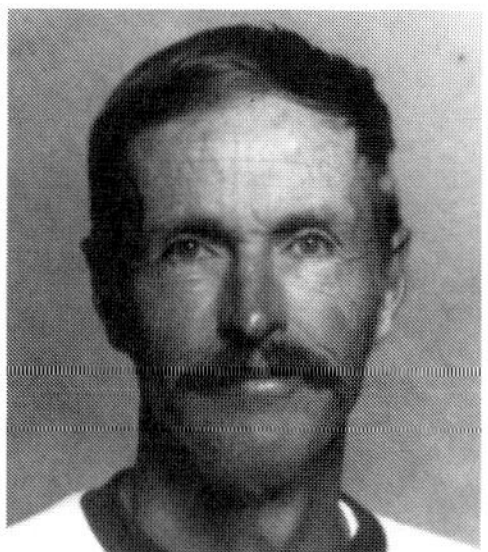

Graduated Flight Officer (blue pickle) and sent to Will Rogers Field Oklahoma City to P-38s. Standing 6'6" (surely the tallest pilot in WWII) he crammed himself into the cockpit and even took a piggy back ride in a P-38. This 20-year-old 2nd Lieutenant flew photo reconnaissance from San Severo, Italy (32nd PRS, 15th AF). Flying up the Adriatic Sea and across the Alps to targets in Germany—sometimes with escort of six P-38s or P-51s, sometimes alone—he was attacked by and evaded the deadly ME 262 German jets on missions to Munich, Stuttgart, Linz, Vienna, etc. Returned to U.S. and graduated from USC. Married to Joan 37 years, daughter Kathryn; now retired after 30 years of teaching. Retired from AF Reserve as Major.

JOSEPH M. FORSTER, was born Oct. 25, 1919 at Gainesville, FL.

His pilot training consisted of PT-17s at Thunderbird II, Scottsdale, AZ, BT-13s at Pecos, TX and AT-9s and P-322s at Williams Field, AZ.

After graduating in Class 43-F, June 22, 1943 he moved on to Olympia, WA, to receive operational training in the P-38 part of which was with the 55th Fighter Gp. In October 1943 he was assigned to fly P-38s with the 432nd Fighter Sdn., 475th Fighter Gp. (Satan's Angels), 5th Air Force, at Dobodura, New Guinea.

On Oct. 14, 1944 Lt. Forster demonstrated one of the many superior capabilities of the P-38, by flying it on single engine from Balikpapen, Borneo to the Halmaheras - a flight course of over 900 miles.

The 475th Fighter Group was assigned the

mission of escorting B-24s at 22,000 feet over the target while they blasted the oil refineries at Balikpapen.

After torching a Zeke closing in on a squadron mate, Forster was latched by another Zeke. The attack came from clouds above by a vertical dive on the group leader, but without hits. The Zeke swiftly pulled up into a loop with a high side perch at three o'clock. Forster turned into the attacker with all the "G's" he could stand, but to no avail. A horrendous explosion in the left engine dictated that the situation had indeed become serious. Pressing the turn as tight as possible he rolled into a split-S with 2600 RPM and 54" HG. When the aircraft and controls started bucking from compressibility, he did a half roll straight down and started a pull out. The dive flaps saved the day. Salt water spray off the waves clouded the windshield as he leveled out. Oil pressure was zero, and coolant temperatures was rising. So the only thing to do was feather the prop. Luckily the Zeke's ailerons froze at high speeds and prevented it from rolling. Otherwise he could have followed for the kill.

The next four hours and 20 minutes on the return route were spent criss-crossing the Celebes to avoid Jap Naval bases. An excellent radar steer to the Halmaheras saved the day again when he would have missed the island completely. Flying time eight hours 20 minutes.

Forster's tally record was: nine confirmed, three probables, one damaged and three lost to squadron mates by cutting cards!!!

After a 30 year career in the U.S. Air Force, Forster and his wife settled in Arizona where he is currently a Supervisor of Building Inspections in Phoenix.

QUENTIN A. FUHS, was born July 9, 1919 in the small midwest town of Kirkville, IA.

Signed up for the flying cadets on Jan. 5, 1942 at Ft. Des Moines, IA. Shipped out for Bakersfield, CA on the same day. Arrived at the preflight reception center at Minter Field, Bakersfield, CA on Jan. 7, 1942.

Sent to Ryan School of Aeronautics for Primary flight training Jan. 23, 1942.

Completed primary training March 23 and was sent back to Minter Field for Basic training in BT-13s. Completed Basic training May 27, 1942 and then on to Luke Field in Arizona for his Advanced training in AT-6s. Graduated with Class of 42-G and was commissioned a 2nd Lieutenant. on July 26, 1942. His first assignment was with Headquarters A.A.F., Washington, D.C., Aug. 23, 1942.

From Washington he was assigned to the 1st Electronics Experimental Sdn. which was just being formed at Logan Airport, Boston, MA. He arrived in Boston Sept. 9, 1942, where he would continue to fly for the M.I.T. Radiator Lab until January 1945 at which time he received orders for overseas.

He was assigned to the 13th Air Force, 18th Fighter Gp., 70th Sdn. He joined the 70th Sdn. on the Island of Mindoro in the Philippines on

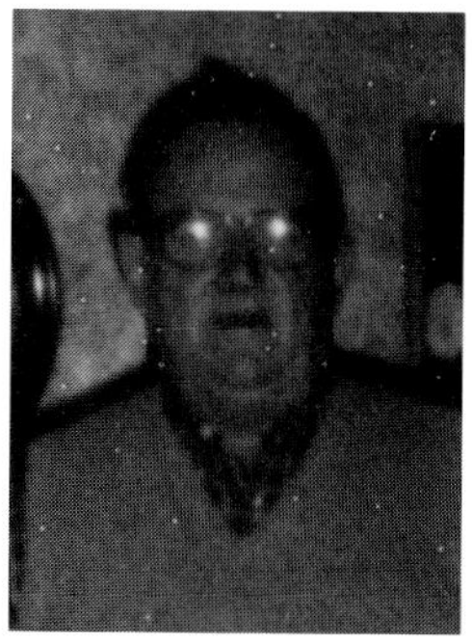

Mar. 26, 1945. After several missions the 70th moved from Mindoro to Zamboanga on the Island of Mindoro. From Zamboanga he was one of the six pilots who flew one of the longest fighter missions of the Pacific War.

They hit Java for the first time since early 1942. On Aug. 1, 1945 the six P-38s left Zamboanga at 6:15 A.M. on a fighter sweep over Java. They refueled at Menggar Airstrip on Borneo both on the way down and back. They hit Soerabaga Naval base at approximately 13:30 and damaged a sub. Then on to the air base where they damaged several Jap planes. They encountered heavy flak here but no air opposition. They carried their drop tanks until they were empty, which gave them two hours over Java. During this time they destroyed several cars, trucks and oil refineries. They also destroyed 37 locomotives. This mission lasted 12 hours and 30 minutes, and was flown over 2,600 miles.

On September 20 he took over command of the 12th Fighter Sdn. which was moved to Tacloban on Leyte. On November 2 he received his orders to go home, arriving in the States on January 9.

He spent 29 years active and in the reserves and is now retired in Oskaloosa, IA. Discharged as a Lieutenant Colonel.

FLOYD H. FULKERSON, born in Little Rock AR, Apr. 6, 1921, entered the Army Air Corps Sept. 29, 1942, but made only what he calls "truck runs" with the 498th Bombardment Sdn. in Australia until January of 1944. His superior aviation skills secured the rare reassignment from the "truck runs" to the 431st Fighter Sdn., and before 1944 ended, Fulkerson was established as a master of the P-38. His first kill occurred on Feb. 3, 1944; an assist Apr. 3, 1944; an assist July 27, 1944; a second kill on Dec. 7, 1944 (taking a Kamakazi into the ocean "in complete disregard for the heavy anti-aircraft fire from friendly ships", and winning a Silver Star from the Navy); three other planes destroyed, two crippled, on Dec. 15, 1944 (witnessed by Maj. Richard Bong); two more kills occurred on Dec. 25, 1944 (resulting in the Distinguished Flying Cross award and a crash landing in a Philippine jungle). Ignoring a broken back, and while behind enemy lines on Luzon, Fulkerson engaged in guerrilla warfare and blew up a major Japanese railroad supply route on Jan. 8, 1945 (for which he received the

Bronze Star from the United States Army, ground forces).

He flew wing for Charles Lindberg in mid-1944. He flew wing for Richard Bong, the Ace of Aces. He flew wing for Thomas B. McGuire. He received from each branch of the United States Armed Forces one of the highest awards give for courage and valor.

On Jan. 7, 1945, Tom McGuire, closing rapidly on Dick Bong for the title "Ace of Aces," was killed in combat in the southwest Pacific Theater. At that time, he owed Floyd H. Fulkerson several hundred dollars from pilot camp gamesmanship. That same savvy has made Floyd H. Fulkerson one of the most successful land developers in the history of Little Rock, AR, where he lives and works today.

CLYDE W. GABLER, was born in LaFargeville, NY, and raised in Royal Oak, MI. He was a member of the 319th Cavalry Reg but was transferred in June 1941 to the 1st Pursuit Gp. at Selfridge Field as a ground officer, being assigned duties of Sdn. Adj. Sdn, Supply and Intelligence Officer. The group was renamed the First Fighter Gp. and Gabler stayed with it for three years, in England, N. Africa and Italy, becoming the Group Intelligence Officer. He commanded one-third of the ground echelon of the First's Squadrons coming ashore during the North African landings. Loyalty to their pilots by the ground support people was an outstanding characteristic in the P-38 units.

During his tour of duty with the First Fighter Gp., Gabler participated in the briefing and debriefing of over 1000 missions. One memorable experience was flying "piggy back" in the P-38 with the Group Commander, Col. Bob Richard.

Gabler was privileged to know and serve such P-38 pilots as Jack Ilfrey, Fred Kahn, Horace Haines, Ralph Garman, Bucky Harris, Dave Stentz, Jim Alford, and Bert Frohling.

After retirement in 1966 in Royal Oak, MI, Gabler in 1978 moved to Williamsburg, FL, near Orlando, with his wife Kathleen, where he is now, July, 1988, President of the Valencia Water Control District.

JAMES P. GALLAGHER, was born in Baltimore, May 28, 1920. Upon graduation from Loyola College he immediately "swore in" the AAF June 1942. Later attended cadet training in communications, graduating Mar. 4, 1943, commissioned a 2nd Lieutenant. Received further training in Fighter Squadrons before being assigned to the 33rd Fighter Control Sdn. with which he went overseas in October 1943 to New Guinea.

He joined the 49th F.G. at Biak to make the Leyte Invasion arriving Oct. 24, 1944 becoming the 7th F.S. Communications Officer and a "hands on" P-38 man. The fierce Leyte Campaign won for the 49ers a Presidential Unit Citation. He was a part of the Guard of Honor for the presentation to Dick Bong the Congres-

sional Medal by MacArthur, one of many memorable events.

His 49ers with their Lightnings were to be the first AAF Combat Unit in the Invasion of Japan. Fate, thankfully, made this a peaceful occupation. Gallagher was separated from the AAF as a Captain in March 1945. As a Civilian, he became a stockbroker.

WILLIAM GARDNER, was born in Gorham, NH Mar. 3, 1920. He graduated from the University of New Hampshire in 1941 with a BSME and a reserve commission in the Coast Artillery. He started work with the Wright Aeronautical Corporation; was called to active duty Nov. 3, 1941 at Ft. Hancock, NJ defending New York harbor. He transferred to the Air Corps Apr. 25, 1942 and graduated from Spence Field, Oct. 20, 1942. After a tour with the School of Applied Tactics at Orlando, FL he flew 172 missions from New Guinea to the Philippines. He became Operations Officer of the 8th Fighter Gp. November 1944. He had eight confirmed Aerial victories and his decorations included the Silver Star, the DFC and one cluster and the Air Medal and eight clusters. He brought the first land based fighter into the Philippines Oct. 27, 1941 with one engine shot out of his P-38 while strafing an enemy cruiser contesting the American beach-head. Following discharge from the Service he was chief test pilot for the Curtis Propeller Division.

In 1948 he joined Sandia National Laboratories and became Director of Development Testing. He retired in 1984. His wife, Marilyn, a WAVE in WWII, came from Concord, NH - they have four children.

DALE W. GARRETT, was born in Worley, ID on Dec. 13, 1917. He joined the U.S. Army in January, 1940 and completed aircraft maintenance instruction at Curtis Wright Training School at Glendale, CA in November, 1941. When WWII broke out, he was assigned to the 343rd Fighter Gp., 54th Fighter Sdn. of the 11th Air Force. He served in the Aleutian Islands as a Crew Chief on P-38s from 1942 until 1944.

He stayed in the service after the war and was later assigned to MATS. He serviced planes in West Germany for the Berlin Airlift and was stationed in Japan from 1950 to 1952 servicing planes for the Korean War.

He retired from the USAF in 1962 as a T/

Sergeant. After retiring, he worked for Boeing in Seattle and the City of Spokane. He was married and had two daughters. He died on Oct. 19, 1986.

LOUIS E. GARRETT, was born Nov. 11, 1923 in Denton, GA; the first of nine children born to Ethridge and Lula Garrett.

He enlisted in the USAAF on Dec. 17, 1941 and completed his basic training and a portion of his A & E Maintenance schooling at Keesler Field, MS near the bustling city of Biloxi. He was one of the first group of maintenance trainees sent to the Lockheed factory plant in Burbank, CA preparatory to expanding the P-38 pilot training program.

He was shipped to SUMAC, on Aug. 28, 1942 and arrived at Newcastle, Australia after a scenic flight from Hamilton Field with stops along the way at little spots of land that he never read nor heard of during his school years in Jacksonville, FL.

He joined the 35th Fighter Gp., 39th Sdn. at Amberly Field near Brisbane, Australia and later moved up to Townsville where the pilots where checked out and the Group moved up to New Guinea later that year.

In 1943, he was selected to help form the 475th Fighter Gp. which was forming in Australia and after assemblying the planes as they were taken off the boats and checking the pilots out at nearby Amberly Field, New Guinea became home ground again. He participated in many of the campaigns throughout the islands and the Philippines. He was aboard LST 1014 awaiting off-loading in Buckner Bay when Japan capitulated on Aug. 13, 1945.

He was discharged at Camp Blanding, FL on Oct. 5, 1945. He reenlisted in the USAF Sept. 1, 1948 in his wartime grade of T/Sergeant. He married a hometown girl, Betty Butler, in August 1948 and they have one daughter.

He subsequently served 14 years in SAC, two years in USAFE where participated in Exercise Delawar while in Turkey. In June 1962 he was assigned to George AFB, CA where he served until his retirement except for a short 12 months in SEA (Thailand) and 14 months in Iran with JUSMAG.

He retired from the 35th Fighter Wing as CMSgt on Mar. 1, 1972. His decorations include the Bronze Star Medal, the Joint Service Commendation Medal, Asiatic Pacific Theatre

Medal, Air Force Commendation Medal, National Defense Service Medal, Air Force Good Conduct Medal with clusters and WWII Victory Medal. He was Student Commander of Class 59C at Westover AFB where he attended the Non Commissioned Officers Academy.

CMSgt. Garrett, USAF (Ret) retired from full time agent status with The Prudential Insurance Company where he has served his community and friends since completing his distinguished career with the United States Air Force and now resides with his devoted wife of 40 years near their grandchildren in Victorville, CA in the great Mojave Desert where he can watch the USAF busily training for the tasks that have kept America great.

WILLIAM M. GASKILL, was born on a farm in Deming, NM on the 24th of May 1921. He joined Troop E of the 111th Cavalry of the New Mexico National Guard on Jan. 26, 1939. He then enlisted as an aviation cadet on Apr. 16, 1942.

He received his wings and was married to Elvena Ford, went overseas and joined the 68th Fighter Sdn., 347th Ftr. Gp. of the 13th Air Force, in the South Pacific. He learned combat flying under Maj. Chandler Worley, Maj. Street and Maj. Shapiro. He learned the importance of a good attitude and disciplined flying.

Gaskill flew 87 missions in P-39s and 90 mission in P-38s. They strafed, dive bombed, skip bombed and glide bombed and destroyed many barges, gunboats, gun positions, shore batteries, ammunition and fuel dumps. Some of the larger missions were lead by Col. Westbrook and Col. Dusard.

Gaskill in time became a 1st Lieutenant, a flight leader and was also appointed Assistant Test Pilot of the 68th Fighter Sdn. and for many months worked together with the Engineering Officer, Capt. Jarrell.

He received the Soldiers Medal and the Air Medal with ten Oak Leaf Clusters and was honorably discharged on July 14, 1945. He is a farmer in Wilcox, AZ. He and Elvena have four children and 14 grandchildren.

DON GELHAUS, attended Gonzaga University in Spokane, WA, where he was enrolled in the Civilian Pilot Training program. He received a private pilots license in early 1942.

In June of 1943 he enlisted in the USAAF, then on to Aviation Cadet's and was in one of the classes to fly T-6s in basic and advanced, graduating from Williams Field Class of 44-C.

At Victorville and Chico, CA, he flew P-39s and then P-63s. Then to Santa Rosa, CA for P-38 training. In September 1944 he left the States for the ETO and more P-38 training at Shrewsberry, England. He was assigned to the 367th Fighter Gp., 393rd Sdn. in November of 1944. Most of the 367th missions were strafing and dive bombing. In February, 1945 the 367th changed to P-47s. He destroyed four planes and

damaged two. In December of 1945 he was separated from the service.

Don married Genevieve Olson in April of 1946, and after several years of college, he returned to the family wheat farm at Plaza, WA, where they raised two sons. He is still getting in some flight time in their Piper Cherokee 180.

MILTON GETKER, Curtiss-Wright Propeller Division, Field Service Rep., Patterson Field, OH. Accelerated flight service test XP-38, P-38. August 15, 1941 to Mar. 30, 1942 (USAAC) Don Young, Milo Bercham and Jack Bartz, there also from Lockheed.

Curtiss Propeller USAAC Tech. Rep 5th AF Air Service Command, Hq. Brisbane, Australia and Townsville Air Depot Aug. 1, 1942 - Propeller OVHI. Shops set up for P-38 and other aircraft - 39th Fighter Sdn. and 35th Fighter Gp.

Curtiss Propeller Tech. Rep. 5th AF assigned to 39th Fighter Sdn. Port Moresby, New Guinea and other bases as moved OIV, Dec. 8, 1942 - June 30, 1944. Contacted other P-38 Sdn. in area as formed.

Hollandia and Biak Island 1944 contacted and assisted engineer and maintenance with propeller problems. Procuring spare parts, finished in P.I. Feb. 22, 1945 assisted service squadrons P-38 repair.

LOUIS MICHAEL GILLET, was born in San Antonio, TX on Feb. 28, 1920, oldest son of Mary and Louis F. Gillet. Graduated Abraham Lincoln High School, Los Angeles, CA 1938. Enlisted U.S. Air Force Oct. 12, 1940, Ft. MacArthur, CA. For Salt Lake City (Air Force used same runways as civilian airlines) known as the Salt Lake City air-dome lease. (88th Recon., B-17s and B-18s) and HQ and HQ Sdn. 7th Bombardment Gp. (Ser. # 19049457, R.A. Tech./Sergeant)

Sent to Curtiss-Wright Tech. Institute, Glendale, CA; completed a 24 week course for "airplane mechanics." Graduated Jan. 8, 1942 and transferred from Glendale to the 37th Fighter Sdn., 55th Gp. Stationed on a small civilian field at Tumwater, WA, a few miles from Olympia. At this time we had P-43s. A short time later our squadron started replacing the P-43s with P-38s. We trained pilots in P-38s for approximately one year. At this time the government sent our squadron, the 37th unattached, to North Africa. Sailed Feb. 8, 1943 and arrived Feb. 21, 1943. A short time later we put oour first P-38s together (they were shipped across on the decks of ships with the wings and propellers removed). We were assigned to the 14th Fighter Gp. consisting of the 48th, 49th, and 50th. We took the 50th's place, so the 14th Fighter Gp. consisting of the 37th, 48th and 49th Sdns. went into combat. We, the 37th, lost our first ship and pilot, Lt. Toth, over Tunis North Africa. We were in both the 12th and 15th Air Force in North Africa.

Foreign service one year, ten months, eighteen days; continental service three years, zero months, eighteen days. Sailed from Italy Aug. 20, 1944; arrived U.S. Sept. 1, 1944. Assigned 247th Army Air Forces Base Unit, Salina, KS (B-29s). Separated Sept. 18, 1945 from Ft. MacArthur, CA.

Met and married one Roselle T. Smith while stationed in Salina, KS prior to being discharged. She was in WAC working the flight line as a mechanic. We have been married 42 years and have two sons, Louis Jr. and Randy.

Upon being discharged from the Air Force in 1945 I spent most of my working years in law enforcement. Los Angeles Police Dept. Nov. 24, 1947 to Dec. 4, 1967 (retired); Campus-Police Officer for University Southern California at Los Angeles Dec. 14, 1970 to Aug. 25, 1973; City of Indio, CA, Animal-Control Officer Sept. 4, 1973 to June 9, 1975; and Riverside Sheriff's (California) Dept. as a Superior Court Bailiff June 9, 1975 to July 29, 1983.

Retired, moved to Nevada and built a home on 13 plus acres near Doyle, CA.

CY GLADEN, Lt. Col. USAFR Ret., was born Nov. 4, 1920 in Guthrie, MN. He joined the Army in 1940 and found himself stationed in Hawaii where he was soon transferred to the Air Corps and trained as a mechanic. In 1941 Corporal Cy Gladen was accepted into flight training, but was still at Hickman Field as Corporal of the Guards at the time of the Japanese attack on Pearl Harbor.

He returned to the States and graduated with the Class of 42-G, Moore Field, TX in August 1942.

2nd Lt. Gladen soon found himself back in Hawaii flying P-40s in the 72nd Fighter Sdn., 15th Fighter Gp. He arrived at Guadalcanal in April of 1943 as a replacement pilot with the 44th Fighter Sdn., "Vampires" serving as a wingman and element leader flying P-40Fs and Ms until September 1943 at which time the squadron converted to P-38s and newly promoted Capt. Gladen took command of the squadron's 6th flight consisting of Lieutenants Marty Glommen, "Red" Fox and Bob Krohn. He flew 30 more missions before the end of his tour. Cy Gladen flew a total of 142 combat missions totalling 320 hours and received the DFC and AM with 19 OLCs. He was credited with the destruction of seven aircraft in air combat, two Zeros on June 16, 1943 and five Vals on Aug. 31, 1943 to received the coveted title of "Ace."

Although he is quick to downplay his own exploits he talks very highly of his flying mates, in particular of 1st Lt. Robert E. Fox who was Cy's wingman on many occasions including Aug. 31, 1943 which Cy vividly recalls. "After downing five Vals off South New Georgia I was low, slow, out of ammo, and couldn't see because my P-40 was covered with oil from the Vals. Before I knew it I had a Zero on my tail shooting darts through my Warhawk. Good ole "Red" Fox flamed him and saved my day. Besides my five Vals, "Red" downed a Val and that Zero."

It wasn't all glory, though, as 1st Lt. Robert Fox was killed in his P-38H on Dec. 23, 1943 when an engine exploded on takeoff from Treasury.

Robert Fox hailed from Waco, TX and joined the Air Corps in 1942. Besides the AM with six OLCs he received the DFC and Purple Heart posthumously.

BURL E. GLASS, JR., was born Dec. 30, 1918 in Alanreed, TX, a very small town in the Texas Panhandle. He finished high school at Lefors, TX in 1936, and graduated from the West Texas State University, Canyon, TX, with a B.S. degree in 1940. He applied for U.S. Army Air Corps Flying Cadet Program and began training in February 1941. Primary and Basic flight training was completed at Cal Aero Academy, Ontario, CA and advanced training at Stockton Field, CA. He graduated on Sept. 26, 1941, Flying Class 41-G.

His first assignment was with the 37th Pursuit Sdn., 55th Pursuit Gp., Portland, OR, flying P-43s. The 37th Sdn. moved to Tumwater Air Field, Olympia, WA in December 1941 and converted to P-38s in February 1942. In July 1942 he was assigned to the 330th Ftr. Sdn., 329th Fighter Gp., a P-38 training group. In March and April 1943 he and several other pilots flew several P-38s across the South Atlantic to North Africa, delivering them to Casablanca. On Aug. 1, 1943 he was assigned as commander of the newly organized 429th Ftr. Sdn., 474th Ftr. Gp. In February or early March 1944 the 474th Gp. went to England, and in August 1944 moved to France, working its way to Belgium. He returned state side in October 1944, having flown 65 missions totalling 198 hours.

His decorations include the DFC, Air Medal with tenOa k Leaf Clusters, the French Croix de Guerre with silver star, Presidential Unit Citation and other campaign type medals.

He retired from the USAF on Jan. 31, 1968 as Lieutenant Colonel.

FRANCIS R. GOLDSBERRY, (Dick) Col., USAF, (ret), born in Athens, OH, March 1922. Entered Aviation Cadet training at Santa Ana Army Air Base, CA in November 1942. Graduated from Williams Field, AZ, Class 43-I, October 1943. Completed P-38 RTV at North Island, NAS, CA in February 1944. Joined 364th Fighter Gp., 383rd Ftr. Sdn., Honnington, England in April 1944. Flew 68 missions, 270 combat hours in P-38 and P-51s. Completed tour in December 1944. Assigned to Ontario AAB, CA instructing in P-38s until WWII ended in August 1945.

Subsequently, served in 36th Ftr. Gp., Panama Canal Zone, 1946-47. Group converted to P-80s in July 1947, transferring to Furstendfeldbruck, Germany in July 1948.

Flew combat tour in F-80s, 8th Fighter Bomber Gp., 22nd Ftr. Sdn., K-13, Korea in 1952. Flew FAC missions in Vietnam, 1966, while assigned as DASC. Director IV Corps.

Varied and interesting flying and staff assignments over period of 31 years service, including four years Pentagon duty 1967-1971. During career flew P-38, P-51, P-80, F-84, F-86, F-94, F-100 and various other single and multi-engine aircraft.

Retired in Phoenix, AZ in July of 1973. *(See photos, page 140)*

PHIL GOLDSTEIN, born in Baltimore, MD. Played French Horn in Symphony Orchestras.

Inducted in Army June 6, 1941. Assigned to Co "D", 12th Inf. He became an Aviation Cadet October, 1942, class of 43G. Pre-flight Santa Ana, CA, Sdn. 39, Primary - Oxnard, CA, Basic - Lemore, CA, Advance - Williams, AZ.

After graduation he was assigned to Muroc Field, CA.

He landed in Algiers in November, 1943, and was assigned to American-British and French unit in Berteaux.

Involved in crash of C-47, as a passenger, on take off. The cause of the accident was due to ice on the wing and fog in the cockpit.

He joined the 49th Sdn., 14th Ftr. Gp., 15th A.F. Jan. 19, 1964.

Returned to U.S. to Santa Ana, AFB, CA.

Decorations: Distinguished Flying Cross, Air Medal with Clusters.

Victories: Officially two in Air four on the ground. Unofficially eight victories.

Certificate of Valor from Gen. Nathan Twining.

HECTOR GONZALEZ, born Feb. 28, 1914 in Nogales, AZ. Family moved to California 1915. Graduated from L.A. High in 1934.

Enlisted with Royal Canadian Air Force in Vancouver, BC on Oct. 8, 1941.

Attained rank of Sgt./Pilot September 1941.

Assigned to #19 Advanced Flying Unit in Dalcross, Scotland October 1942.

Assigned to #1 Beam Approach School, Watch Field, England December 1942.

Transferred from RCAF to U.S. Army Air Force Feb. 23, 1943, rank of Flight/Officer.

Posted to #11 C.C.R.C. Bovingdon, England to #326 Bombing Sdn., 8th Air Force.

Transferred from #11 CCRC to HQ 322 Bomb Gp. May 1943 - 8th Air Force.

Transferred to the 13th Photo Sdn. A.A.F. Station #234 June 1943, Oxfordshire, England (8th Air Force).

Commissioned 2nd Lieutenant Aug. 30, 1943.

Attained rank of Captain August 1944.

Awards: D.F.C. Air Medal with five Oak Leaf Clusters. Distinguished Unit Badge.

Flew 52 Combat Missions in a F-5 Photo Recon. Plane (P-38) in the European Theatre. 13th and 27th Photo Recon. Sdn. - 8th Air Force, 7th Photo Recon. Gp. 8th Air Force.

Relieved from active duty Dec. 11, 1945.

HOWARD L. GOSS, born June 3, 1919 in Bird City, KS. Graduated from Alhambra High School in 1938. Entered the AAF in 1943, Commissioned 2nd Lieutenant Class of 44D at Williams Field, AZ.

Received P-38 training at Santa Maria, CA, P-40 and P-51 training in India. Was assigned to 449th Fighter Sdn., 51 GP, 14th AF May 1945. Had five missions in French Indo China. After VJ Day he was transferred to 2nd Combat Cargo Sdn. assigned to pick up American Personnel throughout China. Separated from service Apr. 24, 1946 as a 1st Lieutenant.

After leaving the Air Force he returned to the construction trade as a Job Supt. Was married to June Jordan in April 1958. Has a daughter Patricia A. (Goss) Vaitaitis and a granddaughter Vilija Jo Vaitaitis.

Retired from Steed Bros. Construction Co. in 1983. He and his wife enjoy traveling and the retired life in Arcadia, CA.

GERALD A. GREENE, was born Mar. 5, 1925 in Brooklyn, NY, he graduated with class 44G in August, 1944. He was assigned to Photo Recon. and joined the 25th Photo Recon. Sdn. of the 5th Air Force in January, 1945. He won the Air Medal in June 1945. He stayed with them for the remainder of the War, including occupation duty in Chofu, Japan. He left for home in January, 1946.

Gerald stayed in the Air Force Reserve, and retired from the Reserve with the rank of Major He became a financial advisor in 1961 and owns his own business. He is married and has two children and two grandchildren. He was elected to the Board of Directors of the National P-38 Association in May, 1987, and appointed General Chairman for their First National Convention, to be held in Dayton, OH in September 1989. He is currently President of the Dayton Chapter of the Retired Officers Association beginning his second year in office. *(See photos, page 140)*

TRAVIS M. GREENWOOD, a native of Ft. Worth, TX, joined the Army Air Corp in 1942. He graduated from two-engine school in 1943 at Williams AFB, AZ and served with the 8th AF during WWII, where he logged 380 combat hours in the P-38 and P-51.

During his career, Travis logged over 5000 hours in conventional and fighter aircraft. He retired from the AF as a Lieutenant Colonel in 1965. In 1946, he joined the 1st Ftr. Gp. March AFB, then moved to the 14th Ftr. Gp. Dow AFB, ME. He was in the 36th Fighter Gp. Furstenfeldbruck, Germany 1948-1951, the 318th Ftr. Sdn. at McChord AFB, Thule, Greenland, and Presque Isle AFB from 1951-1957. EADF Headquarters Stewart AFB NY on a Tactical Evaluation team for three yeas, 68th FIS K I Sawyer, MI 1960-1962. His last duty was with the 38th Missile Wing at Sembach, Germany.

He married Gleneva Sullivan from Ft. Worth in 1946. They have one son, Travis Jr. and now reside in Austin, TX where Travis taught in the Austin School System for 20 years.

HAROLD F. GRENNING, JR., (Bud), was born in Chicago, Nov. 5, 1922, so was 19 years old when the attack on Pearl Harbor got us into WWII.

He enlisted in the Army Air Corps May 31, 1942 and started training in California in November, leading to graduation in Class 43-K in December at Williams Field, AZ, as a 2nd Lieutenant P-38 Pilot.

Assigned to the 434th Fighter Sdn. of the 479th Fighter Gp., he went with them to active combat operations in England in May, 1944.

His outfit, along with other P-38 Groups, provided the low level cover and protection for the invasion forces for the first couple of weeks of the Normandy invasion. After that, it was both high altitude bomber escort, and low level ground targets for the rest of the year.

He became Flight Leader of "D" Flight in mid-August and flew a total of 67 combat missions, getting credit for the destruction of one ME-109 in the air and one each HE-111 and JU-88 on the ground.

After returning stateside in January, he was assigned to the Ferrying Command and picked up and delivered brand new P-63s, P-47s as well as some "real junkers" to various places in the States. The best flight of these was up the AlCan Highway to deliver a P-38 to the 11th Air Force in Anchorage, AK.

His decorations consist of; Distinguished Flying Cross, Air Medal with seven Oak Leaf Clusters, ETO Ribbon with Air Offensive Europe Star, Normandy Star and Germany Star.

He is enjoying retirement in Glenview, IL, with Peggy, his wife of 43 years, his children and ten grandchildren.

JOSEPH H. GRIFFIN, was born in Pauls Valley, OK Nov. 18, 1917. Joined the Army Air Corps in March 1941. Trained at San Diego, Moffet and Mather and was assigned to 51st Pursuit Gp. at March AFB, CA. Checked out in P-36 and P-40 and moved with the 16th Pursuit Sdn. to Mines Fields Los Angeles for coastal patrol at the outbreak of WWII. Transferred to 14th Pursuit Gp. Checked out in P-38 and flew anti-sub patrol out of North Island, San Diego, CA.

Transferred overseas and was attached to the American Volunteer Gp. in China June 17, 1942. Original cadre for the 23rd Fighter Gp. 75th Fighter Sdn. Flew 75 combat missions, in P-40 and P-43 and destroyed two zeroes and one I-99 bomber.

Returned to the U.S. in May 1943 and activated the 393rd Fighter Sdn. in the 367th Fighter Gp. Trained the squadron in P-39 fighters and deployed with them to the 9th Air Force in England in March 1944. The Group was equipped with P-38s for combat. Participated in the invasion of Europe, Cherburg breakthrough. Flew 75 combat missions in P-38s destroying one ME-109, three FW-190 in aerial combat and one JU-88 on the ground.

Received the Silver Star, Legion of Merit with Oak Leaf Cluster, Distinguished Flying Cross with two Oak Leaf Clusters, air medal with 15 Oak Leaf Clusters and the Belgium Croix de Guerre.

Now resides at 2633 Guilford Lane, Oklahoma City, OK 73120.

ROLAND KARRH GRISHAM, born Mar. 3, 1922 at Dekalb, TX. Parents - Oliver E. Grisham and Bettie (Karrh) Grisham. Attended school at Rural Union School, Texarkana Junior High and Texarkana Senior High, Magnolia A&M College, Fresno State and the University of California College of Pharmacy - B.S. 1949.

In January of 1940 he joined the Arkansas National Guard, 153rd Inf., Co., D. Federalized in December of 1940 and sent to Camp Joe T. Robinson in Little Rock, AR. August of 1941 sent to Ft. Raymond, AK. CTD University of Montana, Missoula, Pre-flight Santa Ana, CA. Primary Flight - Santa Maria, CA, Basic Flight - Williams Field, Chandler, AZ, class of 44 - E. June of 1944 Ajo, AZ air to ground gunnery. July of 1944 Victorville, CA P-39 Indoctrination. August of 1944, Chico, CA P-38 Flight training. March of 1945 New Guinea Replacement Depot.

In New Guinea he flew missions to partially secured islands and ferried war weary P-38s to their final resting ground on Biak Island. Also, flew new P-38s to various units as replacements, at Clark Field and Kunming, China. Plus, he flew a couple of missions as a gunner on A-26s into western New Guinea. July of 1945 permanent assignment to 475th FG, 431st FS at Lingayen Gulf. Flew ground support missions on the Philippines. In August of 1945 the unit went to Ie Shima off Okinawa. January of 1946 separated from the service as a 1st Lieutenant.

In December of 1944 he married Gladys Mae Bergthold of Reedley, CA. They have three children: Anita Lorraine Grisham - an office manager for a law firm, Susan Ilene Grisham, - employed by the County of Kern Medical Department, and David Leroy Grisham - a Vietnam veteran of 200 missions in Heuy helicopter with two Purple Hearts, 48 Air Medals, Bronze Star and two Army Commendation Medals with V device. David is a security guard and college student. Roland and Gladys live in Shafter, Kern County and he works as Chief Pharmacist for Thrifty Drug Store in Wasco, CA.

CARLYLE W. HACKBARTH, was born in Syracuse, NY. He was sworn into the U.S. Army Air Corps there on July 3, 1942 and reported to the Reception Center at Ft. Niagara on July 17. After basic training at Keesler, Field, Biloxi, MI, he attended the Photo School at Lowry Field, Denver, CO. Upon completion of his training as a camera repairman, he was assigned to the 10th Photographic Sdn. at Peterson Field, Colorado Springs, in November 1942. Due to a shortage of Squadron Supply personnel, he became Chief Clerk of that department. When the overseas squadrons were formed in early 1943, he was Chief Clerk of the 27th Photographic Sdn. first, then the 26th Photographic Sdn., with which he served two years in the Pacific area. His responsibilities included requisitioning and record-keeping of regular clothing and supplies, preparation of bills of lading, ship diagrams for loading and logistics report for all squadron base moves, and direction of loading and unloading crews on these moves. In July 1945, he became Supply Sergeant, a position he held until returning to the U.S. in November of that year.

The 26th Photographic Sdn. was based in New Guinea, Biak, Luzon and Okinawa before serving in the occupation of Korea at the end of WW II. It relied totally on P-38s in distinguishing itself earning eight Battle Stars and two Presidential Citations in the various Pacific campaigns.

J.C. HAISLIP, was born in Sherman, TX on Aug. 25, 1917.

Entered the Army in October, 1940, as a member of the 132nd Field Arty., 36th Div. where he attained the rank of Sergeant. Accepted for Pilot Training at Goodfellow Field, San Angelo, TX in October, 1942.

He went through Single Engine Flight School, graduating from Craig Field, AL, Class of 43-G. After 85 hours of Tactical Training in the P-47 and P-40 at Page Field, FL he was assigned to the 9th Fighter Sdn., 49th Gp. at Dobodura, New Guinea. He flew 96 combat missions in P-47 until Squadron was equipped with P-38G15 in April, 1944. Was Engineering Officer April, 1944 to October, 1944, when he was made Operations Officer. Completed 57 Combat Missions in P-38 and 96C/M in P-47 for total of 456 combat hours. Attained rank of Captain. Was credited with one (1) confirmed Jap plane in aerial combat. Received two (2) DFC, Eleven (11) Air Medals and Unit received two (2) Presidential Citations.

Following return to Stats was assigned to Luke Field, AZ until release in January, 1946.

He worked as Sales Representative for a Construction Machinery Company until September, 1950, when he was recalled to Active Duty by the Corp of Engineers as Base Equipment Officer at Ft. Leonard Wood, MO. Released October, 1950.

Returned to Dallas with Equipment Company until retirement in 1969. After retirement he was a teacher at Eastfield College in the Science Div. until 1976. He now operates a remodeling business in East Texas Lake Area.

CHESTER A. HALLBERG, flew 74 combat missions with the 20th Fighter Gp. during WWII.

A native of Turlock, CA, Hallberg enlisted in 1942. After pre-flight, he trained at Santa Maria, Lemoore, and Williams Field graduating with the class of 43-E.

He received tactical training at Muroc and Glendale and was then assigned to the 20th F.G. at March Field. The 20th F.G. was sent to England in August of 1943 and began operations on December 28, 1943.

Hallberg was credited with one ME-109 destroyed, 1 1/2 ME-110s destroyed, and one ME-110 damaged. Decorations included the Air Medal with six clusters, DFC with one cluster, the ETO ribbons with three battle stars. He achieved the rank of Captain.

After leaving the military in 1945, Hallberg eventually became a Baptist minister. He and his wife, Mary, have two sons and seven grandchildren. He is pastor of the Calvary Baptist Church in Hollister, CA.

ED HAMILTON, (Hambone), (16053303) born in Pekin, IL on May 14, 1913. Enlisted U.S.A.F. Jan. 23, 1942. Boot training Keesler Field, MS. Base Sdn. Eglin Field, FL. Activated P-38 combat group May 23, 1943. Graduated Lockheed Tech. Training School Burbank, CA. The 429th "Retail Gang" made active Aug. 1, 1943 at Army Airdrome, Glendale, CA. Metropolitan Airport Van Nuys, CA Oct. 10, 1943. Lomita flight strip Jan. 4, 1944. Santa Maria, CA Feb. 6, 1944 - Feb. 15, 1944. Cross country to Camp Miles Standish, MA Feb. 15, 1944 - Feb. 20, 1944 North Atlantic to Glasgow Scotland Feb. 26 1944. Warmwell, Southern England, March 12 - Aug. 6, 1944 Normandy Beachhead, UT, Nevilly, France. Aug. 6, 1944 - Sept. 1, 1944 St. Marceau, near Lemans, France Sept. 2, - Sept. 12, 1944. Peronne, France Sept. 12, 1944 - Oct. 2, 1944 Florennes, Belgium Oct. 2, 1944 - Mar. 30, 1945. Strassfeld, near Euskirchen Germany Mar. 30, 1945 - Apr. 25, 1945. Langansalza Germany Apr. 25 - June 15, 1945 Schweinfurt Germany June 15, 1945 - Sept. 10, 1945. Laharve France Sept. 10 - Sept. 20, 1945. Camp Kilmer, NJ U.S.A. Sept. 26, 1945. Separation U.S.A.F., Ft. Sheridan, IL Oct. 20, 1945.

Promoted to T/Sergeant Ed Hamilton, Crew Chief, P-38 April 1945 with 429th Fighter Sdn. E.T.O. Awarded Bronze Star Medal May 24, 1945 9th Air Force in E.T.O. Employed 33 years Caterpillar Tractor Co, Peoria, IL. Now retired. Enjoy traveling, fishing and hunting.

FERDINAND HANSON, at age 24 Ferdinand C. Hanson was inducted into 40th Div. California National Guard in March 1941. Transferred Regular Army and assigned to March Field, CA. Saw first P-38s and B-19 landed there in summer of 1941. Attended Air Mechanic School in Lincoln, NE, 1941-1942. Assigned North Island Naval Air Station as P-38 Crew Chief. Transferred WCTC, Santa Ana, CA, as Air Cadet 43F. After graduation assigned Twin Engine Fighters, sent to 5th Air Force, SWP 475th Gp., 432nd Sdn.

Saw air action over Arawe, Buna, Rabaul, Wewak, Celebes. No confirmed kills, two probables. "Shot and missed.." Discharged as 1st Lieutenant 1945.

He, in 1945, let the bad guys get another shot at him - joined Los Angeles Police Dept. Retired in 1966.

Worked later as Research Mechanic repairing Black Boxes for the Air Force's most secret aircraft. Retired on Soc. Sec. in 1982. Married to Eva in 1942, three daughters, one son.

ROBERT HARGIS, born Sept. 13, 1922, in Monticello, AR, joined Co. "H", 153rd Inf. AANG on July 17, 1940, in Arkadelphia, AR. In September of 1940, he was then transferred to Btry. "B" 206th CA (AA) which was located in Monticello, AR. Basic training for this regiment took place until August of 1941 at Ft. Bliss, TX.

After being attacked by the Japanese on June 3, 1942, Mr. Hargis then applied for the AC program and arrived in Santa Ana, CA, in October of 1942 for preflight pilot training. Primary training, flying a Ryan P-22, was done in Dos Palos, CA.

During basic pilot training in Chico, CA, he flew a BT-9 and at Williams Field, AZ, flew an AT-6. Advanced training was done in an AT9 and RP-322.

On July 28, 1943, with a 43-G, Mr. Hargis graduated as a commissioned Second Lieutenant. Operational training was done in Muroc and Ontario, CA in a P-38. Departing the U.S. in October 1943, he then arrived in Kurmitola, India, to serve with the 459th Fighter Sdn., 80th Fighter Gp., 10th AF, where he would fly the P-38H. From here in March 1944, he then went to Chittagong. All of the missions flown at this time were against the Japanese over the surrounding vicinities of Burma. Around January of 1945, another move was made to Rhumkapalong, and he returned to the U.S. in April of 1945.

Mr. Hargis' tour of duty included the following: 79 combat missions for 271 combat hours and 2 1/2 aerial kills. His rank in the military constituted a promotion to First Lieutenant in June 1944 and Captain in December 1944. Mr. Hargis retired from the reserves in September of 1982 as Lieutenant Colonel.

Married in 1950, he has four children and was widowed in 1980.

HAROLD T. HARPER, joined the Army Air Corps as an Aviation Cadet at Minter Field, CA in December 1941 at the age of 20. He was born in Bakersfield, CA on Dec. 5, 1921.

Upon graduating with the class of 42-H at Victorville AAB on Aug. 27, 1942, he requested to fly P-38s at Hamilton Field, CA. He was assigned to the 83rd FS, 14th FG and was sent to England in November 1942 on the Queen Elizabeth. After the invasion of North Africa most of the pilots of the 78th FG were sent to reform the 14th FG at Marraletel. He was assigned to the 49th FS and entered combat in early 1943. He met his brother at Telergma who was flying C-47s supplying gasoline and engines for the fighters.

He flew 50 missions escorting B-17s, B-25s, B-26s, and strafing and dive bombing over North Africa, Corsica, Sardinia, Sicily, Italy, and the Island of Pantelleria. He had two ME 109s and several probables along with many destroyed on the ground.

His most memorable mission was the strafing of the Foggia, Italy, German Air Field by 160 P-38s of the 14th, 82nd and 1st FG. Almost 200 German transports, bombers, and fighters were destroyed. He was awarded the DFC, Air Medal with ten OLC and other unit citations. He separated from the service in September 1945 and entered Oregon State University where he graduated in 1947.

He was recalled to active duty in 1950 to fly P-51s at Tyndall AFB. Shortly thereafter, he was sent to Gunter AFB for training in Aviation Physiology. Lowry and Mather AFB were his next stop. After 33 months of active duty he returned to his profession of research biologist with the California Department of Fish and Game. He kept active in the AF Reserve Program and retired from there in 1967 as a Lieutenant Colonel. He has since retired from State Service after 37 years. Wife Elinor and he lives in Sacramento, CA.

He would like to praise the crew chiefs and their assistants who didn't get the glory and the medals of the pilots. Especially, Michael Whiteford, his crew chief who kept his P-38 in top shape and never had to abort a mission due to mechanical problems.

FRANCIS H. HARRIS, was born Mar. 15, 1918 in Epworth, IA. He joined the U.S. Army Air Corps as a Flying Cadet September 1940. His Primary Training was at Parks Air College, IL; Basic, Gunter Field, AL; Advanced, Maxwell Field, AL.

Commissioned, Apr. 25, 1941. Assigned to the First Pursuit Gp., Selfridge Field, MI. Flew P-35s, 36s, 43s, and on July 18, 1941 the YP-38s. Flight leader for the Southeastern Maneuvers September-November 1941, flying P-43s, in the 71st Fighter Sdn.

The squadron flew to San Diego in December 1941. On May 20, 1942 the First Pursuit Gp. left California for its historical flight to England, fully equipped with P-38s.

Harris flew with the Group (now the First Fighter Gp.) on its first mission over France. On Sept. 7, 1942, he bailed out of his burning P-38 over England and landed without injury. On Nov. 14, 1942 the First Fighter Gp. made the second longest P-38 over-water flight when it was airborne eight hours, 25 minutes to fly to Oran to join the invasion of North Africa, Harris was squadron leader for 25 missions, and, by Mar. 1, 1943, had completed his first 50 missions, one ME-109 confirmed.

He returned to California in May 1943, assigned to the 329th Overseas Replacement Training Gp. as Operations Officer. He returned to Italy and the First Fighter Gp. in May 1944 and received orders to take command of the 94th Fighter Sdn. on July 20th, 1944. He led the squadron in the invasion of Southern France (to Corsica) on Aug. 15, 1944.

He was promoted to Lieutenant Colonel Sept. 4, 1944. The First Fighter Gp. was chosen to escort President Roosevelt to Yalta early in 1945. Lt. Col. Harris was given Top Secret orders to travel and make arrangements for the supply and billeting of the pilots on this mission.

On Feb. 18, 1945, Harris led the first flight to a rendezvous with the President's Cruiser, 60 miles West of Gibraltar. Tae First Fighter Gp. escorted the President to Yalta and return.

Returned to combat until Apr. 14, 1945, leading missions to Lentz and Regensburg, and escorted B-17s and B-24s on many more missions. He finished the second tour with 92 missions and over 325 combat hours. He was promoted to Colonel May 30, 1950.

Decorations: DFC; Air Medal with one Silver and two Bronze Oak Leaves; Croix de Guerre avec Palm. Theater decorations included the American, World War II Victory, E.T.O. with Silver and Bronze Star, Pre-Pearl Harbor, Armed Forces Reserve.

HAROLD W. HARRIS, a native of Tennessee, he signed up for the Aviation Cadet Program in Nashville, TN on Aug. 11, 1942 and began training in February 1943 at Montgomery Field, AL. He graduated from Napier Field, Dothan, AL, Class of 43-J.

After fighting training in P-47s at Dale Mabry Field, Tallahassee, FL he was sent overseas to the Southwest Pacific Theatre where he joined the 49th Fighter Gp. in New Guinea and flew P-40s. The group changed to P 38s in September 1944 and was the first land based fighter group to be based in the Philippines following U.S. invasion.

While stationed in Leyte he was credited with four enemy planes destroyed in aerial combat, including two on one mission for which he was awarded the DFC. After flying 120 combat missions he was transferred to the States and assigned as an instructor pilot in P-51s in Sarasota, FL.

He was relieved from active duty as a reserve officer in January 1946 and is now a retired Lieutenant Colonel living in Shalimar, FL.

JAMES GARRY HARRIS, was born in Rich (near Clarksdale), MS, on Dec. 12, 1921. He was inducted in the U.S. Army with the National Guard in 1940. In 1942 he was accepted in the U.S. Air Force and he began training in Santa Ana, CA. In 1943 he had Primary at Oxnard, CA; Basic at Gardner Field, Taft, CA; and Advanced at Williams Field, Chandler, AZ, where he graduated Class 43-F. Later the same year, he trained at Edwards Air Force Base, CA, and McChord Air Force Base, Olympia, WA, before being sent to Kurmitola, India, to help form the 459th Fighter Sdn., flying the P-38 Lightning. While overseas, he was a Flight

Leader and flew 60 missions (183:30 combat hours) and had three victories in the air and three damaged on the ground.

His decorations included the DFC, OLC, Air Medal OLC, and 10th Air Force Unit Citation.

J.G. Harris left Chittagong, India, on Jan. 27, 1945, via a Liberty boat from Casablanca to New York City and was honorably discharged from Santa Ana June 8, 1945, as a Captain. From 1951-1952 he trained pilots in the AT6 at the Greenville, MS, Air Force Base during the Korean War. In 1980 he retired from cotton and soybean farming. He presently owns a pecan orchard in Marks, MS.

JACK R. HARROW, was born in Ramsey, NJ and entered the USAAC as a cadet in March 1942, and earned his wings at Craig Field (SE) Selma, AL in March 1943, transitioning in P-40s and P-51s. He was sent to North Africa September 1943, checking out in the P-38 at Telergma. That was quite exciting since it was the first twin engine airplane he had ever flown. He then joined the 37th Fighter Sdn., 14th Fighter Gp. in Tunisia. Flew 50 missions with the 37th, mostly out of Foggia, Italy. Most memorable of which were five bomber escort missions over the Ploesti oil fields. Some of Luftwaffe's finest ME-109 pilots were deployed there to protect those fields. Lt. Harrow learned how to get the most out of his "Lightning" during the dog fights over that target. During a head-on pass with an ME-109 he closed so fast that neither he nor the 109 pilot had time to shoot and came so close that the two pilots could have swapped scarfs. Later he claimed one ME-109 kill and one probable and several on the ground during fighter sweep missions.

After flying most all of the early jet fighters during his career he retired in December 1964 as a Lieutenant Colonel in Denver, CO. Then worked for United Airlines as an instructor, retiring from there in 1984. Now lives in Daytona Beach, FL.

JOHN SHERMAN HASCALL, was born Jan. 1, 1918, in Marquette, MI. He graduated from Graveraet High School in 1935 and from Michigan College of Mining and Technology (now Michigan Tech. University) in 1939. He worked as a mining geologist in Colorado Springs.

Graduating from Army Advanced Flying School in Chandler, AZ, on May 20, 1943, he arrived in Greenock, Scotland, on Aug. 25, 1943, and was then sent to King's Cliffe, Northhamptonshire in East Anglia (England). He was assigned to the 20th Fighter Gp., flying P-38s and was in the 77th Fighter Sdn. He was shot down over Wenneperveen, Holland, bailed out succesfully but landed in a river where his chute dragged him under. He was buried in Wenneperveen Cemetery, Holland.

WILLIAM L. HASTY, USAF (ret.), was born in Kershaw Co., SC, June 16, 1918, and moved to Bishopville, SC, in 1936. He volunteered for the Army Air Corps in January, 1943, and began flying school in October 1942. He graduated from Williams Field, AZ, P-38s on July 28, 1943.

After P-38 RTU Training at Muroc Army Air Field, and Sandiago Naval Air Station, he joined the famous 475th P-38 Fighter Gp. in Dobo Dura, New Guinea. His assignment to the 433rd Fighter Sdn. was highlighted by bomber escort missions, fighter sweeps to northwest New Guinea, and to the Admirality Islands, and fighter support for the recapture of New Britain.

On June 5, 1944, Bill was shot down over Balo, New Guinea, by AA fire while escorting B-25s. He successfully bailed out of his new P-38 J-15-LO, and was captured by the Japanese and became a P.O.W. for the rest of WWII. After liberation on Sept. 5, 1945, he returned home to various assignments in transport aircraft and staff assignments at the Pentagon, Ninth Air Force, and U.S. Strike Command. After 26 years of service, Bill retired from the Air Force at McDill AFB, FL, on Nov. 1, 1967.

HERBERT B. HATCH, (Stab), 1st Lieutenant, was born May 23, 1918 in St. Paul, MN, where his father was attending O.C.S. at Ft. Snelling. He was accepted into Aviation Cadet Training in September 1942, for Preflight at Santa Ana; Primary at Visalia, CA; Basic at Lancaster, PA; and graduated from Williams Field, Chandler, AZ, in December 1943.

He became a member of the 71st Fighter Sdn., 1st Fighter Gp., at Foggia, Italy, in March 1944. He flew 62 combat missions during his tour, the most memorable of which was on June 10, 1944 to Ploesti, Romania. On that mission the 71st lost eight of their 16 38s while he was credited with five confirmed victories, one probable and two damaged, becoming one of the few who became an ace in one mission.

He now resides in LaConner, WA, with his wife of 50 years. He still retains his pilot's license and his love of flying at age 70.

EDWARD J. HAVLIK, MSgt, USAF (Ret'd), was born at Lawton, OK on Mar. 28, 1920. He enlisted in the Army Air Corps on Dec. 2, 1940, and received basic training at Kelly Field, TX. After basic training, he was assigned to the 76th Bombardment Sdn., 42nd Bombardment Gp., then located at Ft. Douglas, UT, and moved with the Group to Gowen Field, ID and McChord Field, WA. After leaving the 42nd, he was assigned to the 337th Fighter Sdn., 329th Fighter Gp., located at the Grand Central Air Terminal (Cal-Aero Tech), Glendale, CA in 1942-1943, as Chief Clerk in the Engineering department and maintained a visual status display, technical order files, and maintenance files on approximately 25 P-38s. After leaving the 337th, he was transferred to the 20th Fighter Gp., March Field, CA for a short stay. Overseas requirements caused his movement to the Fresno, CA Staging Area for subsequent relocation at Adak and Attu in the Aleutians, via Ft. Lawton, Seattle, WA, on the Liberty Ship George Flavel in 1943 to 1945. On return, he served at these military installations until retirement in 1960: Ft. Logan, CO, Ft. Douglas, UT, Ft. Sam Houston, TX, Lowry AFB, CO, Scott AFB, IL, Gulfport (TTAF) and Keesler AFB, MS; Wheelus Field, (N.A.), Dhahran Air Field (S. Arabia), Ellsworth AFB, SD, and Amarillo AFB.

Ed was awarded "The Air Force Outstanding Unit Award", while with the 28th Bombardment Wing, Ellsworth AFB, SD, and is also a graduate of the 15th AF NCO Academy, Class 58D, March AFB, CA.

His most remembered events while with the P-38 Sdn., were a "belly" landing, without pilot injury; a P-38 taxiing over the cab of a parked oil truck, due to brake failure, and the accidental death of a buddy, by a released tow target and cable, near the flight line area. The shrieking whistling sounds of the P-38 in flight - the posture and cockpit location are also unforgettable features of the P-38.

Ed is married to the former Shirley M. Thompson, of Perryton, TX, and a son, age 20, Dana D. Havlik, was assigned to the 388th Tactical Fighter Wing, Hill AFB, UT, as an F-16 Aircraft Armament Specialist, in December 1987.

Ed retired from the USAF on Dec. 31, 1960, while serving with the 4128th Strategic Wing (SAC) (B-52) Combat Operations Div., at Amarillo AFB, TX, after 20 years service. He also retired from the U.S. Postal Service on May 2, 1987, after 18 years of employment.

RAYMOND S. HAWKS, was born near Versailles, MO on Apr. 3, 1923. He enlisted in the Air Force and entered military service in April 1943. Primary Flight Training was at Wickenburg, AZ. Basic Training at Gardner Field, Bakersfield, CA; Advanced at Luke Field, AZ where he checked out in the P-40 and received his Wings in Class E on May 23, 1944. Advanced Fighter Training was at Chico, CA in P-63 King Cobra and the final hours in new P-38 L Models.

In Europe he was assigned to the 367th Fighter Gp., 394th Sdn. near Reims, France in December 1944. At this time the P-38 was being phased out of the European Theater and he was a bit disappointed to find only the older "J" Models in service. He was discharged in April 1945.

Ray is now retired after 17 years with General Motors and 24 years in automotive sales.

FREDERICK C. HAYNER, ASN 12035014, enlisted: U.S. Army Air Corps, Dec. 10, 1941.

Attended Aircraft Mechanics School, Kessler Field, Biloxi, MS.

Shipped to Santa Monica, CA. Attended Special School to study the P-38 at Lockheed Aircraft, Burbank, CA.

Assigned to 55th Fighter Gp., 38th Fighter Sdn. (then called Pursuit Gp.) McChord Field, WA (May 1942).

Assigned to 329th Fighter Gp., 332nd Fighter Sdn. Paine Field, WA. Became the squadron painter.

The 332nd Fighter Sdn. under the command of Capt. Joseph A. Shurlmistras moved to Orange County Airport, Santa Ana, CA (August 1942). Stayed as a training outfit until about March 1944.

Assigned to 479th Fighter Gp., 434th Fighter Sdn., Lomita Flight Strip. Moved then to Santa Maria staging area.

Shipped out to Wattisham, England April, 1944. Aircraft P-38s later to P-51s.

Returned to Ft. Dix by Aircraft carrier (Lake Champlain) October 1945.

Main duties Aircraft I.D. plus lots of nose art and victory symbols.

Highest rank S/Sergeant.

Occupation since WWII, Layout Artist and Screen Printer.

JAMES E. HAZLETT, (Gene), was born in Whittier, CA and raised in Long Beach, CA. He graduated from Poly High School in 1942. Since WWII was already in process he decided to take the aptitude and physical tests to qualify for selection as an Aviation Cadet. He was called to active duty on January of 1943 and was selected to be a Cadet in class 44C. He graduated from Williams Air Field in Arizona and was selected for fighter pilot training. He flew the P-39, P-63, and P-38 fighter planes in tactical training in California. In 1944 he joined the 49th Fighter Sdn., 14th Fighter Gp., which was then headquartered in Foggia, Italy. He completed 33 missions in P-38 fighters and was still with the 49th on V.E. Day.

His missions were mainly fighter escort for B-17 and B-24 bombers, dive bombing, strafing, and a couple of F-5 photo escort missions. He participated in several missions involving the shooting of ground targets such as trains, planes, and other strategic targets. He saw some of his buddies shot down by groundfire, but his plane was only hit twice by groundfire, and in both cases the damage was not serious enough to prevent his return to home base in Foggia. Gene obtained his Commercial Pilots License after the war but decided on a career in Electronics. Gene retired from his career in Electronics in 1986 and resides in Los Alamitos, CA with his lovely wife, Zita.

WILLIAM WIGG HAZZARD, JR. born Birmingham, AL 1916. BS U of GA '38. Alabama Forest Service until Private 31st Inf. Div., March 1941. Aviation Cadet, August 1941 - March 1942, at Pine Bluff, AK; Randolph & Brooks Field, TX. Joined 8th Photo Sdn. (P-38s) at March Field, CA, May 1942 as 2nd Lieutenant pilot. Overseas with 8th to Australia, July 1942. Later transferred to 95th Bomb Gp. Awarded DFC, PH, AM (eight OLCs), After U.S. training assignments, ended WWII as Major, subsequently Lieutenant Colonel in AL NG.

Died, Pittsburgh, PA 1977, leaving wife Jaquelin and three daughters. You will best remember Bill's spirit by the first and last stanzas of a poem he wrote on Dec. 8, 1941.

When God spent six days making the earth
He must have spent full five
To build this land - America
That freedom here might thrive
****
And though the fight is not our choice
We did not set the stage
Pray God that I may live or die
For our Great Heritage.

JOHN W. HEDENBERG, was born in San Antonio, TX, Dec. 9, 1922. He entered Cal Tech in September 1940, and worked on the P-38 final assembly line at Lockheed during the summer of 1941 as a bucker and riveter. After Pearl Harbor, he left his sophomore year at Cal Tech to enlist as a Private in the old Army Air Corps on Jan. 3, 1942, and applied for the Aviation Cadet training a few weeks later. Receiving his wings and commission May 20, 1943, in the class of 43-E, he spent the next year in, successively, the 329th FG and the 364th FG at Van Nuys, CA then the 479th FG at Palmdale, CA before going to the 8th Air Force with the 479th in May of 1944.

During John's combat tour he was awarded the Distinguished Flying Cross and the Air Medal with five Oak Leaf Clusters. He was returned to the U.S. in November 1944 and was separated from the service as a 1st Lieutenant in September of 1945. He returned at once to Cal Tech, graduating with a BS in Applied Physics in 1948. He worked as an engineer in the rocketry and computer fields for the next 15 years, then returned to school at UCLA for a doctorate in psychology (granted in 1969). He has worked since then as a Clinical Psychologist in private practice, in Santa Monica, CA.

EARL C. HEDLUND, was born and reared in Nebraska, graduate of University of Nebraska 1938 with ROTC Commission in Field Artillery. Masters Degree University of Illinois, 1939. Continued graduate study until called to active duty August 1941. Flying school August 1942, trained for 40 hours in P-38 when assigned Alaska/Aleutian Islands October 1942 to 54th Fighter Sdn. at Adak. Flew 67 fighter, bomber and strafing missions from Adak and Amchitka—supported ground forces in battle of Attu. He really appreciated the single engine capability of the P-38 when he lost an engine while strafing at Attu and returned 500 miles over cold, northern waters on single engine on his very first combat mission! Promoted to Captain, awarded DFC and Air Medal; returned to Z.I. July 1943.

Assigned Squadron Commander newly activated P-38 Sdn., 428th of 474th Fighter Gp. — trained in California. To England February 1944. Flew from Southern England on escort and interdiction missions until moved to Normandy after D-Day. Continued support of ground forces and interdiction missions and moved toward Germany with ground forces. Left 428th Sdn. as Commander to Operations Officer 474th Gp. — promoted Major — then to Lieutenant Colonel and Deputy, then Group Commander. Flew 103 missions in Europe. Shot down by ground fire April 1945—prisoner—escaped. Returned just as war ended. The 474th Fighter Gp. was the last unit flying P-38s in the theater.

Hedlund returned to U.S. December 1946; completed his Ph.D degree at University of Illinois and assigned in logistics/transportation sales in Hq. USAF; Far East Air Force, Japan during Korean operation. Back to Hq. USAF. Promoted to Colonel 1951; Brig. General 1960. Assigned Deputy Ogden Air Material Area, Commander, Warner Robins Air Material Area as Maj. General. In 1966 assigned Director Defense Logistics Agency, Secretary of Defense as Lt. General. He continued flying and flew most of the fighters up to F-4 and F-111. In 1971 assigned U.S. Military Representative to the Central Treaty Organization, Ankara, Turkey. Retired as Lieutenant General October 1973 after 35 years service. Awarded the DSC, DSM, Legion of Merit, DFC, Air Medal, Purple Heart, British DFC, French Croix de Guerre, Belgium Fouriguerre.

Married Eleanor Ruth Neff in October 1948. Six children; all finished college and on their own. Five grandchildren. Since retirement worked as consultant and active in volunteer work. Hobbies are gardening, golf, hunting and fishing.

I loved the Air Force and what it stood for and accomplished. I loved the feel and power and majesty of the P-38!

ARTHUR W. HEIDEN, born, schooled and entered service from Keya Paha Co., NE. Aviation Cadet exam, Omaha, NE June 1942. Aeronautical University, Chicago, till November 1942 (aviation mechanic). Santa Ana preflight, Visalia/Dinuba primary, Merced Basic, and graduated Class 43-H Williams Field at Phoenix August 1943. Assigned P-38 RTU's at Murock Dry Lake, Salinas, and North Island NAS. Rush ETO assignment from San Diego on Christmas Day 1943.

Arrived Glasgow on the Aquitania. Orientated at Achem and Gox Hill then assigned 20th FG 79th FS on Feb. 11, 1944. Combat Ops in P-38s until July then P-51D-5 until Aug. 7, 1944. Ground crew were Sgt's Pyles, Birdwell, and Fink. Max Pyles was awarded Bronze Star for crewing the first P-38 to exceed 300 combat hours without an abort. Unit Citation on April 8, 1944, DFC with cluster, and Air Medal with four clusters. Lucky Lady name used on aircraft.

Returned to ZI and assigned Fighter Training Instructor at Tallahassee. Attended Fighter Gunnery & Instructors School Pinellas. Flew P-40N and P-51D aircraft these stations. Joined Caterpillar Club after P-40 midair. December 1944, detached to Long Beach Ferry Command for rush P-51 deliveries during the Battle of the Bulge. Flew P-40s in Victory Parades, major U.S. cities, to honor returning ETO General Officers after V-E Day.

Assigned Far East Theater fall of 45 for occupation of Japan. Duties included 8th FG Operations and Assistant A-3 Combat Operations, Fifth Fighter Command. P-38Ls and P-51Ds. Returned US and separated from service June of 1947. *(See photo, page 140)*

GEORGE HENDRIX, (Dan), was born in Piqua, OH Feb. 3, 1924 and was raised in Tarzana, CA.

He enlisted in August, 1942 and completed pilot training at Williams AFB, AZ in December 1943. He was an original member of the 434th Ftr. Sdn., 479th Gp. which was formed at Glendale Airport, CA in December, 1943. The Group was sent to Whattisham, England in May 1944. He flew 67 combat missions and destroyed four enemy aircraft. He also flew 50 combat missions in Korea in F-94s and over 200 missions in RF-4C Phantoms in Southeast Asia.

He graduated from UCLA and the USAF Experimental Test Pilot School and served as a Test pilot for ten years. He retired from the Air Force in 1973 after 31 years of service, with the rank of Colonel.

He then spent 11 years with Rockwell International and retired as their Director of Flight Test in 1984. He was married in 1946 and has two daughters and two granddaughters.

CHARLES W. HERRALL, was born in Yakima, WA on Apr. 1, 1920 and lived and raised in Portland, OR. He enlisted in Aviation Cadet program at Santa Ana Army Air Base, CA and graduated and commissioned as single and twin engine pilot at Williams Field, AZ, Class of 43-B.

He was then sent to Muroc, CA (now Edwards Air Force Base) to check out in P-38s in March of 1943, and by April of 1943 was assigned to the North African theatre, landing in Oran, Algeria, and assigned to the 37th Fighter Sdn., 14th Fighter Gp.

He assisted in chasing Gen. Rommel out of Africa, covered the invasion of Sicily and again the invasion of Salerno, Italy. His longest single combat mission was eight hours plus, escorting B-24s from North Africa to the Brenner Pass in Austria and Italy and back. He had three single engine returns, the longest of these being after a dog fight over Yugoslavia which involved crossing the Adriatic Sea, the southern tip of Italy, then across the Mediterranean Ocean back to his base at Tunis, North Africa. He is sold on the P-38.

His decorations include the Air Medal with seven Oak Leaf Clusters and the North African Campaign Medal with two Bronze Stars.

GENE HERSEY, started flying in 1941 after receiving a scholarship under the Civilian Pilot Training program. Just before Pearl Harbor he received his Private Pilot's license. He tried to enlist in the USAAF but did not have the required college credits. He was on his way to Canada to join the RCAF when he took a college equivalency test and was accepted as an Aviation Cadet in the USAAF. From March of 1942 to February of 1943 he went through Primary, Basic and received his wings as a twin engine pilot from Williams Army Air Force Base on Feb. 6, 1943. Checked out in P-38s at Muroc Dry Lake. (Now Edwards AFB).

Arrived in North Africa May 19, 1943 and was assigned to the 15th Photo Sdn. 3rd Photo Gp. Because he had no photo recce training he was checked out in the F5, flew two high altitude practice missions and flew his first real mission in early July of 1943. After several missions was promoted to 1st Lieutenant in September 1943. Seven months later was promoted to Captain. Flew total of 188 hours of photo recce missions. After returning to U.S., he held various administrative positions and was discharged as a Major in late 1946. Highest rank held was Lieutenant Colonel in the Air Force Reserve. Worst memories: being shot at over Ploesti, chased by FW 190s and the bitter cold at 32,000 feet.

HARRY CLARK HIGGINS, was born in San Diego in 1918. He graduated from the University of California in 1949 thanks to the GI Bill of Rights. He recently retired from Boeing after a career in aerodynamics.

His Draft Board tried to get him in July, 1941 for going on strike at North American Aviation but he wiggled out of that one and volunteered as a Cadet in the Army Air Corps in February 1942.

He learned to fly in the class of 42-J at Visalia, Merced and Luke where he was commissioned, rated and married Oct. 30, 1942. He checked out in P-38s at Muroc AAF and flew as a trainee out of Grand Central Air Terminal in Glendale before they had rules about buzzing in LA.

With five of his classmates he ferried P-38s to the 54th FS in the Aleutians arriving in March, 1943. Here they harassed the Japanese on Kiska and Attu until the enemy was defeated and withdrew from this largely forgotten invasion of the USA. During this period the 54th moved from Adak to Amchitka and Shemya and finally came to rest on Attu, the end of the line.

He was credited with destroying a Betty and spent the rest of the war flying the gravy train in southern California tooling around in P-38s, P-51Ds and P-80s. He was probably unique in WWII being a fighter pilot with a caterpillar, two DFCs and no Air Medal!

CLARENCE W. HOFFMAN, entered military service June 15, 1941 as a ROTC 2nd Lieutenant in the 25th Inf. at Ft. Huachuca, AZ. As 1st Lieutenant, transferred to the Army Air Force for pilot training and graduated Class 43-K (December 1943) at Luke Field flying the AT-6.

After instructing in BT-13s for 1 1/2 classes, subsequent training in P-39s and P-38s, he was assigned overseas to the 15th Air Force in Italy. Reported Nov. 10, 1944 to the 49th Fighter Sdn., 14th Fighter Gp., north of Foggia. Hoffman participated in 35 missions in the P-38 escorting B-17s, B-24s; on fighter sweeps, dive bombing, etc. Campaigns were Balkans, Rhineland, North Apennines, Po Valley and Rome Arno.

Hoffman returned home to Reserve status. Employed by Phillips Petroleum Co. at Bartlesville, OK and Borger, TX as an engineer for 37 years. He and Mary Jane Murray married in 1948 and they have two sons.

ROBERT HOLBURY, assistant Group Operations Officer, volunteered to fly a mission, even though all others had been cancelled due to bad weather. He flew an F-5 #608 he headed toward the heavily defended 12 mile sector and began his camera run at Merzig north along the Saar to the Moselle, then made a left turn down the Moselle to complete the mission.

Holbury said the mission required his complete attention so he could follow the river while at the same time not hit any obstacles. He was chased and shot at by Jerries. He discovered that his left coolant gauge began creeping past the red line and a glance in the rear view mirror showed a stream of white vapor behind him. The Jerries continued to chase him and Holbury continued on his photo recon. mission.

After circling the field at about 300 feet and visibility at about one mile, he landed the plane.

The plane was a virtual wreck, the stabilizer was shot off, large holes were in the horizontal stabilizer and the left rudder was just hanging on - but he had brought back 212 very revealing photos that indicated that there was no imminent threat from the Panzer units across the Saar and that redeployment of U.S. forces to the area was not required. Holbury was awarded the Distinguished Service Cross for this mission.

BERKLEY E. HOLLISTER, (B.E.), was born Nov. 12, 1920 in Los Angeles, CA. In July of 1940 he was employed by Lockheed as a final assembler in Burbank, where repeated "buzzings" by Milo Burcham in the YP-38 increased his already great desire to fly a P-38. He left Lockheed to enlist in July of 1942 and the dream came true at Williams Field, AZ, graduating in 43-K class, December 1943. He was assigned to the 434th F.S., 479th F.G. taking O.T.U. training at Lomita Flight Strip, CA (now Torrance Airport). The group arrived in England April 1944, and flew P-38s most of the tour before changing to P-51s.

He flew 67 Combat Missions and was awarded the Air Medal with eight Oak Leaf Clusters and the D.F.C. With the tour completed, he returned to the U.S. in January 1945, and was separated July 1945. He is now retired and living in San Juan Capistrano, CA.

RAYMOND C. HOLMAN, was born Apr. 12, 1921 in Williamsburg, KY, graduated from Ft. Lauderdale High School, Ft. Lauderdale, FL in 1940 after which he took courses at the University of Miami and the University of Florida and an I.C.S. course in engineering.

He enlisted in the U.S. Army Air Force on July 23, 1941 and was honorably discharged Sept. 5, 1945. Ray served with the 49th Fighter Gp., 9th FS and had foreign service for 2 1/2 years from January 1942 thru June 1944 in Australia and New Guinea. He served in the U.S. for 1 1/2 years. Ray was a Staff Sergeant when discharged from military duty as an airplane power plant mechanic.

He started working as a meter reader and worked up to Water and Wastewater Distribution Superintendent and Wastewater Operations Supt. for City of Fort Lauderdale for 30 years. Ray also formed and directed Utilities Services Inc., a company that maintained wastewater pumping stations and made pressure taps. He retired from the City of Ft. Lauderdale in 1975 and sold Utilities Services Inc. in 1981. Ray died on Apr. 26, 1983 from complications of diabetes.

Ray was married to Doris Herman in Ft. Lauderdale, FL in August, 1949. They have a son and daughter and three grandchildren.

HARVEY HONEYCUTT, born Modesto, CA Jan. 19, 1922. Graduated from Alameda High School in 1939. Drafted into Service November 1942. Requested and reluctantly received permission to enter flight training with Class 43-J, Preflight at Santa Ana - Primary at 29 Palms - Basic at Lemoore - and Advanced at Luke Field.

Graduated 2nd Lieutenant. On to Oroville for O.T.U. training in P-39. After approximately 95 flight hours, our O.T.U. program was cancelled, the A.F. needing mechanics in South Pacific and replacement pilots in Europe.

On to Orange County Airport (across highway from SAAAB and Preflight) for 25 to 30 hours in P-38s prior to overseas shipment. By convoy to Casablanca - slight delay enroute to Bari, Italy in C-47s. A Capt. Schwab, C.O. of the 154th Weather Recce Sdn., requested a couple of P-38 pilots throw their gear out for assignment to his outfit. After three missions and a slight disagreement with Gen. Bourne, Oper. Officer of the 15th A.F., proceeded to Triolo, Italy for assignments to 37th FS - 14th F.G. - Up thru the ranks flying #2 in various flights to tail-end Charlie to element leader-flight leader, section leader, Squadron and Group leader. Served as Operations Officer - Completed 66 missions - Attained rank of Captain. Decorations include seven Air Medals, two DFC's, one Soldier's Medal and the French Croix De Guerre with Palme.

At completion of missions, returned Stateside for reassignment. Ended Military career as Engineering Test Pilot at Chico and Paine Field, separating from Service at Portland, OR, in October 1945.

Having married my wife, Beverlee, in March 1945, we moved to Southern California where I worked construction, raised five children, four of which survive, and retired to Oregon in 1982 where I smile quite a lot. Always a cribbage player, in retirement I've sharpened my game to where I am now the World's Greatest. (If there's any cribbage players out there, I'm only joking).

VERNE E. HOOKER, was born in Great Falls, MT on Nov. 12, 1921. He volunteered for military service Mar. 12, 1942 and was sent to England in April of 1942. In December of 1942 he was one of few selected to return to the U.S. for pilot training. He graduated pilot training at Chandler Field, AZ in December 1943, Class of 43-K.

He was assigned to the 435th Fighter Sdn. of the 479th Fighter Gp. at Oxnard, CA for overseas training in P-38. Subsequently, the group was stationed at Mottisham, England in May 1944. He was squadron operations officer March 9 to Dec. 2, 1945. Held the rank of Captain and two aerial combat confirmed and three and one-half destroyed on the ground. Received Air Medal with 12 Clusters, DFC, with one Cluster, and Croix de Guerre.

He was separated from service on Jan. 28, 1946. Graduated from Colorado School of Mines in 1950 with a degree of Geological Engineer. He became an authority on the measurement of stresses in the earth's crust and retired as a research supervisor from the U.S. Bureau of Mines November 1981.

His wife's name is Virginia and they have two daughters and a son.

BRUCE HUBBARD, was born Jan. 30, 1924. Attended University of Washington; signed for Cadet program October 1942; called March 1943. Went through Shepherd Field, TX; Oklahoma A&M at Stillwater; Classification and Cadet Centers at San Antonio, TX. Primary flight training was at Stamford, TX; Basic, Perrin Field, TX; Single-engine Advance, Eagle Pass, TX, Class 44-D. Aerial gunnery and P-40 transition at Eagle Pass. O.T.U. at Peterson Field, Colorado Springs (P-40s) and held over as instructor. In November 1944 was assigned to 18th Fighter Sdn. (P-40s), 343rd Gp., 11th A.F., on Attu Island, AK. Transition to P-38s on Attu. The "plan" was to use our P-38Ls to attack the northern Kurile Islands.

Only missions were scrambles after incoming radar signals lacking I.F.F. response. The Navy, of course, and, of course, no "kills." I became good subject for a lot of kidding for half "credit" for a Japanese balloon. Was R.O.P.A.'d into retired reserves as Captain in 1955. Presently employed by City of Seattle as a Housing/Zoning Inspector.

Married another native Washingtonian, Jacquelyn Douglas, in 1950. Have four daughters and two grandchildren.

W. GENE HULL, Brig. Gen. joined the 8th Ftr. Gp. Met them in Manila. Did train in California with some of them before they left for the Pacific. I stayed for some extra gunnery.

My instructor at Santa Maria in 1944 was

Capt. Donald M. Leighton from Wells, NV. When I joined the 8th in Manila, Capt. Leighton was the Squadron Commander of the 35th Fighter Sdn. and I went on to Ie Shima with the squadron. Leighton later became Major.

I went on to Japan with the squadron. I was squadron commander of the 36th FS when we made the move to Japan. Flew with the 8th FG during my time in Japan. Ferried the P-38s that we had to Kimpo Strip (outside of Seoul) in early 1946. Not many left in the outfit with P-38 time then as we were converting to P-51s. I made several trips to Korea taking the 38s to Kimpo. Had to do our own maintenance as all P-38 mechanics had gone home. Understand that when the Korean War started, the P-38s lined up at Kimpo were bull-dozed off the runways and destroyed to make room for the planes that were going to be used in the Korean War.

I stayed in the reserve and served 37 years retiring several years ago as Brig. General.

LOWELL, RICHARD, and GERALD HUNTER, from Pimento, IN enlisted in the U.S. Army Air Corp in June of 1941. Basic training at Jefferson Barracks, St. Louis, MO and then to Chanute Field, IL for basic aircraft mechanics school. All were assigned to the 49th FS, 14th FG at Hamilton Field, near San Francisco, CA then on to Ft. Dix, NJ. While at Ft. Dix, waiting for the convoy to be organized, the squadron took up a collection and purchased musical instruments for the three brothers, who were known as the Hoosier Harmonies. Richard played the bass fiddle, Lowell played the guitar, and Gerald played the violin. While in Shrewsbury, England the brothers played for officers' dances.

From England the squadron went on to North Africa docking at Oran. From the base at Constantine, the P-38s went on long-range escort missions using the newly-developed, jettisonable, 150 gallon fuel tanks.

The brothers said that they lived in tents in the desert and that it was very hot during the day and cool at night. After about a year in North Africa, the Hunter brothers went to Foggia, Italy. At this time, Lowell was promoted to Staff Sergeant and became a crew chief.

After four years in the service, the Hunter brothers had accumulated enough points to go home. They were discharged at Camp Atternury, IN on June 25, 1945. They all went to work in a glass factory in Terre Haute, IN. Lowell attended Purdue University and earned a degree in vocational agriculture in 1949 and taught school for the next 36 years.

He and his brother still get together to play and sing.

JULES J. HYMEL, Lt. Col., A0-800545, Pilot Class 43-C was active in all aspects of the Army and U.S. Air Force, Active and Reserve, with total service of 40 years and eight months and nine days.

Jules was born in a small town, Luling, LA. 70070, the oldest of four children, with the normal likes of a boy with desires to fly and know mechanic and become a Pilot.

On Mar. 25, 1942 he was excepted into the Aviation Cadet Program. After attending Preflight at Maxwell Field, Primary at Dorr Aero Teck, Inc., Basic at Greenville Flying, Advanced at Napier Field and Fighter Training at Page Field.

In July 1943 it was overseas to Africa where he was assigned to the 82nd Fighter Gp., 95th Fighter Sdn., with no P-38 time. After a Cockpit Check, was assigned a P-38 to get his transition time. Ha! Ha! Perfection was achieved on my first job, running the still in Africa.

Flying 20 missions, plus other assignments, such as slow timing engine and taking new pilot training. It was during one of those slow timing trips, he was intercepted on the return from enemy territory, he was escorted to the field by four Spitfires, then all hell broke with Headquarters and Commanding Officer, (RECON MISSION). On one of those weather missions we returned through two fronts, low on gas he had to land and did so in a field with wheels down, (Story too long), but gas was dropped from a C-47 then he took off, upon returning to our field he told the Commander that the Germans had left that area the night before.

On his 20th mission, escorting bomber to Udine M/Y in Northern Italy the war came to a sudden end, he had plane trouble and had to bail out, (Story too Long), was turned over to the Germans and "FOR YOU THE WAR IS OVER", which was a statement to all Prisoners of which he remained for 501 days in Barth, Stalag I, until liberated by the Russians near the end of the war.

Upon his return to the USA at the end of the war and with the end of the war in Japan, he was relieved from active duty, he remained very active in the Reserve, in the New Orleans area. After about 27 years he was transferred to the standby reserve at the rank of Lieutenant Colonel. On Dec. 4, 1982 he was retired from the U.S.A.F.

In civilian life he did plenty in school, government and for the good for one's community, plus managing several businesses. He will be retired from all except Asst. Civil Defense Director of St. Charles Parish, which is a part time job.

Lt. Col. Jules J. Hymel, Pilot, Class 43-C, was active in all phases of the Army and U.S. Air Force, active and reserve, but he P-38 indeed was my favorite period. "HEARS TO A GREAT PLANE THE P-38."

JACK ILFREY, born July 31, 1920 in Houston, TX. Graduated from high school there and went on to Texas A&M where he learned to fly in the first Civilian Pilot Training Program in 1939. Had secondary C.P.T.P. University of Houston, 1940, while working for the Hughes Tool Co. at night. Entered Army Air Corps as an Aviation Cadet April 1941. Graduated 41 I Luke Field, AZ first war time class Dec. 12, 1941. Assigned to the 94th Pursuit Sdn. 1st Pursuit Gp. flying P38 D&Es in defense of the southern California coast.

In late Spring of 1942 the First Fighter Gp. was equipped with new P38Fs and flew to Dow Field, ME to stage with the 97th Bomb Gp. in preparation for the Bolero Mission, first mass flight of fighter planes to England.

On July 4, 1942 the 94th "Hat In The Ring" Ftr. Sdn. took off on first leg of the mission, four P38s flying wing on a B17, Presque Isle, ME to Goose Bay Labrador. July 6 Goose Bay to Blue West I, Greenland. July 15 B.W.I. to Reykjavik, Iceland.

It was on this day that six of the 94ths P38s and two B17s, low on fuel, went down on Greenland's icecap. All crews were rescued but the eight aircraft are still there to this day.

July 26, saw most of the 94th land at Kirton in Lindsey, Lincolnshire, England to be stationed with the famous Polish 303 Koscinsko Sdn., who taught them many tricks of the trade.

On September 1 the 1st Ftr. Gp. made the first all American pilots and aircraft sweep over Northern France for the fledgling 8th A.F.

Other missions were to follow; however, on November 15 they all took off from Chivenor, in the Lands End area on Operation Torch, the invasion of N. Africa where they were to suddenly find out what it was all about.

Shortly out Ilfrey lost a belly tank but calculated he could make Gibraltar. Calculations proved wrong and he had to land in Lisbon Portugal, short on fuel, where he was immediately told he and the P38 would be interned. However; he conned the Portugese out of some gas, fired up and made an unauthorized take off. By the time he got to Gibraltar an international incident had flared up.

Jack was awarded commendation letter from Chief of Staff, U.S. Army, dated February 3, 1943 for being one of the first aces in 12th A.F. Mediterranean Theatre, fifth plane shot down day after Christmas 1942.

After 208 combat hours, 72 missions, he was relieved from combat duty, returned to the States and instructed in P38s and P47s.

In April 1944 Jack went back to the E.T.O. and became Squadron Commander of the 79th Ftr. Sdn. 20th Ftr. Gp. flying P38s. On May 24 he was credited with two ME109s in a hairy dog fight near Berlin. One of the 109s collided with his P38 and ripped off 4-1/2 to 5 feet of his right wing. But he was able to return to England.

On June 13 after successfully dive bombing a railway bridge over the Lorie River near Angers, Jack was shot down while strafing a train near there, some 200 miles below the front lines. He evaded and was back in England in four days.

In late July the 20th converted to P51Ds and went on Frantic VI, Shuttle Mission to Russia, Italy and back to England.

The last memorable incident to happen was on November 20, 1944 when he landed behind enemy lines, near the front, around Maastrick, Holland and successfully picked up his wing man for a short ride to Brussels. After 70 missions, 320 hours of combat flying he was reassigned to the States where he later became a Troop Commander at McChord A.F.B. In two tours, completed a total of 142 missions, 528 combat hours.

Decorations: Silver Star, Distinguished Flying Cross w/five Oak Leaf Clusters and Air Medal 12 Oak Leaf Clusters. Tally records: eight kills, all air. Presently a retired bank officer after 30 years at Alamo Nat'l. Bank, San Antonio. Historian 20th Ftr. Gp. Assoc. Editor, "King's Cliffe Remembered" (their 8th A.F. base in England) newsletter.

ARCHIE JACKSON, (Gil), was born in Kansas, and was raised on a farm near Louisburg, KS. He joined the RCAF in 1941, transferred to the USAAF December 1941, and graduated from Luke Field. He checked out in a P-38 at Muroc, CA, went to England November 1942, then to Africa January 1943, he joined the 94th *Hat-in-the-Ring* Sdn. Presented a silver *Hat-in-the-Ring* by Eddie Rickenbacker April 1943, as were 21 pilots of the 94th. He was a wing man, element, flight, and squadron leader for the 94th. Later group leader of the 1st Fighter Gp. Missions were North Africa, Sardinia, Sicily, Pantelleria, and Italy. He was Operations Officer of the 94th, and CO before returning to the U.S. He had 74 Sorties. On return to U.S. he attended Matagorda fixed gunnery school, was Gunnery Officer, and instructor pilot (P-38), Ontario, CA AFB. Attended Meterology school, and Nuclear Chemistry School, Berkley, CA. Was on the team of Chemist who discovered Russia's first Atomic Blast.

He retired from the Air Force as Lieutenant Colonel. In civilian life, was a Chemist Sunkist Growers, Ontario, CA. Then Chief Chemist, and Director of Manufacturing "Hawaiian Punch" in Fullerton, CA. Retired from "Hawaiian Punch" to South Laguna, CA 1984.

CALVIN G. JACKSON, born Lorenzo, ID Aug. 18, 1918. Has identical twin brother. Joined Army Air Force Dec. 27, 1941, Seattle. Basic Training Sheppard Field. Assigned to Radio School, on completion joined 48th Fighter Sdn. March Field. Our P-38 Sdn. sent to England Aug. 6, 1942. He was transferred from communications to set up welding shop, due to his welding experience. Went in invasion North Africa at Oran, October 1942.

Welded many parts P-38s and A-20s at Youks-les-bains, Telergma. Our squadron served in Algeria, Tunisia, Sicily, Italy. P-38s flew cover for Infantry, tanks, also escort for heavy bombers.

Medals include: Good Conduct, Air Offensive Europe, African, Middle Eastern Service, Distinguished Unit Citation, 15 Battle Citations. Squadron attached to 8th, 12th, 15th Air Force. Discharged T/Sergeant June 30, 1945, Seattle.

Thirty years Insurance Adjuster/Examiner. Retired. Lives in Seattle with wife of 41 years. Have motorhome and travels. Proud to have served his country.

NORMAN W. JACKSON, was born near Carter, MT Apr. 14, 1920. He enlisted in the Army Air Corps in January, 1942. Attended pre-flight at Williams Field, AZ, primary flying school at Ryan School of Aeronautics, Lindbergh Field in San Diego. Basic at Merced and Advanced at Williams Field, graduating Aug. 27, 1942 in class of 42-H. Checked in P-38 at McCord Field, WA with 55th Fighter Gp. and assigned to 37th Sdn. at Olympia. Bailed out after mid-air collision on Dec. 2, 1942. Flew to North Africa via Florida, Brazil, Asencion Island the British Gold Coast on Pan American arriving on Christmas Eve. Joined the 14th Fighter Gp., 49th Sdn. Jan. 14, 1943. Transferred to 82nd Fighter Gp., 96th Sdn. when the 14th Gp. reorganized. Shot down and crash landed on 7th mission March 15.

Assigned to Fighter Training Center near Casablanca March to September. Rejoined 14th Gp., 49th Sdn. at Kairouan, Tunisia on September 23. Moved to Foggia, Italy in December. Flew 50th combat mission to Ploesti May 5, 1944. Received Silver Star, Distinguished Flying Cross, Air Medal with nine clusters and Presidential Unit Citation.

After returning to the States, he attended Armament and Chemical Warfare school at Buckley Field, CO and was Armament Officer at Ontario, CA when separated from the Service as Captain in September 1945.

EVERETT JACOBSON, was born in Oakland, NE on June 16, 1917, and raised on a farm. He joined the Army Jan. 7, 1942. In May at Long Beach, CA, after training at Keesler Field, MS and Indianapolis, IN, he was assigned to the 94th Fighter Sdn. of the famous "Hat-in-the-Ring" Sdn. of the 1st Fighter Gp.

He left for England June 4, 1942 where their pilots trained for combat in the sleek and powerful P-38 fighter planes. October 17, 1942 his squadron was given a special assignment—the Invasion of North Africa. Soon after he was qualified to be Crew Chief on P-38, his job for the rest of the war.

The Battles and Campaigns he participated in were: Algeria, French Morocco, Tunisia, Sicily, Naples, Foggia, Rome Arno, Southern France, Rhineland, Air Combat Balkans, Northern France, Northern Appenines and Air Offensive Europe.

With the war ending in Europe, May 8, 1945, found Everett back in the States June 21, 1945. He was discharged June 30, 1945. Everett went back to farming and at 71 is active, feeding cattle and hogs and grain farming.

JACK S. JENKINS, born in Comanche Co., TX Nov. 17, 1914. He graduated from Texas Tech University with a degree in mechanical engineering and joined the U.S. Army Air Corps as a flying cadet in March 1938.

After graduation from the advanced flying school at Kelly Field on Feb. 1, 1939 he was assigned to the 20th Pursuit Gp. at Barksdale Field, LA flying P-36s.

Soon after Pearl Harbor he commanded the 37th Fighter Sdn. of the newly formed 55th Fighter Gp. at Olympia, WA flying P-43s and P-38s.

As a lieutenant colonel and deputy group commander he accompanied the 55th Fighter Gp. with P-38s to England in September 1943 to escort heavy bombers over Germany.

In February 1944 he was promoted to Colonel and commander of the 55th Fighter Gp. He led the 55th in a P-38 fighter sweep on Mar. 3, 1944 to become the first American fighter unit to penetrate the skies over Berlin.

On Apr. 10, 1944 after both engines of his P-38 TEXAS RANGER were crippled by flak while strafing a German airfield near Paris, France he crash landed, was seriously injured, captured by the Luftwaffe, and held as a P.O.W. until the end of the war.

He was credited with two enemy aircraft destroyed, one probably destroyed and one damaged. His decorations include the Purple Heart, Legion of Merit, Distinguished Flying Cross, and Air Medal with three Oak Leaf Clusters.

Before retiring in 1968 with 30 years military service he held several assignments relating to fighter aircraft operations some of which were: Commander, 49th Fighter Wing, Commander 35th Fighter Wing, and Commander 50th Tactical Fighter Wing equipped with F-100 aircraft.

No other aircraft he has flown has approached the thrill and satisfaction of the P-38.

However, he believes it could have been improved if equipped with Merlin engines.

Jenkins resides in San Antonio, TX with his wife Julie.

DARREL L. JOHNSON, Minnesota veteran flight instructor and examiner, has been aloft for nearly four years; putting it differently, he has been airborne an average of two hours a day, 365 days a year, for 46 years! That's nearly 33,000 hours logged as a U.S. Army Air Force and civilian pilot and flight instructor. Johnson has been both teacher and tester to generations of Rochester-area pilots.

Born in Barron, WI, Johnson became an early prototypic "airport kid" - his father was the second employee of Minneapolis's Wold-Chamberlain Field, and the family actually lived on the airport. Soon after high school he found his way back to the field, working as a lineman for McGinnis Aviation. He was paid half in cash and half in J-3 Cub time and lessons. This permitted him to get his private license in 1941.

War changed many plans in those days, and Johnson was no exception; he entered the Army Air Force (AAF) in 1942, first attending Armament school, then transferring to Glider school, where he flew TG-3 and CG gliders. The glider pilot program was closed out as he finished advanced training in 1943. This led him to Aviation Cadet School and two years as an Army flight instructor in Texas. He finished his AAF career patrolling the Panama Canal and coastal waters in a P-38. He relates with sadness being ordered with his squadron to fly their P-38s to a depot, to be loaded aboard ship and dumped at sea, with crated spare engines and parts. (Warbird lovers, weep!) Johnson did manage to keep the control yoke from his P-38 as a momento. During his AAF time, Johnson recalls flying many of the finest aircraft of the day, including the AT-6, PT-19, BT-13, P-40, P-51, P-38, C-45 Beech 18, UC-78 bamboo bomber, C-47, and various light aircraft. He described the P-38 as "the finest airplane I ever flew,"despite having to learn to fly it without the benefit of dual instruction, and having to teach himself engine-out procedures in an actual emergency at pattern altitude!

When the armistice sent thousands of pilots home, Johnson found his way to Rochester, MN in 1946, to obtain his civilian instrument and instructor's ratings at the recently-founded Gopher Aviation. After completing that training, he stayed on at Gopher as a charter pilot and instructor for 25 years, until the company was dismantled. Subsequently, Johnson has been an independent instructor and charter pilot, but closely identified with Neuman Air Charter of Rochester Municipal Airport, run by his former Gopher Aviation colleague, Andrew C. Neuman. Johnson has flown many air charter missions in Neuman's Bonanzas.

Perhaps Johnson's most lasting impression on the Southeastern Minnesota aviation community has been as a Designated Flight Examiner, for which he was licensed by the FAA in 1950. Many have quaked in their boots as the steely-eyed, firm jawed judge in the right seat set out impossible tasks, mis-set their radios, distracted their thoughts, and pulled their throttles!

Johnson has trained over 2,000 military and civilian pilots.

Johnson says that he flies now "just enough to keep me happy," and considers himself semi-retired. When the day of actual retirement comes, he will be missed as one of the finest, steadiest soldiers of general aviation. The thousands of passengers and patients safely transported, students trained and retrained, and pilots tested, all owe Darrel L. Johnson a great debt. He never got rich - in fact, never owned an airplane - but he continues to give his best. One can only wonder anxiously, who will take his place?

HERBERT E. JOHNSON, (Herb), born Dec. 22, 1915; South Dakota. Entered Flying Cadets November 1940; graduated July 1941. Assigned to 14th Fighter Gp. March Field.

On maneuvers the morning of Dec. 6, 1941 he landed at Salinas CA to find wild eyed ground crews shouting war, war!

The Group converted from P-40s to P-38s during the spring of 1942 and flew to England in July by way of Greenland and Iceland. In October the 14th Gp. flew to North Africa via Gibraltar for combat operations in Algeria-Tunisia. By February 1943 all assigned aircraft were lost to combat so pilots were rotated to the U.S. Maj. Johnson returned to England in October 1943 and flew with the 20th Gp., first as 77th Sdn., C.O. and later as Deputy Group Commander.

Johnson flew 108 missions during the two combat tours. He was credited with five a/c destroyed two probables and two damaged. Herb likes to remember that the big fighter brought him back from combat missions on one engine four times. Lt. Col. Johnson was assigned to 8th Fighter Command Operations in August 1944 and was rotated to the U.S. in February 1945.

After a brief interlude as a civilian he accepted a regular commission and served 27 years in the Air Force, retiring in 1967. Now retired again from city administration he lives in Ventura, CA. He and his wife Connie, have four children and enjoy bridge and public service activities. Herb is an avid golfer and sailor.

ALVIN H. JONES, (Jack), was born Feb. 23, 1923 at Horseshoe Bend, ID. He graduated as a pilot with class 44E at Williams Field, AZ.

In May 1945 he was assigned to 7th Sdn., 49th Ftr. Gp. as a P-38 pilot.

Most of his 30 ground support missions were completed in "LAP DOG", Sdn. #14. He made his last P-38 flight in LAP DOG Jan. 11, 1946, when the Group switched to P-51s at Atsugi, Japan.

He was discharged in 1946, and recalled in 1952. He retired in 1969 as a Lieutenant Colonel. During his 20 years of active duty he flew the P-39, P-63, P-51, C-47, B-25, T-29, T-33, EC-121, C-121, C-131 and his favorite the P-38. He logged over 7000 hours including 896 combat hours.

His decorations include the DFC, Air Medal with 17 Oak Leaf Clusters and Vietnam Gallantry Cross with Silver Star.

LINCOLN D. JONES, born Dec. 4, 1921, Los Angeles, CA.

Military: Commissioned 2nd Lieutenant (Luke Field, AZ September 1942 - may have been youngest commissioned pilot). Sent to Africa Christmas Day 1942 with a group of "replacement" P-38 pilots. Joined the 49th Sdn. of the 14th Fighter Gp. and later (after the remaining pilots were sent home), was transferred to the 96th Sdn., 82nd Gp. (end of January 1943). Flew 50 combat missions in North Africa. Destroyed FW-190, ME-110, and a ME-109 (half credit). Awarded DFC and nine Oak Leaf Clusters. Returned to USA in July 1943 and stationed in Salinas and then Santa Maria, CA as P-38 combat instructor and as a Base Training Aids Officer. Was (medically) discharged at end of WWII.

Professional: After WWII obtained both BS (1951) and MS (1956) degrees in Electrical Engineering from University of Arizona; taught there for two years. Moved to San Jose State University in 1956 and have been teaching to present time (except for a year off to obtain degree of EE at Stanford, 1964).

Family: Married Muriel (Midge) Nott (of Orcutt, CA) on Sept. 7, 1944. Three grown children and three grandchildren.

ANDREW ADAM KAHANAK, (Sept. 2, 1924) enlisted Dec. 3, 1942 for Aviation Cadet Flight Training. He completed 'Boot' and Aviation Student Training prior to cadet graduation as a pilot. He flew P-39s and P-63s before shipment (1944) to 1st Ftr. Gp., 71st Ftr. Sdn. for P-38 Combat, Foggia, Italy. Missions were Bomber Escort and Divebombing Rail Bridges in Germany, Austria and Yugoslavia mountains.

Occupation Germany (1945-1946).

He flew KB-29 and KC-97 Tankers as Aircraft Commander, instructor and Test Pilot,

Sac, (1951-1959). Missions flown in Newfoundland, Labrador, Greenland and England.

He served a three year tour with 2nd Comm. Gp., 8th Radio Relay Sdn., at Dreux and Eureux Air Bases, France with spouse and three children.

He was reassigned to Sac, Fairchild, AFB, WI. (1962), as Personnel Officer (CBPO).

He retired as Lieutenant Colonel, Aug. 1, 1967, from Minot AFB, North Dakota.

Awarded Air Medal, Commendation Medal, EAME (four Major Air Campaigns), Senior Pilot Wings.

GILBERT V. KAMARA, born Feb. 25, 1919 in Astoria, OR, was drafted June 23, 1941, and commissioned in the Ordnance Corp., January, 1943. Trained at Pendleton Field, OR and Thermal Air Base, CA. Went to India with the 1819th Ordnance Co., 68th Service Gp., arriving in Bombay, Dec. 26, 1943 and was stationed at Jorhat.

He transferred to the 459th Fighter Sdn., May, 1944, replacing LeRoy Delaney, the Ordnance-Transportation Officer. During this time, the P 38s were used as bombers, destroying Burma targets with 1000 lb. or Napalm bombs. Several bombardiers were assigned for the converted P-38s that had the 20MM and 50MM guns removed from the nose so the bombardiers could have a seat during the bombing runs.

He returned with the squadron, October 1945, and was released when the squadron was deactivated. Gil finished college at Oregon State and today appraises real estate in Astoria, OR.

DORSEY G. KAMMERER, born in Spirit Lake, ID - Mar. 27, 1912.

Went to Lockheed from Vultee in 1936 and worked on the first pressurized cabin. After completion of this cabin Flight Test Group was sent to begin work on the mockup of the first P-38. The mockup was made of cardboard and wood and the instrument panel was made of photos cut out and glued on a simulated panel. Testing on the first tricycle landing gear was begun on the taxi strip. After completion of the mockup - design drawings were developed and the craft was assembled of handbeat parts of aluminum.

The first aircraft was completed in 1938. After completion it was disassembled and loaded into two trucks that delivered it to a Secret hanger at Riverside (March Field) and reassembled. After completion Kelsey flew it to Dayton, OH. Dorsey was flying United back to Dayton when he heard the report than an experimental plane had crashed in New York. After landing at Dayton he boarded a flight for New York for the crash sight. The plane had been dismantled by the Army and there was so much damage to the frame that Dorsey and Ernie Beale had the job of reassembling the aircraft and segregating the components to be sent back to Burbank.

Dorsey was then sent with the first production aircraft with Don Young to Dayton, OH to begin the flight test program. The first of three P-38s were flown by military pilots. After one year of flight testing the first production aircraft was delivered to Selfridge Field in Michigan where the Air Force 1st Fighter Gp. served as test pilots. Dorsey taught Air Force crews and pilots how to service and handle the P-38. The full complement of 100 craft were divided between the 71st, 94th and 27th sdns.

After completing the assignment at Selfridge Field Dorsey left on a cross country trip with the 1st Fighter Gp. At landing in Texas they learned of the attack on Pearl Harbor. They were attached to the 94th Sdn. in Glendale and the 27th Sdn. in San Diego as well as the 71st at Santa Barbara to guard the west coast of the US.

The 1st then left for England - the first group sent to Europe. The squadron was assigned all new airplanes and sent to Bangor, ME where new color lights, radio frequencies, etc. were devised for the 100 aircraft. An additional 150 Lockheed employees joined the pioneers and worked in two 12 hours shifts to complete the modifications.

On completion the crews left for Europe with spare parts and five gallon cans of fuel loaded in C47s. In Maine they had landed on fresh dirt runways - serviced the equipment and flew on to Greenland. From Greenland they flew to Iceland where they were held for weather. On takeoff they were approached by a German Navigational Bomber. Some pilots were so excited on takeoff that they forgot to load their extra parts. One pilot, emerging from the clouds, came face-to-face with a German Bomber and shot it down to score the first U.S. Air Victory in WWII.

The planes landed back in Iceland and celebrated the victory where Douglas Fairbanks, Jr. led the celebration. They then left Iceland for Preswick, Scotland and on arrival were met at 12,000 feet by three aircraft. With only 45 caliber handguns for defense, the group was relieved to learn that the aircraft were British Air Force escort planes sent to fly them in.

Dorsey remained with the 1st Fighter Sdn. in England until 1942 when Kelly Johnson called him back to Burbank to work with a small group of mechanics and engineers to develop a new aircraft in a group which came to be known as the "Skunk Works."

Dorsey retired in 1967 as Division Manager of Advanced Design Projects for Lockheed.

TROY KEITH, was born in Little Rock, AR on Sept. 20, 1911. He entered the Army Aviation Cadet program and graduated from Kelly Field, TX in 1936. He was assigned to the 20th Pursuit Gp., Barksdale Field, LA.

The 20th Pursuit Gp. was transferred to the West Coast in 1939 and in January 1941 the 14th Fighter Gp. cadre was formed under the command of then, Lt. Keith. The 14th Fighter Gp. was activated on July 1, 1941 at March Field, CA. After Pearl Harbor the 14th Fighter was re-equipped with P-38E aircraft and alerted for overseas. In June 1942 the 14th Fighter Gp. departed for England via the North Atlantic and participated in four missions over France in September 1942 becoming the first P-38s over Europe. In October 1942 the 14th Fighter Gp. became a part of the Invasion of North Africa and remained in the M.T.O. for the duration.

Col. Keith assumed command of the 14th Fighter Gp. in January 1943 and returned to the United States from Italy in early 1944. One confirmed victory. Decorations included the Silver Star, DFC and the Air Medal with eight Oak Leaf Clusters. He retired from the Air Force in 1962 in the grade of Colonel.

ROBERT BURRELL KELLY, was born in Tulsa, OK and raised in Bellflower, CA. After completing the Advanced CPT Course at LBJC he entered the Army Air Corps Feb. 11, 1941 at March Field, CA. Training completed as a Flying Cadet at Stockton, CA he received his Wings and a commission of 2nd Lieutenant assigned to the 38th Pursuit Sdn. 55th Gp. Portland, OR. The start of WWII moved the Group to McCord Field, WA and transition from Republic P43 to Lockheed P38E aircraft. A new assignment took him to England with flight training in the British Spitfire while the 93rd Fighter Sdn. 81st Gp. awaited outfitting with P400 aircraft to be flown to North Africa. At Casablanca, Morocco he was back in P38H aircraft assigned to the 37th Fighter Sdn. 14th Gp. joining them at Telergma, Algeria moving shortly to St. Marie Du Zit, Tunisia where he later became Squadron Operations Officer. On completing 50 combat missions he returned to the USA and assigned to the 331st Fighter Training Sdn., Ontario, CA (a P38 CTU) instructing pilots headed for combat until War's end.

A civilian for three years, he was recalled for Korea with subsequent assignment to ARDC serving 13 years at several locations, retiring as a Lieutenant Colonel in 1965.

STANLEY T. KELTON, was born in San Francisco, CA on Oct. 27, 1921. He entered active duty on Feb. 23, 1943 and was assigned to Santa Maria, CA for aviation cadet training. Then to Tex Rankin Academy, CA for primary, Marana, AZ, Basic, Williams Field, AZ advanced.

He was sent to Guadalcanal to join the 44th Ftr. Sdn., 18th Ftr. Gp., 13th Air Force for combat in the P-38.

Lt. Kelton participated in campaigns including the Northern Solomons, The Bismarck Archipelago, Ryukus, New Guinea, Luzon, Southern Philippines, Western Pacific China. He holds the Asiatic Pacific Medal, Philippine Liberation Ribbon with two Bronze Stars and the Air Medal with two Oak Leaf Clusters. He also participated in 98 combat missions which totaled over 340 combat hours.

Mr. Kelton is now retired after spending 40 years in the printing industry and lives in San Carlos, CA with his wife Bonnie.

JACK KELTY, was born at Fort Douglas, UT and lived at many U.S. Army Posts until his father retired from the service at Fort Sherman, Panama Canal Zone in 1936.

Jack entered military service in December 1942. He was assigned to the 84th Inf. Div., then in training at Camp Howze, Gainesville, TX. He transferred to the Army Air Corps in May 1943. On attaining his commission upon graduation from flight training, he was assigned to Photo Recon training at Coffeyville Air Base, KS and subsequently served with the 25th Photo Recon Sdn., 5th Air Force in the Philippines, Okinawa and Japan.

WALLY KENNEDY, enlisted in Army Air Corps December 1942 at the age of 18 years...was called to active duty February 1943 at BTC #8 (Fresno County Fairgrounds). Basic training at Utah State University, then to Santa Ana Army Air Base Pre-flight school. Received primary training at Blythe, CA; basic at Marana, AZ; Advanced and Transition at Williams Field, AZ. Trained as single engine fighter pilot. Transferred into twin-engine fighters (P-38). Sent to Photo-reconnaissance School Coffeyville, KS, before going overseas in 1944.

Flew 46 missions in unarmed Lockheed P-38 Lightning from New Guinea, Dutch East Indies, Philippines, Okinawa, China, and Japan with 8th Photo-Recon. Sdn. Received Battle Stars for campaigns over New Guinea, Southern Philippines, Luzon, Western Pacific, Air Offense of Japan. Philippines Liberation with a Bronze Star, Air Medal with Oak Leaf Cluster, Asiatic Pacific Service Medal, American Theater Service Medal, and World War II Victory Medal.

Aircraft flown included: Stearman, BT-13, AT-6, AT-9, UC-78, AT-11, L-5, RP-322, B-25, C-47, and P-38. Returned home to graduate from Fresno State University. Entered into the wholesale lumber business with brother-were partners for 35 years as Kennedy Lumber Company.

Remained in Air Force Reserves 24 years serving at Castle Air Force Base, Merced; Hamilton Air Force Base, San Rafael; McClellan Field, Sacramento; and Fresno Air National Guard Terminal. Now retired with the Rank of Lieutenant Colonel.

RICHARD F. KENNEY, Colonel (Retired) USAF, born Mar. 2, 1920 Santa Cruz, CA; s. Lieutenant (USN) Patrick John and Doris (Frusher) Kenney; attended San Diego State University 1940-41; University of Nevada 1968; Sierra Col. 1972. Commissioned May 1942, Class 42E Stockton Field, CA. Deployed with 82 Fighter Gp. 95th Sdn. to England in September 1942 and on the Invasion of North Africa November 1942. Awarded the DFC for skipbombing a ship on Apr. 28, 1943 and destroying an ME-109 that attacked him as he pulled up from the bombing run. On May 5, he shot down three aircraft on a mission over the Mediterranean Sea. On June 15 he was shot down by ground fire over Marsala Sicily while strafing and sent to Stalag Luft 3 in Germany.

Making a career in the AF he was commander of the 23599 Tactical training Sdn., Nellis AFB (F-86),comdr., 67th Tactical Fighter Sdn., Korea (F-86), comdr., 721st Fighter Day Sdn., TAC, F-100, Comdr., 451st Fighter Sdn., TAC, F-100, comdr 416 Tactical Fighter Sdn., F-84/100, Japan, 1960. He retired from Myrtle Beach, SC where he was Dep. Cdr. of Operations for the 354th Tactical Fighter Wing.

As a principal the Squaw Valley Nordic Center, a member of the North Tahoe search and rescue, president of the Lake Tahoe Ski Club and an official for the United States Ski Association as well as a Technical Delegate of the International Federation of Skiing he was active in ski sport until 1983 when he moved to Coronado, CA, where he resides.

EMIL A. KIELBASA, was born in Cranbury, NJ Mar. 19, 1919, and raised in Milford, DE. He joined the Army Air Corps in June 1941 right out of college, having passed the Flying Cadet Physical exam in May 1941. He was a member of 42-D, graduating from Moore Field, Mission, TX Apr. 28, 1942 where he received his Wings and Commission. He and many of his classmates were assigned to Hamilton Field for Transition into P-38s. His squadron, the 49th received brand new P-38s less than three months later and flew across country, across the North Atlantic into England and November 1942 saw the Squadron fly seven hours and 45 minutes into combat at ORAN, NORTH AFRICA, on D-DAY plus one, being the first P-38 Sdn. to enter combat in WWII. After having fulfilled mission requirements and with two Oak Leaf Clusters, Emil returned Stateside West Coast to train combat P-38 pilots until the

war ended. Emil's love of flying kept him active for 39 years before retiring as Major in 1979.

Before retiring Emil flew choppers and assorted aircraft for the Ohio National Guard and earned a Masters Degree at Ohio State. He retired to California in 1979. remaining active in his old P-38 49th Ftr. Sdn. and maintaining an active interest in the National P-38 Association from Pollock Pines, CA where the 3500 foot altitudes suits him fine. One of his sons is a career Air Force officer with rank of Lieutenant Colonel in the European HQ Command; and, another son is a career chopper pilot with the Tennessee Guard, having served a one year tour in Vietnam combat. One of Emil's daughters soloed in a civilian 150 at the age of 16. His family of eight children and 16 grandchildren keep him busy to say the least.

DONALD DAVID KIENHOLZ, was born in Edwall, WA on June 7, 1923. On Mar. 10, 1943, Donald graduated from Luke Field; class of 1943-C. His overseas assignment was to the 94th FS, 1st FG, (North Africa-Italy), which he joined at Mateur. He was credited with six official aerial victories, all of which were scored in the P-38.

Donald remained on active duty for 27 years; participating in the Korean and Vietnam conflicts. He retired as Major, Oct. 1, 1965. His last assignment was as a B-52 Command Pilot, and Check airman. His decorations include: the Distinguished Flying Cross, the Air Medal, Presidential Unit Citation, and various campaign and theater ribbons.

After retirement, Donald flew for World Airways, was a FAA flight examiner, flight instructor, and corporate pilot. Sadly, Donald was killed in an aircraft accident at Walla Walla, WA, Dec. 27, 1973.

RICHARD D. KIICK, (Dick), enlisted in the Army Air Corps - from Sheffield, AL - on Nov. 25, 1940. He graduated from flying school at Maxwell, AL (Class 41-E) in July 1941.

Flight instructor duty followed graduation at Maxwell, Turner Field, GA and Randolph (CIS). P-38 RTU was at Santa Marie, CA. He was then assigned to the 475th Fighter Gp. in the Pacific theatre, commanded by Col. Chas. MacDonald.

On his first mission, he flew wing for the group commander on the first fighter escorted bomber mission to Balikpapen, Borneo. After flying 50 P-38 missions, he was selected by the Comdr. 5th Fighter Command to command the 421st Night Fighter Sdn. He was 24 years old.

Dick left the service in 1946 but was recalled for the Berlin Airlift in 1948. He commanded the 12th Troop Carrier Sdn. of the 60th TCGp at Wiesbaden and made 25 C-54 flights into Tempelhof.

Subsequent duty included two assignments to the Pentagon - one on the Air Staff and one on the Joint Staff. He retired in 1967. Decorations include The Distinguished Flying Cross (two enemy aircraft destroyed on one mission) and the Air Medal.

Dick and his family are permanently residing in Honolulu, HI.

CHARLES W. KING, three years ROTC and Civilian Pilot Training led Charles W. King to the Air Corps in October 1940. His graduation assignment to the 39th Pursuit Sdn. lasted 2 1/2 and provided the beginning and apex of 30 years flying. The Squadron was the first to fly the P-39 Airacobra. After combat baptism in June/July 1942 it got General Kenney's first P-38s. Early in 1943 with lightnings it scored 47 victories in 13 days, provided much of the fighter action in the Bismark Sea Battle and played a major role in chasing the Japanese from their forward New Guinea bases. The 39th was the first in the Army to tally 100 victories and during this time King advanced from 2nd Lieutenant to Captain.

August thru November the Squadron did its share of the 5th Air Forces neutralization of Wewak and Rabual. King took command of the Squadron and became its 12th and last P-38 ace before he left as Major in December 1943.

Before he returned to the Pacific early in 1945 King commanded a squadron of the first US Jet Group gaining early experience in the Bell P-59 and the XP-80.

Stateside King flew the F-86A with the 56th Fighter Gp. as part of the newly organized Air Defense Command. The Korean War dispersed his 63rd Sdn. to Oscoda where he was also the Base Commander. In 1951 with ten years of commissioned service he was promoted to Colonel and moved to operations of Hq. ADC. 1952 brought Korean combat opportunity with assignment as Deputy Commander of the 4th Wing. He flew 66 sorties in the F-86E against North Koreans, Chinese, and Russians. He was the Wing Commander when the Wings 300th MIG was destroyed and then he was again in Hq. 5th AF both as a staff officer and as Liaison Officer to the 7th Fleet aboard the Battleship Missouri.

King returned to the all weather Air Defense Command as an Air Division Deputy for Operations and as a Wing Commander. He captained two F-86D rocketry teams and one of the F-102 teams in the first William Tell missile competition. After three years as an advisor in Spain and two years in a SAGE Air Division he ended up his career with five years operations staff duty at North American Air Defense Command.

Tally: five victories, three probables, five damaged.

Aerial Combat: 267 sorties. First victory came in the first action of the Bismark Sea Battle. The fourth and fifth were a double over Rabual New Britian. His last combat success was the damage of two MIG-15s.

ROY KING

ROBERT E. KIRTLEY, was born in Spartanburg, SC on July 26, 1917. He was graduated from Wofford College in 1939 when he entered the Air Corps as a flying cadet. He graduated from Kelly in the Class of 40-A as a 2nd Lieutenant. He was then stationed at Selfridge Field, MI as a pilot in the 1st Pursuit Gp. Was one of the first six pilots to conduct service test on the first three hand made P-38s at Wright-Patterson. A flight leader in the 27th Pursuit Sdn., he trained the first P-38 pilots in the Squadron.

At the start of WW-II he accompanied the 1st Pursuit Gp. to the West Coast. When the 1st Group went to England, he became the first Squadron C.O. of the 95th Fighter Sdn., 82nd Fighter Gp., which was manned by Flying Sergeants. This group went to England in September 1942. The group flew to North Africa in December 1942. Enroute, Major Kirtley claimed the first victory for the groups, a JU-88.On this tour, he destroyed a total of four German aircraft. On his return he was Operations Officer of the P-47 Combat Training Gp. at Dove, DE.

From 1948-1950 he was Group C.O. 49th Fighter Gp. in Japan. During the Korean War he formed and ran the Joint Operations Center for 5th AF.

His further assignments were with the Air Staff, JCS and Weapons System Evaluation Group. In 1960 he was appointed a Harvard Fellow.

Col. Kirtley retired after 30 years of service and became president of Sernco, a systems analysis company.

Decorations include the Legion of Merit, Bronze Star, DFC, AM (11 Oak Leaf Clusters).

Currently he resides in Williamsburg, VA.

LOUIS R. KITTEL, born and raised at Casselton, ND; majored in Engineering at North Dakota State College and became a charter member in the Carl Ben Fielson Flying Club (Curtiss Pesher and Curtiss OX-5). Entered Army Air Corps flying cadet program at Santa Maria, CA, Class of 40A. Graduated from Kelly in March 1940. First assignment was Selfridge Field, MI with the 31st Pursuit Gp. Flying P-35s and P-40s during this time, he was assigned TDY to Patterson Field, OH for duty as service test pilot on P-40 aircraft. When the war broke out on December 7 he was with the 50th Pursuit Gp. in Meridian, MS and shipped overseas to the Pacific with the 12th Sdn. and a compliment of P-39 aircraft.

As stagings moved south he wound up on New Caledonia where he checked out in P-38s, subsequently moving to Guadalcanal. On Apr. 18, 1943 he participated in a 16 ship (P-38) strike resulting in the shooting down of Admiral Yamamoto—a classic example of a perfect mission.

While at Guadalcanal the observed he shortcomings of the existing night fighter operations. He coordinated and inaugurated the first night fighter operation with the P-38 working with ground search lights. The first night of application resulted in two Betty aircraft being shot down.

He was assigned as Squadron Commander to the 70th Sdn. and flew combat missions with the 12th, 44th and 339th (P-38/P-39/P-40).

He is credited with one Zero and three Betty Bombers shot down; plus a Zero probable.

Upon the war ending, his career switched to aircraft maintenance. A tour in Alaska resulted in a F-80 aircraft being recovered from the wilderness by building and installing skiis which allowed it to be flown back to Elmendorf. His aircraft recovery crew was responsible for many millions of dollars worth of aircraft being returned to full flight conditions.

His hobbies are hunting and fishing which are bountiful in the great state of Oregon.

An interesting P-38: Have you ever heard of a P-38 carrying a load of four 1000 pound bombs? A young Lieutenant by the name of Tucker designed four special bomb racks for a P-38 while on Guadalcanal. He successfully flew one mission but the aircraft was lost on a routine dog fight some days later. Tucker belonged to the 12th Sdn. and I have no idea of his status today.

Another humorous event: While stationed at Chico A.F.B., CA where we were training P-38 pilots, we were alerted to the fact that the Japs were sending huge paper balloons with incendiary devices attached. A short time later on a nice sunny day, a huge, white object was sighted and Ops. called for an intercept. I climbed into the standby P-38 and charged off into the blue. At first I thought this was a sure thing, but as I approached 35,000 feet I became skeptical as to the possibility. About then Ops. called and says, "Come on down, you are chasing the star Venus." Ironically, two other west coast AF bases had also scrambled P-38s. What you see and what you think you see can really be different in many ways.

WILLIAM A. KLEIST, (Bill), Entered Armed Forces January 1943. Graduated and commissioned Jackson, MS AFB May 1944 Transition P-38 Chico, CA AFB October 1944 through April 1945. Joined 5th A.F. - 8th Ftr. Gp., -36th Fighter Sdn.

July 1945-Mindoro, Philippines Separated from Service June 1946.

LESLIE E. KNAPP, was born in Aurora, IA. He joined the Army in 1942. As an aviation cadet he was assigned to Thunderbird Field, Marana Field, and graduated from Williams Field, AZ in the class of 44-D. He flew P-39s at Victorville, CA before being assigned to Santa Maria, CA in P-38 RTU. He was assigned to the 37th Fighter Sdn., 14th Fighter Gp., in Foggia, Italy and flew 26 combat missions.

He was recalled to active duty during the Korean War, serving in Japan. In 1968 he flew 350 combat missions as a Forward Air Controller in Vietnam.

In 1961 he was the last active duty Air Force pilot to fly an Air Force P-38. He flew the P-38 from Kelly AFB, TX to Wright -Patterson AFB, OH where it is now on display at the Air Force Museum in Dayton, OH.

He retired as a Command Pilot with 18 decorations in San Antonio, TX in 1972.

FREDRIC L. KOHN (formerly, Fredric Kohn) completed 50 missions with the 71st Ftr. Sdn./ 1st Ftr. Gp., in North Africa flying P-38s; he destroyed seven enemy aircraft, was shot down, captured and escaped from a prison camp in Sicily.

He was awarded the DFC and Air Medal/nine Oak Leaf Clusters. He served as HQ AAF Project Officer for P-38 spin, stall, and bailout techniques. Promoted to Chief, Safety Education Div., Office of Flying Safety, he wrote and illustrated manuals on how to fly fighter aircraft including the P-38, P-51, and P-47. He test flew America's first jet, the XP-80, developed power steering for aircraft, and became a major at 23.

Since the war, he's been an inventor of labor saving devices, founder of the Fredric Arnold Company, Arnold-Blau Corp., and Photo Engineering International. He is a painter, sculptor, film and TV actor, and author of DOORKNOB FIVE TWO and KOHN'S WAR, books about his war time experiences.

FREDERICK J. KREMMER, was born in Minneapolis, MN on Jan. 8, 1919.

On Aug. 21, 1942 he enlisted in the USAAF and was called to active duty on Jan. 3, 1943 and sent to Jefferson Barracks, St. Louis, MO, for basic training and then to Stevens Point, WI, eight weeks later to participate in the College Training Unit which included ten hours' flight time in J-3 Cubs. From there he entered the flight training pipe line in the Western Flight Training Command through Santa Ana, Classification Center, CA, to Thunderbird Field in Phoenix, AZ, Marana Field, Marana, AZ, and Williams Field, Chandler, AZ. At that time Williams Field had P-38s (322) without turbo chargers but had lost so many he stayed in AT-6s, graduated and received his Wings on Apr. 15, 1944. From there he was sent to Santa Maria, CA to train in P-38-Js and Ls for 120 hours. Finally received orders for overseas duty and arrived in Foggia, Italy in January 1945 and assigned to the 15th AF, 14th Fighter Gp., 37th Fighter Sdn. The German Air Force was practically ineffective although on several missions were attack while escorting the bomber force ME-109s and ME-262s but Lt. Kremmer's 23 missions were primarily directed to dive bombing bridges, marshalling yards, ball bearing factories, strafing trains, electric sub stations, anything that moved, until cessation of hostilities on May 7, 1945.

Returned to the United States in September, joined the Air Force Reserve in Minneapolis, MN and was recalled to active duty in 1951 for the Korean hostilities , but that's another story.

WALTER F. KUEMMERLE, born Germania, NJ Aug. 16, 1919. Aviation Cadet Oct. 29, 1942. Preflight Santa Ana, CA. Primary Dos Palas, CA. Basic Chico, CA. Advanced Williams, AZ.

Graduated August 1943, class 43-H. Served ETO, 20th F.G. from Dec. 27, 1943 to Oct. 17, 1944. Flew 62 missions. Flew second shuttle run from England to Russia to Italy and return.

Six days after invasion, June 12, 1944 our flight of P-38s was strafing enemy lines. Lts. Rader, Kendrick, Fiebelkorn, and Kuemmerle were all shot down. Three were returned to base and Lt. Kendrick was a POW. Decoration Air Medal with three Oak Leaf Clusters. DFC with an Oak Leaf Cluster. Unit Citation and ETO Medal.

After discharge went in business and now retired.

WARD ALBERT KUENTZEL, (Lt.), was born in the small farming town of Delano, CA located in the central valley. He grew-up and was educated in this small town. He enlisted in the Army Air Corp. February 1942.

He took his primary flight training at Rankin Aeronautical Academy, Tulare, CA just 30 miles from home. After he finished his basic flight training in California. He was sent to Luke Field, AZ for his advanced training. He was being trained as a fighter pilot. At that time he knew that he was going to fly either a P-47, P-39 or his choice the beautiful P-38. He was sent to Williams Field, and was given three hours dual instruction time in a twin engine AT-9. He knew then that he was going to fly the P-38. At Williams Field he wrote me a letter that he got to sit in a P-38, "it was love at first sight". He took his P-38 training at Murdoc Field, CA, finishing his P-38 training November 1942.

He embarked from New York to England where he was then assigned to the 82nd Fighter Gp. He and others flew from England to Africa early in December. His first combat mission was from Telergma Africa on a Bomber Escort Christmas Day, 1942.

Lt. Ward Kuentzel was a Flight Commander of his P-38 Sdn. in North Africa, making him at 21 one of the youngest in the service. He holds the Distinguished Flying Cross, the Air Medal with 21 Oak Leaf Clusters and the Battle Medal with four bronze stars.

Lt. Kuentzel had many air battles including 22 successful bombing missions in Sicily. He was pounding the Axis with skip bombs dropped from his beautiful P-38. On one occasion he was responsible for the blowing up of an ammunition dump in Pantelleria in which nine planes on the ground were also destroyed. He very nearly lost his life on the mission as a result of the blast.

In an air encounter he accounted for two German planes in period of three minutes. He knocked down a Junkers 88 and a Focke Wulf 190 got on his tail. At first he believed it to be one of the planes in his squadron until the ship's tracers started coming in too close. He quickly out-maneuvered the Nazi by executing a tight loop and sent him down in flames. Lt. Kuentzel has five Messerschmitt 109s to his credit, two of them being members of Herman Goering's famous "yellow nose" group. Lt. Kuentzel's victories include the five Messerschmitt 109s the Focke Wulf 190, the Junkers 88, two 202 Maschi's, one Caproni 2007 bomber, one Italian transport, and 22 probables.

He was a personal friend of General Jimmie Doolittle and flew as the General's escort when he made his trip to London conference preceding the invasion of Sicily.

After 265 combat flying hours, his tour of duty in Africa was over. He was to fly his P-38 from Casablanca Africa to the east coast of the United States via Canarias Isles, Brazil, and Cuba. When he was about seven miles out to sea, his right engine caught fire. He radioed a mayday for help. Lucky for him there was a British Destroyer below. The Destroyer told him to bail out and they would pluck him from the sea. He rolled the P-38 over and bailed out. He said that he cried when he saw his Beautiful P-38 go into the sea. Uncle Sam had ideas for this brave young hero. He was to use his trusted P-38 for War Bond Tours around the country. He did indeed make many Bond Tours when he eventually got home.

His next duty came as a P-38 Instructor, instructing combat flying. He instructed at Salinas, Muroc, Glendale, and Van Nuys, CA.

He was then sent back to England to participate in the D-Day invasion of Europe. On June 19, 1944 he was to lead a flight of P-38s to meet with B-17s returning from their bombing mission deep into France. The weather was very bad, and the P-38s had to climb to 32,000 ft. trying to get out of the overcast. As my brother and his Squadron broke clear of the overcast, he started to wander aimlessly in the sky. He then began losing altitude with both of his wingmen following. His wingmen tried to contact him by their radios to no avail. One of his wingmen saw that it was fruitless to follow, climbed back-up to the squadron. Ward was using an English leather flying helmet. He had adapted the American oxygen mask to it. It is suspected that he did not get a good fit of the mask causing asphyxia. My brother Ward and his faithful wingman crashed in France, killing both pilots.

It has been 44 years since his death, and I still miss my brother and always will.

JOSEPH A. KUHN, born Roxborough, PA, July 15, 1923. Entered Army Air Force Mar. 19, 1943, trained Santa Ana preflight, Ryan Field, primary, Marana basic, and Williams Field advanced. Graduated flying school 44A, Jan. 7, 1944. P-38 training Ontario, CA, January-May 1944. Assigned 401st Sdn., 370th Gp., 9th AF at Andover, England. Flew 79 missions (60 in P-38 and 19 in P-51). Airplane-"Pennsy Lady." One FW-190 destroyed.

Flew combat missions during Korean and Vietnam Wars, 7000 flying hours C-141, C-131, C-124, C-82, C-47, C-54, C-45, SA-16, H-21, H-19, H-5, R-4, AT-6, P-51, P-38, AT-9, BT-13, and PT-22 aircraft. Vietnam 13 months Nha Trang and Saigon as Dir. of Maint. 7th AF. Retired colonel, 1974, 31-1/2 years.

Spent three years Operations Director, Iran Aircraft Industry, Tehran, with Northrop and 11 years with Lockheed/Burbank L-1011 Commercial Support /Contracts. Member AFA, Daedalians, TROA, Air Rescue Association, and now President, P-38 National Association. Awards LOM (two), Air Medal w/13 Clusters, AFCM (two). Married to the former Nancy Coleman and has seven children and nine grandchildren. He resides in Valencia, CA.

THOMAS J. LANPHIER, JR. , Capt. US Army Air Force. He was educated at Stanford University and received his flying wings a month before Pearl Harbor.

He was a bonafide ace who by war's end would fly 112 combat missions, sink an enemy destroyer and shoot down 17 Japanese planes. On Apr. 8, 1943, he bagged the largest single prize of the entire war in the Pacific. After American Intelligence had cracked the Japanese code they intercepted a secret message: Admiral Isoroku Yamamoto was to fly on April 8 from Rabaul, New Britain, to Kahili on Bougainville in the Solomon Islands. Yamamoto was the man behind the attack on Pearl Harbor.

Lanphier and the rest of the P-38 pilots reached the area specified and saw Yamamoto's Mitsubishi bomber escorted by 12 Zeros. The admiral's plane tried to elude the Americans but after a steady stream of ammunition on the Zeros the Betty bomber was sited and destroyed by Lanphier. (Some also say that Rex Barber shares in this aerial victory). Regardless of who shot him down, it was a great day for the Allies.

After the war Lanphier was an editor at the Idaho Statesman at Boise for several years and later was a vice president in the Convair Div. of General Dynamics Corporation.

He died on Nov. 26, 1987 at La Jolla, CA where he had been a cancer patient for nearly a year.

STANLEY W. LAU, was born and raised in Honolulu, HI on Jan. 18, 1923 and was a student at the University of Hawaii when Pearl Harbor was attacked. Within hours after the attack, his ROTC unit was immediately activated.

In November 1942, he transferred to the Army Air Corps as an aviation cadet. His flight training was completed at Cal-Aero Academy, Ontario, and Taft Field, CA and Luke Field, Phoenix, AZ in the Class 43-J. His first fighter assignment was in P-39, Redding, CA where he was trained as a "skip-bombing" specialist. However, just weeks before being sent to the South Pacific, his unit was deactivated and he was sent to Santa Ana for his operational training in P-38. In March 1944, he was sent overseas where he joined the 27th Fighter Sdn., 1st Fighter Gp. in Foggia, Italy. On his 6th mission, he shot down his first German FW-190 over Southern France. By his 50th mission, he had a record of four confirmed kills; three probable and two destroyed on the ground. He was awarded the DFC and the Air Medal with five clusters. Nicknamed "Poco", he was the only Chinese-American fighter pilot in Southern European Theater.

After his European tour, he transferred to the Pacific to fly P-47s and P-51s. His love of flying fighter aircraft saw him turning down an offer to fly commercially. He remained on active duty and retired as Lieutenant Colonel in July 1964.

After his retirement, he taught in the public schools in Hawaii and ended up as a State School Dept. planner before retiring again in December 1984.

DICK LEE, born in St. Louis on July 18, 1921 entered the Army Air Corp in March 1942. After Pre-flight, Primary and Basic Training in California he graduated from Luke Field, AZ in Class 43-C in March 1943.

Lt. Lee checked out in P-38s at Muroc Air Base and Santa Ana before shipment to North Africa in May 1943. He joined the 94th "Hat in the Ring" Sdn. of the 1st Fighter Gp. While with the 94th Lt. Lee became an ace, destroying three ME-109s, one FW-190, one Macchi 202, plus an ME-109 probable and two damaged.

Upon return to the U.S. Capt. Lee instructed in P-38s at Van Nuys and Chico, CA. He left the Army Air Force in September 1945 and immediately joined Trans World Airlines. He retired as a senior Captain in July 1981 after 36 years.

Dick and Esther Rae Lee have three sons and are retired in Scottsdale, AZ.

G.F. LEINWEBER, (Jerry), was born Sept. 18, 1920 in Richards, TX. He graduated from Houston High Schools in 1938 and entered University of Houston School of Engineering. After two years at University of Houston he joined the Air Corps in September 1941 as an aviation cadet.

After completing pilot training in April 1942 he was assigned to 55th Fighter Gp. at Paine Field, WA. He logged 600 hours in P-38s before going overseas and over a 1,000 hours total in P-38s after completing 57 combat missions in England. After returning to States he was assigned to training command until his discharge from Air Force.

He was recalled to active duty for Korean War but saw no combat. After Korean War in 1953 he entered the oil field supply and manufacturing business. He retired in 1982 after serving from salesman to C.E.O. and chairman. Currently living in Kerrville, TX (Texas' Best Kept Secret) pursuing golf instead of enemy aircraft. *(See photos, page 140)*

ENOCH P. LEMON, was born in Ripley, WV, Jan. 12, 1920, was raised in Charleston and Du Pont, WV. Took physical exam for Aviation Cadets Dec. 8, 1941 at Huntington, WV. Enlisted as Private Army Air Corps Mar. 29, 1942. First soloed in Aranica Chief sea plane at Clark's Flying School, Kanawha River.

Upon receiving Cadet call he was sent to Santa Ana, CA. Preflight; training at Ontario, CA. (Chino Field). Took Basic flight training at Lancaster, CA. Applied for P-38 training—sent to Luke Field, Phoenix, AZ. Got two weeks twin engine flight at Williams Field. He graduated from Luke as 2nd Lieutenant Feb. 6, 1943.

Lemon was assigned Operational Training Unit in P-39s, Ft. Myers, FL. On May 18, 1943 left for North Africa, arriving Casablanca and the Fighter Training Center at Ber Rechid. He transferred to P-38s and got two hours flying time. Joined the 14th Ftr. Gp. - 37th Ftr. Sdn. P-38s on Aug. 31, 1943 at Ste, Marie Du-Zit.

On 2nd mission he was flying Red 2 on "Soupy" Campbells wing, to dive bomb "Montocvino Revella". While walking to aircraft, Soupy said "Lemon if you want to get back stick on my wing". At debriefing, several pilots wanted to know, "since when do we dive bomb in tight formation?". The cat was out of the bag, for Lemon took Soupy at his word.

He flew 52 combat missions with the 37th being credited with three confirmed, two probables, two damaged, two shot up on the ground. Caused a "yo-yo" practicing ME-109 to crash and explode in Yugoslavia. No credit—didn't fire at him; just scared him into the ground. That was on Apr., 24, 1944 on mission to Bucharest. He served 17 years on ADC Assignments, retiring as Lieutenant Colonel on May 31, 1963.

HENRY K. LENNOX, (Hank), born Jan. 1, 1925, Jamaica (Queens County), NY Entered Aviation Cadet Training Santa Ana, CA, September 1943. Training bases included: primary - Wickenburg, AZ; basic - Marana, AZ; advanced - Williams Field, AZ. Assigned to the 9th Photo Recon, Sdn., 8th Photo Recon Gp. Theater of Operation was China-Burma-India, based at Myitkyina, Burma. Missions consisted of photo reconnaissance - 50 missions total.

Retired from the USAF in November of 1945 with the rank of 1st Lieutenant. Currently employed as the president of a computer sales agency. Married Sally Ann Slack on Dec. 13, 1946 in Utica, NY. Have four children.

LEROY L. LETTE, (Lee), from Oklahoma City, OK , a Cavalry Officer in June 1942 and Class 43-H graduate completed OTU at Muroc, CA before joining the 97th Sdn., 82nd FG in Italy in March 1944.

Until July 1944 he flew 51 escort and dive-bombing missions with four confirmed and one probable aerial victories and four destroyed on ground.

Out in May 1945 he rejoined the AAF and flew F-80s and P-51s with the 8th FW and 35th FW in Japan, completing a second combat tour in September 1950.

He was awarded the DFC and Air Medals (15 Oak Leaf Cluster).

He retired (disability) from Office for Advanced Studies (USAF) in 1957 after plowing up a Pennsylvania pasture in a T-33 three years earlier.

Retirement has been as an Aerospace and automotive (air bag) engineer until 1975. He resides in Mendocino, CA with his wife, Betty where he teaches High School and directs a choir.

JOHN D. LEWIS, was born in Westport, CA June 23, 1918. John was raised in the Central Valley of California, in Waterford and Denair.

In 1941 he enlisted and was a mechanic at Stockton Field. He took his pre-flight training at Santa Ana in 1942. His primary was at Cal-Aero Flight Academy in Ontario, CA. Basic Ffight training was at Lancaster, and his advanced training was at Williams Field - Chandler, AZ. John graduated with the class of 43H.

John's overseas training was at Santa Ana, Muroc and Lomita Flight Strip. He arrived at Casa-blanca on Dec. 15, 1943. Then was assigned to the 49th Fighter Sdn., 14th Fighter Gp., located near Foggia, Italy. John flew 53 missions escorting bombers in P-38s from January 1944 to July 1944.

After the service he was a farmer and a self-employed businessman. Now retired, John still enjoys flying his own small airplane. He lives, with his wife, Diane, in Coulterville, CA.

EVERETT S. LINDLEY, was born in Massachusetts and enlisted in Boston as an aviation cadet with the USAAF in 1942. He earned his Wings in 1943 at Craig Army Air Field and was assigned to teach at Courtland Basic Flying School.

He joined the 94th "Hat-in-the-Ring" Sdn., flying his beloved P-38, in August 1944 at Salsola, Italy.

His two most interesting combat missions were—a 20 minute dog fight with a Luftwaffe ME-262 jet fighter in November 1944-and leading the 1st Fighter Gp. (maximum effort), Apr. 15, 1945, on a low-level strike in Bavaria, Germany, when he was shot down.

After active duty with the Army, Navy and Air Force, coupled with reserve service, amounting to 23 years, he retired as a Lieutenant Colonel in Arlington, VA.

ROBERT M. LITTLEFIELD, Lieutenant Colonel, USAF (Ret.). Born in San Francisco, CA, raised in Monterey, CA. Graduated from Monterey High School and attended Salinas Jr. College and University of California at Davis. He learned to fly with the Civilian Pilot Training Program. Flew the P-322 at Williams Field AZ, Cadet Class 43-J. As a 2nd Lieutenant he flew the P-38G-10LO and H in California in an air defense mission with the 360th and 329th Fighter Gps., 4th Fighter Command, 4th Air Force.

He flew his first combat mission in April 194 with the 38th Fighter Sdn., 55th Fighter Gp. 8th Air Force Wormingford, England. He totaled his P-38J-10LO in a crash landing after returning from a strafing mission with his left engine and landing system shot out by flak. Twice wounded, he was shot down on his 59th combat mission by flak while strafing a train in France. After a 400 foot bail-out he was hidden from the Germans by the French Resistance in Chateau Le Matre, near Pavilly, Normandy. In September 1944 he walked out of enemy occupied territory into English/Canadian lines.

A career fighter pilot he flew the Lockheed F 5, T-33 and F-104.

KENNETH B. LLOYD, was born in Hermosa Beach, CA only child of Karl and Marjorie (Oughton) Lloyd, June 4, 1924.

On his 18th birthday was sworn into th

USAAC. Graduated from the P-38 Advance Flight School, Williams Field, AZ, class of 43-J. After tactical flight training in P-39s and P-38s was flown to Nadzab, New Guinea in late March 1944 and assigned to the famous 80th "Headhunters" Fighter Sdn., 8th Fighter Gp. Flew the "Lightning" on 101 missions over New Guinea, Dutch East Indies, and Philippines. In aerial combat had one confirmed and one probable, plus six planes destroyed in strafing missions. Received DFC and Air Medal with four Clusters.

During night strafing mission against Naval Task Force off Mindoro Dec. 26, 1944 his P-38 was hit by water thrown up from flak crumpling the left end of the wing and wrapping the left prop around the engine. He managed to keep his badly damaged P-38 airborne long enough to gain altitude and bailed out over his base. While floating down in his chute was mistaken for enemy paratrooper and shot at by ground troops, but not hit.

June 1945 was rotated home and married Evelyn Ann Brugge. They have three children and four grandchildren. He was assigned to Luke Field, AZ and then to Norton AFB, CA where he was separated from the USAAC in November 1945.

After 32 years with the Orange County Insurance Agency, Santa Ana, CA he and his wife Evelyn retired to their mountain home in Crestline, CA.

RICHARD C. LOIZEAUX, was born in Plainfield, NJ on Aug. 23, 1919. Growing up in Plainfield he spent most of his teen years dedicated to sports, baseball, football competitive swimming and exhibition gymnastics. A very modest amount of time to studies.

In 1939 he went into apprenticeship in the importing business in New York City, concentrating on business management.

He enlisted in the A.F.E.R. Cadet Training on Aug. 3, 1942, however while waiting for call up a freak accident caused a broken back. This inconvenience necessitated a one year delay and a Government waiver before proceeding with Cadet training.

In February 1943 training began at Maxwell Field, AL. Primary in P.T 17s at Decature, AL basic in Vultee Vibrators at Courtland, AL and on to advanced in A.T.6s at the jewel of the S.E. Air Command Craig Field, AL. On graduation he was assigned as an instructor in advanced trainers at Craig Field but one week later applied for transfer for fighters and in three months was in Portland, OR training in P-39s and P-63s. Next stop was Santa Rosa, CA for introduction to the P-38. While at Santa Rosa he was assigned to a Capt. Boggs, a very unassuming pilot home on rotation. No stars, no bars, no ribbons, no talk, but much experience as he was later to learn.

In September of 1944, Loizeaux was sent to Chittagong India as a replacement pilot with the "459th" Fighter Sdn. and shortly thereafter Capt. Boggs now a Major returned to take over the Squadron from Daddy Bear Luhering who was to go stateside to A.J. School.

1944/1945 brought 46 missions over Burma/Thailand, primarily strafing, bombing and top cover in the Irrawaddy and Rangoon area, also the Air Medal, two clusters, Asst. Operations Officer and Silver Bar. He was also privileged to be part of that lonesome Squadron that received The Presidential Citation for 123 Jap planes in 58 days.

As a dedicated civilian he was mustered out in November 1945 married his high school sweetheart, and they had four children.

After 17 years of specializing in business reorganization and consulting, purchased a 600 acre dairy/crop farm in Vermont.

Husband and wife actively operated this farm until 1986 when much of the acreage was sold and they built a retirement home on the 100 acres retained.

STANLEY A. LONG, was born in Munising, MI, on Aug. 20, 1918. He attended schools in Marquette, MI, including Northern State Teachers College before entering military flight training during WWII. He began flying in the Civilian Pilot Training Program while in college and graduated from Randolph and Kelly Fields as an Army Air Corps pilot in September, 1941, (class of 41-G).

He flew a P-38 in the Aleutians from June, 1942, to March, 1943. He was the first pilot in WWII to down an enemy plane, a Japanese bomber (Mavis) in a P-38. Out of a squadron (54th FS, 55th Gp. 343rd Gp., 11th FC) of 31 pilots sent to the Aleutians in 1942, he was one of nine who survived. His mission was to fly cover for bombers and shoot up Japanese gun emplacements. Besides the Mavis, he was credited with a Zero.

Following his tour of combat duty, he was assigned to the 441st Army Air Base Unit as commander of a training class at Metropolitan Airport, Van Nuys, CA. Other assigned stations were Muroc, Salinas, and Chico, CA. He was at Portland, OR, in the 432nd AAF Base Unit, when the war ended and was discharged in September, 1945, with the rank of major.

After WWII, he became owner-operator of Long's Air Activities and manager of Marquette County Airport. He also helped locate the site for K.I. Sawyer AFB.

He was recalled to active duty in 1951 and sent to Korea, where he served as a commander of the 67th Fighter Sdn. - known as the "Red Scarf" Sdn. - flying the F-51D Mustang, and later, the F-86. He was shot down over the Yellow Sea but rescued. The mission of the 67th was interdiction and support for ground troops. He was credited with the destruction of a large hydroelectric plant in North Korea. He flew combat in Korea from mid-March, 1952, to mid-November 1952.

Returning from Korea, he was reassigned to Reno, NE, as an advisor to the Nevada Air National Guard, then, from 1953 to 1956, to Milwaukee with the Wisconsin ANG.

In 1957, he became commander of K.I. Sawyer AFB, and in 1958 he was reassigned to Elmendorf AFB, AK, as chief of NORAD combat operations center.

His last assignment was as chief, combat operations center, Truax AFB, WI, from 1962 to 1966. He retired in 1966 with the rank of colonel.

His awards and decorations included the Distinguished Flying Cross with one oak leaf cluster, Air Medal with four Oak Leaf Clusters, and the Air Force Commendation Medal.

After his AF retirement, he worked as Assistant Tax Assessor for the city of Marquette from 1970 to 1980.

He married Virginia Butler of Marquette on May 1, 1943. He had two sons: David, a software engineer, and Dennis who died in 1969 while involved in a swimming rescue for which he posthumously was awarded the Carnegie Hero Medal and a Valor Award from the U.S. Department of the Interior.

Stanley died Aug. 13, 1987, and is buried in Park Cemetery, Marquette, MI. *(See photos, page 140)*

JOHN R. LUND, was born Feb. 19, 1916 and grew up on a farm southwest of Dodge City, KS. He enlisted in Co. D, 137th Inf. Kansas National Guard in 1935 and graduated from Ft. Hays Kansas State College in 1939.

On Dec. 23, 1940 he was commissioned a 2nd Lieutenant, Infantry in the 35th Div. In January 1943 he was accepted for pilot training in the "Student Officer" program and graduated from Luke Field, AZ in the class of 43-H. Next came "Phase I Fighter-Pilot" training, flying P-39s in Nevada and California. After a period as an instructor in P-39s and P-63s on the west coast, he transitioned to P-38s and joined a former organization (394th F.S., 367th F.G.) in France in September 1944.

On one of his 30 missions in P-38s the right engine was shot out while strafing a column of horse-drawn artillery. On one engine he climbed up through the clouds, made a successful return to the field only to have to make a belly landing as the hydraulic lines had been shot out. In February 1945 his squadron transitioned to P-47s. On March 19 John's P-47 was hit on a dive bombing mission over Germany. He bailed out and was taken prisoner near Hanau. This was his 43rd mission.

His decorations include the D.F.C.; Air Medal with five Oak Leaf Clusters; Purple Heart; E.T.O. Medal with three Battle Stars; Presidential Citation Badge and Belgium Fourragere.

He and his wife Esther live in Oak Ridge, TN.

GEORGE B. LYNCH, JR., born, only child of G. Buford and Elmerita Lynch, July 11, 1922, Benson, AZ. Graduated Tucson High School 1940.

Quit Douglas Aircraft immediately after Pearl Harbor. Enlisted in Army.

Graduated Air Mechanics School, Shepard Field, Wichita Falls. There passed flying cadet tests, then to pre-flight, San Antonio, Primary flight school, Jones Field, Bonham - Class 43-D. Basic, Perrin Field, Sherman, Advanced, Moore Field, Mission, (all in Texas) where flew ten hours in P-40.

Wings, commission, and overseas orders same day.

7th Air Force, in Hawaii - flew P-40s, and P-47s.

Asked for transfer to New Guinea. Joined 35th Fighter Sdn., 8th Gp. 5th Air Force.

He flew 111 combat missions, dive bombing, strafing and escorting in P-40s, last 60 missions in P-38s.

Leading a four ship flight Christmas, 1945 - shot down two of four "Jack" fighters. Wing man and element leader, the others.

Placed inactive as Captain, Retired from Reserve, a Lieutenant Colonel in 1969, with D.F.C. three air medals, six battle stars.

1947, married sweet Bonham girl. One daughter.

J.D. MacARTHUR, (DOUG), graduated from Williams Field, AZ, class of 43-J. He was assigned to the 77th Sdn. 20th Ftr. Gp., 8th AF, stationed at King's Cliffe, England on Apr. 20, 1944.

In July 1944, the group changed from P-38s to P-51s. Completed his tour on Oct. 6, 1944, flying a total of 61 combat missions (27 in P-51s); 270 Hours (139 in P-51s). Claimed one JU-88 destr. on the grd. Was awarded the DFC & the AM w/five Oak Leaf Clusters. Flew five missions as convoy top cover during D-Day Invasion. Flew escort on the last mission from Eng. to Russia to Italy to Eng. Longest combat mission was 7:10 hours in a P-51. His, perhaps, unique record was that he never saw an enemy aircraft in the air or never used the relief-tube.

After returning to the Z of I, he flew P-39s, RP-63s & P-47s at various bases. Separated from the AAC on Oct. 30, 1945, he then attended the University of Washington, receiving a Bachelor of Science degree in Aeronautical Engineering.

After graduation, was employed by Douglas Aircraft Co. until his retirement 33 years later on Dec. 31, 1981. He and his wife, Pat, live in Malibu, CA. They have five children and six grandchildren.

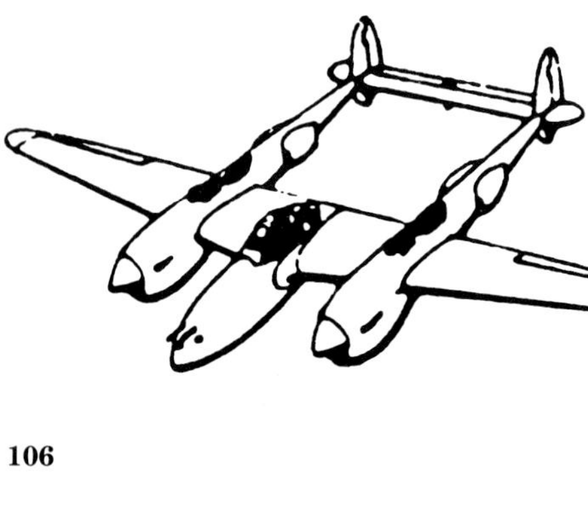

COLIN DAVID Mac DOUGALL, was born July 8, 1920 in Longmont, CO but grew up in Cheyenne, WY. He attended Nebraska University for two years and worked in a bank for a year before Pearl Harbor. Enlisted in the Army Air Corps in March 1942. Graduated from Luke Field, AZ in Class 42-K and went directly to P-38s at Muroc Air Base, CA.

He spent 13 months flying P-38s in the 54th Ftr. Sdn. in the Aleutian Islands from July 1943 to August 1944 after which time he instructed new pilots in P-38s at Santa Rosa, CA. His last year and one-half in the service was in the 94th Ftr. Sdn. of the 1st Ftr. Gp. flying P-59s and P-80s at Santa Maria, CA and March Field, CA.

After the war, he attended University of Southern California and the University of Colorado in Boulder, CO graduating in aeronautical engineering in 1950. He married Colleen Board in Santa Maria, CA in June of 1946. They have two children, Janet and Dick and one granddaughter.

He retired in 1981 after 30 years as a flight test engineer with Lockheed Aircraft Corporation working on the Discoverer (Agena) program and the last 18 years in the "Skunk Works".

JULIAN MarDOCK, July 22, 1919 Tyler, TX, High School, Tyler Junior College, The University of Texas, Cornell University Medical College Primary, Sikeston, Basic Waco, TX, Advanced Lake Charles. O.T.U. Westever Field, checked out in P-47. Fifth flight, engine conked out and made a forced landing in the hills of western Massachusetts. Westover hospital for six months. Grounded for six months and service pilot status for six months. Pulled enough strings to be removed from service pilot status. Sent overseas as a casual officer. Ending up flying the Bamboo Bomber (C-78).

Gen. Brereton was persuaded to transfer this inactive pilot to a recon squadron. After ten hours in F-5 (P-38) began flying combat missions—Flew 65 photo missions plus 35 in air transport in the combat zone. Discharged from active duty with the longest time in grade of any First Lieutenant. My friends and relatives all made railroad tracks while drilling troops in the ZI. C'est le guerre. Having determined that my future was bleak in the Army, went to Cornell Medical School right after the war and graduated in 49.

Became a surgeon and sent five kids to college and they became three doctors, one school teacher and one soldier. MarDock is a true "has been". He had been a geologist before the war, has been a combat pilot, and now has been a surgeon.

KENNETH I. McCLOUD, was certified as a Private Pilot, license #68166-41 through the Civilian Pilot Training Program, Fullerton Jr. College California, on Sept. 6, 1941. After flying just a short time all Civil Aviation was shutdown on the Pacific Coast following Pearl Harbor. In January/February 1942 he received a letter from the CAA threatening Draft into the Infantry if he did not do something with the flying training he was afforded by his Country!

He enlisted in the US Army on Apr. 14, 1942 as a Private with pay grade of $21.00/ month at Grand Central Airport Burbank, CA a P-38 RTU Base, until the received his appointment to Aviation Cadet. He graduated 2nd Lieutenant from William Field, AZ with over 100 hours in AT-6, AT-9, AT-11, AT-17 and P-322 version of the P-38 as a Cadet. He transitioned into P-38s at Edwards AFB (then Muroc AAF) and North Island RTU, with the Navy and Marines, in San Diego,. He assigned for duty with the USAAFISPA 13th Air Force, 347th Gp. 339th Fighter Sdn. Guadalcanal.

On his second tour in the combat zone he was sent forward on the C-47 Courier plane along with about a dozen new pilots to relieve some of the old heads from the 44th "Vampires" Sdn. (the 339th "Sun-Setters" Sdn. was supposed to have relieved them a month earlier) who were over-due for rotation home after completing their three tours of duty! On a mixed-bag mission of questionable tactics and diverse experience McCloud was assigned to a flight led by a Night-Fighter (loaner type) pilot, a brand-new-shiny-gold-bar (1st mission) pilot and a 44th pilot who aborted and returned to Base. Lead circled to drop tanks (instead of moving out) over Rabaul and when the 360 was completed the strike force was out of sight. When the Zeros hit the flight Lead called "Break" and as McCloud rolled HARD left and felt cannon shells hitting home he observed his wing man explode (unusual for the P-38). Continuing that roll carried him down through a thin stratus cloud and he broke in the clear on the back of (less than 100 ft.) one of those Zeros, who had just made their pass, directly under his guns. That unlucky Zero exploded in a flash as the P-38's triggers were mashed with vengeance for that new pilot who never got home!

In a few days the rest of the 339th made it through the weather front with new P-38s, the 44th was sent on R&R, and the missions rolled on into our well-oiled "Tach weave" which credited McCloud with another Zero that almost ran over him on the tail of an F4U (again right in front of his guns) diving into our protective ball to "get that Zero off my back"!

McCloud ended his flying career after 20

years in the cockpit, from the Lockheed P-322 the first to the Lockheed F-104 the last. He retired Colonel USAF Reserve with 31 years (one year enlisted) service.

FREDERIC E. McCOY, was born Feb. 6, 1915 in Darlington, OK (the old Cavalry Remount Station, Ft. Reno). He was educated in Oklahoma City and graduated from Oklahoma State 1939 with a B.S. degree in Education. Soon there-after, he applied for flying cadet training, was accepted and trained at Cal-Aero Flying School, Ontario, CA, then Stockton, CA, class of 41G. He was commissioned 2nd Lieutenant in September 1941. He was stationed at Portland A.F.B. in the 38th Pursuit Sdn., 55th Pursuit Gp.

He went to Alaska in May of 1942 with the 54th Fighter Sdn. equipped with P-38 E's. On his first escorted mission of a Recon. LB 30, to Japanese held Kiska, the two P-38s and bomber were jumped by three zeros. One zero was destroyed by McCoy. The two P-38s and bomber returned safely to their base at Atka. Some time later McCoy returned to Peterson Field CO and took the 25th Photo Reconnaissance Sdn. to the Southwest Pacific in 1943. The squadron was equipped with F5s. McCoy was the first American man to fly an American plane over the Philippines on mission in two years. The photographs showed airdromes, battleships, destroyers, carriers and in which even the number of planes on the apron could be counted. As a result, the Jap 'drome installations on southern Mindanao were wiped out, pattern bombing destroying seven airdromes and more than 200 planes. Col. McCoy was later the commander of the 1st Lockheed U-2 unit in the European area. He retired in November 1968. His decorations include four DFC's, four Air Medals and other awards and service ribbons.

THOMAS G. McGUIRE, born Burlingame, CA, Mar. 5, 1918. Education: Lowell High School, San Francisco, CA, UC Berkeley, UCLA, California B.A. and M.A. in English. Aviation Cadet Training: Santa Ana Air Base, Santa Ana, CA, Class 43-K.

Flight Training Bases: Primary Flight School-Hemet, CA; Basic Flight School-Merced, CA; Advanced Flight School-Ft. Sumner, NM. Promoted to rank of 2nd Lieutenant Dec. 3, 1943. Assigned to Central Instructor's School, Randolph Field, San Antonio TX, for training as an instructor of cadets in twin engine advanced flying schools. After graduating from Randolph Field Advanced Instructor's School I was assigned to Pecos Air Base Advanced Cadet Twin Engine Flying Train-School from February 1944 to December 1944, after flying about 1000 hours. In January 1945 I was was assigned to Luke Field, Phoenix, AZ, for single engine fighter training in P-40 aircraft. In March, 1945 I was assigned to P-38 flight training, 4th Air Force, Chico, CA and Santa Rosa, CA, 434th AAF BU, from March, 1945 to December 1945 when I was discharged from military service. Left service as 1st Lt. due for Captain.

Theater of Operations: Training Command only. Never went overseas.

Type of missions Flown: Training mission only. I flew P-38Ls on aerial and ground gunnery runs, dive and skip bombing runs, fired eight inch rockets from wing racks at ground targets, night flying missions, and patrol missions for Japanese paper balloons. My favorite instructor was the great Pacific P-38 ace, Jay Robbins, who had about 26 kills.

Post-War: Married, 1947. Three children, divorced 1976. From 1952 till 1982 I was a professor of English at Los Angeles Valley College. Co-authored a college English text book, titled *Design*. Flying P-38Ls was one of the high spots of my life.

FRANK J. McINTOSH, was born in San Francisco, CA on Mar. 2, 1920, youngest of six children born to Alexander and Mae McIntosh. Enlisted as an aviation cadet on Jan. 17, 1942. Took primary flight training at King City, CA (PT-22s); basic at Merced, CA (BT-13s); graduated from Luke Field, AZ (AT-6); Class of 42-I on Sept. 29, 1942.

After training in P-39s with the 77th Sdn., 20th Gp. and P-38s with the 38th Sdn., 55th Gp., both in the Northwest, went overseas to Africa. Flew 50 missions with the 27th Sdn., 1st Fighter Gp.; three confirmed victories, six damaged and/or probable claims.

Returned to the States in Sept. 29, 1943. Instructed on P-38s, P-39s and P-63s, again in the Northwest; returned for a second tour with the 402nd Sdn., 370th Gp. in Belgium and Germany. War ended after 11 missions. Returned to the States; discharged in 1947. Received D.F.C. and Air Medal with 13 clusters.

Married for 40 years to Alma Gogel, three children and two grandchildren. Still in San Francisco, retiring after 38 years with the city and county of San Francisco Real Estate Department. Still fly, owner of D.H. Chipmunk.

Wildest moment in the service were landing a P-38, wheels down and on the runway at Pendleton, OR with both engines dead. Also a memorable dog fight on last mission on Aug. 30, 1943 when the group lost 13 out of 43 aircraft, P-38s to 109s led by Steinhoff (now General in Luftwaffe).

MAURICE R. McLARY, was born in Anderson, IN on Mar. 5, 1916. Graduated from Pendelton High School, Pendleton, IN and from Ohio Mechanics Institute (now University of Cincinnati) in Cincinnati, OH in Engineering in 1938. (Sigma Epsilon Chi Fraternity).

Joined Army as Aviation Cadet in August 1941 and graduated as pilot in class of 42-C. Took primary training at Oklahoma Air College, OK, basic at Goodfellow Field, TX and advanced at Foster Field, TX.

Was assigned as instructor at Foster Field for approximately one year before being transferred to 55th Fighter Gp. at McCord Field, WA. Went overseas to England with the 20th Fighter Gp. in August 1943. Progressed through Flight Commander, Operations Officer and Squadron Commander of the 55th Sdn. Also Group Tactical Inspector and 77th Sdn. Commander. Flew 289 hours of combat with one FW-190, one ME-110 and one ME-410 destroyed, two ME-110, two FW-190, probably destroyed in the air with six or seven aircraft probably destroyed or damaged on the ground along with many bridges, locomotives, flak towers, etc. as strafing targets.

Awarded D.F.C. with two Clusters and Air Medal with seven Clusters and Presidential Citation.

Married Lt. Murphy DeWitt, Army Nurse Corps while at McCord Field in May 1943, celebrating 45th this year (1988). One daughter, Kaye.

Returned to private sector in September 1945. Became President of Ingersoll Products Div. of Borg Warner Corp., and then President and C.E.O. of Accurate Mfg. Co., Chicago, IL Retired in 1978.

NORMAN S. McMORROW, JR., was born in Washington, D.C. on Jan. 19, 1922. He graduated from Gonzaga College High School June 21, 1939; was attending George Washington University when Pearl Harbor was attacked. He immediately applied for Aviation Cadet training, was accepted and was sworn-in as Private Air Corps unassigned on Apr. 14, 1942, awaiting call-up for Aviation Cadet classification and training at Nashville, TN on Aug. 28, 1942. He was classified Pilot and began Preflight School Dec. 7, 1942, Santa Anna, CA; Primary training Ryan School, Hemet, CA; Basic, Minter Field, Bakersfield, CA; and graduated from Williams Field, Chandler, AZ, Class 43-I, as a Two Engine Fighter Pilot (P-38).

His first assignment was with the 372nd Fighter Sdn., 360th Fighter Gp., Salinas, CA for 1st phase P-38 operational training. Next was 331st F.S., 329th F.G. at Ellensburg, WA and Ontario, CA for 2nd and 3rd phase P-38 operational training. He was transferred to 29th Photo Recon. Sdn., 2nd Photo Gp., Will Rogers Field, OK for Photo Recon. Pilot training Jan. 19, 1944, to June 2, 1944.

He arrived at Borgo, Corsica on July 26, 1944, for his combat assignment with the 23rd Photo

Recon. Sdn., 3rd Photo Gp. He completed 51 missions in F-5E aircraft, flying from bases at Borgo, Corsica; Le Luc and Valence, France; Malignano and Florence/Peretola, Italy. He was Assistant Aircraft Engineering Officer and Squadron Test Pilot. Promoted to Captain and Leader of "A" Flight. Decorations included Presidential Distinguished Unit Citation: MTO Aug. 28, 1944, for Southern France Campaign, Distinguished Flying Cross, Air Medal with five Oak Leaf Clusters, ETO Medal with seven Battle Stars and American Theater Medal. The 23rd PRS was a composite of British, French (33rd PRS) and American thru the Southern France Campaign and Invasion. The 23rd PRS was awarded the French Croix de Guerre with Silver-Gilt Star: Aug. 15, 1943 to Oct. 15, 1944. The 23rd PRS completed 2000 combat missions during WW II with only the loss of two pilots lives due to enemy action. McMorrow departed for home on Apr. 15, 1944.

He served with the U.S. Air Force until Dec. 24, 1952, attaining the rank of Major. He is Charter Member #97 in the P-38 National Association and a member of the Mediterranean Area Photo Recon. Wing Association. He is now retired and lives in North Springfield VA.

RICHARD A. McNAMARA, was born and raised in Indianapolis, IN. He joined the Air Corp. on May 7, 1941, first being stationed at Ft. Benjamin Harrison, Indianapolis. After Radio School at Scott Field, IL, he was assigned as a Control Tower Operator at Las Vegas Army Air Base Gunnery School. He was then accepted as an Air Force Cadet in September of 1942 and graduated July 28, 1943 as a 2nd Lieutenant from Luke Field, AZ. From here he was sent to the 49th Fighter Gp. 8th Fighter Sdn. at Gusap, New Guinea. At this point, in a P-40, he flew up and down the North Coast of New Guinea between Gusap and Biak Island. Wewak, Aitape, and Hollandia were locations in between.

While at Biak Island, he learned to fly the P-38 and by Oct. 28, 1944 was flying P-38's out of Leyte Island. He achieved the rank of 1st Lieutenant on Apr. 18, 1944.

He contracted yellow jaundice at Leyte and was sent back to Biak Island to recover in the hospital. He was sent home from there and spent four months as an information and education officer during which time he disseminated war information to over 3,000 officers and enlisted men each week.

He was Honorably Discharged July 20, 1945 from Camp Atterbury, IN having served his country 53 months.

Combat Hours: 310; Combat Missions: 140; P-40 160 hours combat; P-38 150 hours combat.

CHUCK AND BETTY MALVEN, Chuck enlisted at Lowry Field, CO in May 1941, and served as a radio operator/mechanic aircrewman aboard several "ancient" aircraft, including the Lockheed A-29, "Hudson." After seeing a demonstration flight of the YP-38, Chuck set a personal goal of eventually flying that gorgeous bird himself. Soon after graduating from pilot training at Williams Field, AZ, he met his goal and flew the Thutty-Eight at Santa Rosa, Santa Maria, and with the 443rd Ftr. Sdn. at Ontario. he was later assigned as Combat Intelligence Officer, Gunnery Officer and Technical Intelligence Officer.

Betty (Hudson) was involved with contracting activities on the P-38 at Lockheed, Burbank when Chuck met her. He then talked P-38s with her until she finally married him in June 1945. The honeymoon is still in progress.

Chuck was recalled in 1950 as a Technical Intelligence Officer and collected a number of goodies in Korea for about five years. He then worked on Intelligence & Reconnaissance Systems, Tactical Systems Analysis, and Technical Planning at several aerospace firms (other than Lockheed), and retired from Northrop Corporation in 1985. After two years of consulting, Chuck and Betty were not called on by Lockheed, so they hung-it-up and are now hangar-flying on a full-time basis.

CHARLES W. MANION, Oct. 2, 1944 at the Redmond Army Air Base, OR, Charles W. Manion (2nd Lieutenant Pilot, U.S. Army Air Corps) sporting a smile of accomplishment, stands with the P-38-L after completing his solo flight. All P-38 pilots will warmly remember this place in time.

Born in Henderson, KY, Manion entered the Army Air Corps Dec. 10, 1942 as a private, attended Aerial photography school, Lowery Field, CO, made PFC (private first class) upon completion. Applied for and was accepted in the West Coast Flying Training Command as a pilot trainee, completing Twin Engine Training school (Advance) Williams Field, AZ, Class 44-E a 2nd Lieutenant.

After aerial gunnery in P-39s, P-38s, Link-trainer, and daylight navigation orientation was assigned to the 7th Sdn. of the 49th Fighter Gp. 5th Air Force, Pacific Theater of operations through February 1946.

Manion's memorable flight: After the Japanese surrender signing Sept. 2, 1945 Gen. MacArthur requested the famous 49th Fighter Gp. to make morning and afternoon flights over the Tokyo area to remind the Japanese we (the United States of America) was there. These two-plane formatin flights continued until diminishing maintenance personnel (due to their returning to the states) was unable to assure continued safe operation of the P-38. Manion's four hour solo flight Dec. 29, 1945 around the Tokyo Bay area doing acrobatics like some stunt jockey in an air show, even buzzing the nearby U.S. Marine base tower. Not realizing this was to be his last affair with a great love in his life, the P-38.

Battle/campaigns: New Guinea, Western Pacific, Luzon, China Defensive, China Offensive.

Decorations/citations: Air Medal, American Theater Service Medal, Asiatic Pacific Service Medal, Philippine Liberation Service Medal w/ one star, Victory Medal.

Separated from active service March 1946. Completed 30 years Air Force Reserve and active duty March 1978 as a Lieutenant Colonel.

Occupation: Owner/managing a Commercial-Advertising Photography Studio D/B/A FOTOgraphics, Inc. Owensboro, KY since 1954.

Married former Jeanette Ward of Henderson, KY Oct. 10, 1959. Children: Elizabeth Ann (1962), David (1964).

BOB MARGISON, was raised in the S. California area. He enlisted in the USAAF at March Field Riverside, CA in December 1936. He was a guard in the hangar where the first P-38 was assembled and flown by Capt. Ben Kelsey. He vowed he would fly the P-38 one day. In advanced flying school he was one of the first Cadets to fly the P-322, which was the British P-38.

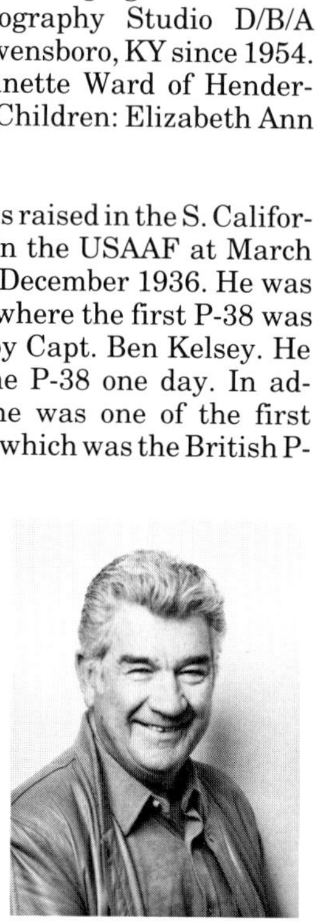

In June 1943 he was assigned to the 37th Fighter Sdn., 14th Fighter Gp. in North Africa under Maj. Carl Rudder, Sdn. C-O. Later flew under Bill Leverett Sdn. C-O and was on one mission where Leverett shot down seven enemy aircraft in 15 minutes. Bob got his first victory then.

After 50 missions and three victories, DFC, Air Medal w/11 Oak Leaf Clusters, he returned to the States in June 1944. He remained in Fighters, flying the first P-80s in the 1st Fighter Gp. at March Field in 1945.

He flew fighters until his retirement in 1963.

He now lives in the Reno/Tahoe area. Still gets a thrill watching Lefty Gardner fly the beautiful white P-38 each year at the Reno Air Races.

LAWRENCE V. MARS, was born and raised in Watertown, MN. Three months after Pearl Harbor, he enlisted in the Army at 21. He received his basic training and graduated from Air Mechanics School at Stockton Army Air Base in 1942. While at Stockton he applied for Aviation Cadet Training and was accepted. Preflight at Santa Ana, Primary at Hemet, basic at Gardner, and he graduated from Williams in Class 43F. Before graduation Larry and some classmates were checked out in RP-322s.

His first assignment was Muroc Army Air Base, and then Lomita Flight Strip. On Aug. 25,

1943, he left for Africa and was assigned to the 37th FS, 14th FG, 12th Air Force, at St. Marie Duziti in Tunisia. In December he moved up to Foggia #7 in Italy and was assigned to the newly activated 15th AF. During the 56 Combat Missions he flew, he advanced to Flight Leader, Squadron Leader, and eventually led the Group on missions. He flew nine missions, escorting "Heavies" in to the Romanian Oil Fields around Ploesti. On one of these missions he had one engine shot out, and fell back into a bomber formation. Two ME-109s attack, and he shot one down, and turned the other away from the crippled bombers. For this he was awarded the Silver Star. He also received the DFC, Air Medal and 11 Clusters, and The Presidential Unit Citation. He had three confirmed victories. Today Larry lives with his wife Judy in Del City, OK. He is now retired.

ONALD JAMES MARTIN, was born and raised in Fresno, CA. He enlisted in the U.S. Army Air Corps in January, 1940 in San Francisco and his first assignment was to Moffat Field, CA.

He was a charter member of the 55th Pursuit Gp. and was a Radio Mechanic (Technical Sergeant) when they received their P-38s at McCord Field, WA.

After a piggy-back ride in a P-38, he entered he Aviation Cadet program as Technical Sergeant, graduating from Williams Field, AZ in he class of 4-E. After training in the P-39 and other aircraft, he first flew the P-38 at Portland Air Base, OR. Fighter Pilot training preceded his transfer to Hamilton Field, CA for assignment to the Pacific Theater.

He was assigned to the 8th Fighter Sdn., 49th 'ighter Gp. and flew over New Guinea, the Philippines, Okinawa and was one of the first J.S. Army Air Force units to occupy Japan as Vorld War II ended.

He returned to the Air Force after a short stay n civilian life and retired from the U.S. Air 'orce in December, 1960. He now lives in Overon, NV after a career as a School Administrator. He and his wife of 46 years have five hildren, 16 grandchildren and one greatrandchild.

DDIE MARTIN, has lived all his life, so far, n Southern California. Eddie was born in 1901, nd learned to fly in 1921. In 1923, he opened the Eddie Martin Airport in Santa Ana and operated the airport and a flying school until 1939, when Orange County condemned the airport land for a road project. The County needed the airport, and tried to lease land for an airport from James Irvine II, owner of the Irvine Ranch upon which the Eddie Martin Airport leased land. Irvine refused to provide land to the county unless they gave Martin Aviation a 17-year exclusive lease on the County airport site. The County agreed. Shortly before the airport operation was taken over by the County as Orange County Airport (1939), Eddie sold the business to his brother, Floyd. In 1941, Orange County Airport was appropriated for the duration of the war.

By the time Eddie began testing for Lockheed, he had not only built the airport into one of the most successful in Southern California, but he owned several ranches and farmed them, flew for both Western Air Express (later Airlines) and American Airways, was a corporate pilot for R.S. Macmillan (owner of Macmillan Petroleum Company and KMPC radio station), and was Sales Manager for MacMillian Ring-Free Motor Oil in the Los Angeles division.

Eddie was considered an "old timer" by the time he started working as a test pilot for Lockheed. He was an acceptance pilot, as well as one of the few pilots allowed to ferry the test pilots back from deliveries.

Today, Eddie continues to take care of ranches he owns in the Imperial Valley, and also cares for his elderly sister, who is in a convalescent hospital. He is also writing an autobiography.

WILLIAM E. MARTIN, born Nov. 14, 1918, Alton, IL. Drafted April 1941, entered Aviation Cadet program and graduated Class 43-E, May 1943. Transitioned P-39s in San Francisco Area. To Orange County Airport for 20 hours P-38 transition. January 1944 to ETO for assignment to the 55th FS, 20th FG. Flew 55 P-38, 12 P-51 combat missions returning to the States December 1944.

Married his present wife, Mary R. (Jean) and assigned to Las Vegas AAF, NV flying P-39s and P-63s (Pinballs). Then to Greenville AAF and Sumter AAF, SC flying P-40s and P-47s. Separated from service December 1945. Joined ILLANG, Springfield, IL flying P-51s. March 1951, with unit, recalled to active duty and assigned to Bergstrom AFB, TX, then to George AFB, CA. May 1952, assigned as Air Liaison Officer with Navy, serving for three years.

Subsequent assignment on Ops/Plans staff of Hqs. 18th AF, 9th AF, 315th Air Div., (Japan) and 19th AF. Retired June 1966.

Decorations are DFC, Air Medal/three OLC, AF Commendation/two 0LC. August 1966 moved to present residence, Myrtle Beach, SC. Here involved with resort motel operations for first eight years, since associated with Myrtle Beach National Golf Club.

BEN A. MASON, was born Dec. 19, 1911 in Nashville, TN. Raised on Cavalry Posts around the U.S. as his father was an Army Officer. Attended U.S. Army Flying School at Randolph and Kelly Fields from October 1933 to October 1934 as a Flying Cadet.

Served one year as a rated Flying Cadet and received commission as 2nd Lieutenant Air Corps Reserve. Received a regular commission while an instructor at Advanced Flying School at Kelly Field in 1938.

Flew a P-38 from California across the South Atlantic to North Africa in 1943. Flew gunnery and dive bombing missions in P-38s for two weeks at Muroc in 1943 while stationed in headquarters Los Angeles Fighter Wing.

Went to Tunis, North Africa toward end of 1943 and was assigned to Operations, MAAF. After the headquarters was moved to Caserta, Italy I got my assignment to the 82nd Fighter Gp. at Foggia, Italy. How I managed this is too long to go into here. I served as Deputy Commander and Commander of the group from March 1944 to August 1944. After returning to the U.S. was Deputy Base C.O. of a P-38 training unit at Van Nuys, CA until it was deactivated. After a tour in Guam, 1946-1947, with the 20th AF, I served at several stations in the Continental Air Command until my retirement in 1963 with the rank of Colonel.

While with the 82nd I flew on the dive bombing mission of Ploesti June 10, 1944, for which the group received its 3rd Distinguished Unit Citation. Flew on shuttle mission to Russia, strafing air fields in Rumania and landing in Poltava. Finished that mission on one engine for three hours. While in Poltava I led a strafing mission on German Air Field at Mielec, Poland. Shot down one ME-109 and one ME-110.

On June 13, 1938 Elizabeth and I had a military wedding at the Randolph Field Chapel. Son William was born Oct. 4, 1940, San Antonio, TX; daughter Sandra was born Sept. 27, 1955, Barksdale AFB, LA.

Decorations: Silver Star, DFC, AM (w/one silver and one bronze Oak Leaf Cluster), AF Commendation Medal, American Defense Service Medal, American Campaign Medal, European-African-Middle Eastern Campaign Medal (w/one silver and three bronze stars), WWII Victory Medal, Army of Occupation Medal (Japan), National Defense Service Medal, AF Longevity Service Award (w/silver

Oak Leaf Cluster and bronze Oak Leaf Cluster), French Croix de Guerre (w/palm) and Distinguished Unit Citation.

FRANCIS CURTIS MASON, son of Emma (Reed) and William C. Mason was born June 13, 1924 in Hamilton, Il.

He worked at Douglas aircraft on the B-17 line when he was called in the draft at the age of 18. After his basic training at Kerns Field, UT and engineering and operations training at Colorado A&M at Ft. Collins, CO he reported to the 394th Fighter Sdn. at Hayward, CA, and was assigned to the operations section until the unit was deactivated in 1945.

After his discharge he worked at North American Aviation, and was in the garage business with an uncle when he reenlisted and was assigned to the base operation and the Form Five Section at Long Beach, CA. From Long Beach he was assigned to the Operation Green House and Atomic Bomb Test in the Marshall Islands. Other bases of assignments were Base Operations at Tachikawa, Japan, SAC Command Post at 22 BW March AFB, CA, 3234BW Mather AFB, CA Lajes Field, Azores. 6007th Recon. GP. Yokota AFB, Japan, he retired at Travis AFB in 1965 with 20 years service and he and his wife, June now live near Turtle Lake, WI.

MORGAN H. MATHEWS, was born in Sparta, WI on May 16, 1920, was raised in Portage, WI and was 21 years old and married when Pearl Harbor was attacked. After convincing his wife and family, he applied for Aviation Cadets and graduated with Class 43D, Wm's Field, AZ. He checked out in P-38s at Muroc, CA and after some training was sent to the 37th Fighter Sdn., 14th Ftr. Gp. near Tunis in August 1943.

As a Pilot he covered the invasions of Sicily and Italy, participated in Missions to Regensburg, Klagenfurt, France, Italy, Austria, Greece and the Ploesti Oil fields. Morgan had two confirmed kills, two probables and one damaged. The Squadron moved to Foggia #7 Dec. 11, 1943. On Jan. 20, 1944 Morgan was assigned as Operations Officer.

After returning to the U.S. in July of 1944 Morgan was assigned as a fighter tactics instructor at Santa Maria, CA. While there he attended an aerial gunnery course at Dagget, CA and finished #1 in his class. His comment "If I had only had this course before going to combat!"

Among his decorations are the DFC, Air Medal with two Silver and one Bronze Clusters, and a Bronze Star.

He is retired and living in Arroyo Grande, CA with is lovely wife (Myrtle) of 47 years. His two daughters each have a boy and girl who are his pride and joy.

JIMMIE MATTERN, born Mar. 8, 1905, in Freeport, IL, the youngest of four children. Parents were Caroline Kennedy and Phillip Mattern.

Took first flight 1924, Army Air Corp, HI. Soloed, 1927 Ryan Flight School, San Diego, CA. Issued FAI License #2919, signed by Orville Wright and carried, by Neil Armstrong, to Tranquility Base, Moon on Apollo 11, December, 1969. The 1920s in OX5 Waco Ten, flew in Hollywood movies "Hell's Angels (one of Howard Hughes first pictures) and in Lilac Time. 1928 was Test Pilot for Eddie Stinson, Wayne, MI. Flew the first Air Freight Line into Mexico in 1928, from Texas to Soto La Marino Mexico. Also carried 1st Airmail Brownsville, TX/Monterey, Mexico. Issue Mexican Pilot's license NO. 76.

Partner, Stinson Distributorship for three states, Ft. Worth, TX. Flew first scheduled Airline from San Angelo, TX 1929, 30 and 31, with J5 Lockheed Vega, Wasp Lockheed Vega and Speedwing Travelair.

1932-made first America to Berlin Flight New York Newfoundland Templehof, July 5-6, establishing eight world records, most being "Firsts", stand for all time. Held Trans. Atlantic speed record of ten hours 50-minutes for eight years.

Flying from Berlin to Moscow, crashed near Minsk, Russia. First American Pilot to be interrogated in the Kremlin.

June 3, 1933, first man to take one monumental task of flying around the World, solo. Took off Floyd Bennett Field, NY. First jump was non stop to Norway, 4200 miles. Time 23.5 hours. First New York non stop flight ever to that Nation. The second leg Moscow/Omsk, Siberia, third, Omsk to Krasnayarsk, fourth to Irkutz, fifth to Khabarask. First man to fly solo across Siberia with no maps, no de-icers, no automatic pilot nor radio. Turn and Bank indicator for instrument flying. Sixth, across the Okotz Sea. Crashed on the Artic Siberian Tundra, 75 miles from Anadyr, Siberia. Three weeks there without food before rescue by Siberian Eskimos. Having covered 11,500 miles of the 15,800 around the World.

1934—Aeronautical Director, Pure Oil Co., produced National Radio Show, "The Diary of Jimmie Mattern". 1936-37—Aeronautical Director Benedume Trees Family of Oil Companies.

P-38 History Jimmie Mattern has more hours in a P-38 than any man living (or dead) 3000 hours. 1939—Joined Lockheed as Experimental Engineering Test Pilot. Tested the first YP-38. In P-38 testing was the first man to reach the Speed of Sound. Became the most experienced on the then Aeronautical Phenomena of Compressibility. Helped engineer Hydraulic Ailerons (later power steering for Automobiles). Developed microphone in Oxygen mask, dive flaps and introduced the first Piggie Back P-38 to the Army Air Corp. Reporting only to the Chief of Staff of the AAC, made 3000 single engine landings in demonstrations to AAC Bases. (They said it couldn't be done). Created an Oxygen Office for all Bases. Although then a civilian, was Chairman of the Board of Safety Command Officers, creating the Flight manual for P-38s and extended to all aircraft models.

Reduced accident rate of P-38s from 12 a day to three a day in three months. Decorated by the AAC and presented with the only "Citation of Appreciation" of its kind and signed by Gen. Hoys S. Vandenberg and Secretary of the Air Force, W. Hoyt Symington. Retired as a full Colonel.

In 1980s, Doctor of Aeronautical Science and enshrined in the Air and Space Hall of Fame.

HONORS AND AWARDS

Aviation & Space Hall of Fame-Oklahoma City. OX5 Aviation Hall of Fame; Smithsonian "Golden Age of Flight". Featured on Cover of Smithsonian 1986 calender. Honorary Fellow of "The Society of Experimental Test Pilots". USAF Citation "Scroll of Appreciation" for P-38 Piggie Back Program, the only one of its kind. Doctor of Aeronautical Science. Colonel, USAFR (Ret.). Elder Statesman of Aviation (National Aeronautical Association). Colonel, Governor's Staff of Texas 1932. Colonel, Governor's Staff of Oklahoma 1932. Texas Ranger, 1937. Honorary Citizen of West Berlin, 1969. Amelia Earhart's "Memory Lane"-Atchison, KS. National President of "OX5 Club of America" (17,000 Pioneer Airmen) 1961. OX5 Aviation Pioneer's "Distinguished Service Award". Honorary Pioneer of Alaska. Member of Igloo #3. Fairbanks, AK. Wings enshrined on the Wall of the "Aviator's Chapel" at the Mission Inn, Riverside, CA. Honorary Cropduster of California (inducted with Jimmy Doolittle). Navy League Award. Honorary Member Navy Fighter Sdn. 151, Miramar, CA. NAS. Honorary Member VX4 Experimental Test Sdn. USN, Point Magu, CA. Honorary Signal Officer USN-Aircraft Carrier "Enterprise". Honorary Member of the "Angeles Antiquers". Received in the "White House Oval Office" by four Presidents-Hoover, Roosevelt, Johnson and Eisenhower. Decorated by six foreign countries. New York Ticker Tape parade 1933. Palm Desert, CA Proclamation—designating May 16, 1987 as "Dorothy and Jimmie Mattern Day". Legion of Honor. Al Malaikah Shrine Temple.

WORLD FLIGHT RECORDS

1932 with single engine 150 MPH Lockheed Vega "First America to Berlin Flight" (New York-Newfoundland non stop to Berlin). Trans-Atlantic Speed Record ten hours 50 minutes. Held for eight years. First America to Berlin Speed Record 29 hours 31 minutes. First Aviators interned and Tried in the Kremlin, Moscow, Russia after having crashed near Minsk, Russia. 1933 Solo World Flight-First ever attempted Solo Flight around the world. This flight attained the following flight records:

First non-Stop Flight New York to Norway (1st leg). Longest Solo over water flight ever made without an automatic pilot. First man to fly the Atlantic Solo after Lindbergh. First flight from New York to Moscow (2nd leg). First man to fly solo across Russia and Siberia (no maps). Crashed on the Siberian Tundra 500 miles from Nome, AK. Was there three weeks without food before being rescued by Siberian Eskimos. While in Siberia-kept a daily diary-as yet never revealed.

Airmail & Airline Pilot-First Air to Air Refueling over Alaska (1920). Operated first scheduled airline from USA to Mexico. Flew first Tri-Motored Ford from Texas to Alaska and return. First non-stop flight from Oakland, CA to Fairbanks, AK. Flew from Fairbanks to North Pole and return to Point Barrow, AK. First non-stop flight from Juneau, AK to Burbank, CA.

C.A. MAUSS, was born Mar. 17, 1919 on a farm near Terril, IA. He enlisted in the Army Air Force Jan. 10, 1942, at Ft. Snelling, MN. Training consisted of basic at Sheppard Field, TX, Radio at Scott Field, IL, and radar at Boca Raton, FL.

He was assigned to the 37th Fighter Sdn. at Tumwater, WA in September 1942. In February 1943 the 37th embarked for Oran, Africa where the Squadron assembled its own P-38s at nearby La Senia airfield. In May the squadron moved to Medouina Air Field near Casablanca, joining the Veteran 14th Fighter Gp. before going into action in July 1943 from Telergma, south of Constantine. After several moves in Africa, Sicily and Libya, the 14th F.G. landed near Naples, Italy on Dec. 7, 1943, settling in at Triola, 14 miles north of Foggia, for the duration.

Mauss attained rank of Tech. Sergeant with flight leader status in keeping the command radios, IFF and Tail Warning Radars Operational. He also assisted in designing and installing a sound system serving all officer and enlisted men's quarters at Triola. He returned July 29, 1945 and was discharged Oct. 2, 1945 at Santa Ana, CA. He earned a BSEE Degree and worked for the Navy at China Lake, CA, then TRW, Bunker-Ramo, and Hughes Aircraft Company, retiring in 1984. He and his wife Roselyn had three children and live in Canoga Park, CA.

GLENN M. MAXWELL, was born in La-Grande, OR. He joined the Army Air Force in 1941, graduating from aviation mechanics' school in 1942, at Keesler Field, MS. Later he

obtained his wings in the class of '43H at Williams Field, AZ.

He later was assigned to the famous "Satans Angels" 475th FG, 5th Air Force, completing 144 combat missions.

ROBERT B. MAXWELL, Mansfield, OH; 1921. Born maverick and prophetic and color-blind and crazy-to-fly. Cheated in A.C. 1942 via CPTP, glider-pilot, etc., until memorized color-tests.

Graduated Luke Field July 1944, sneaked himself and three buddies into fighters, most went to bombers. Trained in old, tricky P-39s for cannon-attack new Tiger-tanks. Next trained massive 180-hours in sweet-flying P-38-Ls. He crashed once, it burned. His analysis: 50% pilot-error. Discovered P-38-L to be prop-fighter could do DOUBLE Immelman any altitude, thanks to special-flaps, lotsa power and NO-TORQUE! Must drop the belly-tanks, first.

Sent to 475th Fighter Gp. in Pacific June 1945 for Japan invasion, but his cocky-self probably saved by atom bombs. Occupied Korea til June 1946; one crash in Stinson L-5 with road-engineer, flipped over: 80% pilot-error? Considering how hot he thought he was, and sometimes was, those crash-blunders yet gall him. Also in Korea, his strange dream of 1939 came true, powerfully strong.

After WW II, he became a Nature-Boy, Vegeterian, Explorer, Fire-tanker pilot, Labor-organizer, Civil-Rights activist, Builder, Inventor and finally a damned Hermit. Along the way he also became a C.O. to Slaughter, War, Poisons & Anti-Nature. Says to his WW II comrades. "Thanks for cover, buddies, and farewell."

PERCY C. MEGGINSON, JR., was born in Shreveport, LA, on Jan. 21, 1923, and at age three moved to Cincinnati, OH.

He enlisted in the USAAF in 1942 and graduated from single engine flight school at Moore Field, TX, in the class of 44B. After training in P-40s, P-39 and P-63s, he entered P-38 transition training at Santa Rosa, CA. Since he had never flown twin engines, he was instructed on his first flight to walk rudders straight ahead to 10,000 feet and then try the wheel. It worked great.

After training, he was assigned to the First Fighter Gp., 27th Sdn., at Foggia, Italy. He flew 13 missions in P-38s received the Air Medal and was made a first lieutenant.

Upon his discharge he returned to college and after graduating as an aeronautical engineer, worked at Wright Field. During the Korean War, he served two years in the USAF as a Captain, pilot and procurement officer.

During 1954 to 1984, he worked for Northrop Corporation in Los Angeles as a contract administrator and manger. He is now retired in Hemet, CA, and flies radio-controlled model airplanes.

HENRY MEIGS, was born June 12, 1921, in New York, NY. Trained in Southeast T.C., commissioned Oct. 7, 1942.

Advanced Training Instructor, Spence A.F.B., GA, until transferred to Glendale, CA for P-38 training. Assigned to 339th F.S., 13th A.F. March 1943-April 1944. Served at Guadalcanal, Northern Solomons and over Rabaul. Shot down three Betty bombers, three Zeros, seven probables, awarded Silver Star, D.F.C., and Air Medal with seven clusters. Discharged as Captain, August 1945.

PAUL MELLINGER, form five shows I flew RP-322s from Williams Field on June 4, 1943 and arrived at Drew Field (Tampa) the next day, June 5, 1943. If we were on any kind of official orders I never got a copy. I think it was a "Special" cross-country "Training" flight and that was what had our C.O. at Williams so agitated when our stay at Tampa kept getting extended way beyond the original understanding. In fact, I was still not completed by mid-July when I was assigned to the Air Force School of Applied Tactics, Orlando, FL. I actually was stationed at a small field at Cross City, FL and transitioned into P-47s (about ten days) and then returned to Williams.

I never flew P-38s in combat-it was all in P-47s. I did fly quite a bit in the RP-322s in the training command at Williams Field, AZ.

Retired from Lockheed in 1988; had been the Field Service Manager on the SR-71 Blackbird at Beale AFB, CA.

ALBERT H. MERYMAN (AL), was born and raised in Winthrop, MA. An early interest in radio, photographing airplanes and building solid and scale flying Cleveland Models led to

acceptance in the U.S. Army Air Corps, Panama on July 27, 1939, when it expanded in July, 1939.

He was assigned to the 29th Pur. Sdn./16th Pur. Gp. at Albrook Field, C.Z. on a P-26A as a mechanic's helper. Then on Feb. 1, 1940, he became a member of the newly-reactivated 43rd Pur. Sdn. (I)/16th Pur. Gp. (I)—the famous pursuit 43rd School Sdn. at Kelly Field, TX, for more than ten years, assigned to the crew of the official squadron commanders P-36A as an air mechanic. Later that year, while on TDY at Rio Hato, airfield, the CO, Maj. W.W. Gross, flew this A/C there bringing him orders to report to the Caribbean Air Force Radio School at Albrook, for training. After graduation, he did radio checks on the new P-40C's—switching to the Allison V1710-33 engine change crew before furlough to the States, May, June, 1942.

Upon returning to the now 43rd Ftr. Sdn., he worked on various jobs on the P-39K. Then transferred to the 53rd Ftr. Sdn./32nd Ftr. Gp./36th Ftr. Gp. (P-39D/P-40C) for return to the States during May, June 1943. Then transferred to the 442nd Ftr. Sdn./321st Ftr. Gp., Westover Field, MA for ground crew combat training and thence to the 485th Ftr. Sdn./370th Ftr. Gp. for OTU as a P-47D2, three mechanic to crew chief.

Then on to England via the Maurtania during February 1944. Thus he became a crew chief of the Lockheed P-38 Lightning, followed by three others, that for a year before transitioning to P-51D Mustangs, proved as exciting as anything during his six years of military service.

After the war, he went to school under the GI Bill and worked in aircraft plants until retiring from Lockheed Aircraft Dec. 31, 1984.

He has been in the Civil Air Patrol/Auxiliary USAF since July, 1948 as Photo Officer and Mission Observer (now with Squadron 35/Gpl/CAWG) and continues his radio, photographing airplanes and Cleveland SF Models.

ALPHONS J. MIKES, was born in Shiner, TX Aug. 19, 1917, son of James and Mary Mikes. He enlisted Mar. 20, 1942, and was assigned to the 358th Inf. Reg., 90th Div., then was appointed Aviation Cadet September 1942 and received pilot flight training graduating at Williams Field, AZ, class 43 G.

Mikes was sent overseas November 1943 to North Africa and joined 82nd Fighter Gp., 96th Sdn. at Foggia, Italy in January 1944. He flew 56 missions consisting of heavy bomber escort, dive bombing and strafing. He participated in P-38 dive bombing of Ploesti oil refineries. He was awarded DFC and Air Medal with seven Oak Leaf Clusters and the group was awarded two Presidential Unit Citations.

Mikes returned stateside September 1944 and served at Portland, OR airbase and Gowen Field, Boise, ID. He was separated from the service as Captain in August 1945 and attended Southern Methodist University. Mikes later engaged in masonry contracting, until his retirement in 1980, in the Shiner area, where he and his wife, Anita, currently reside.

RUSSELL S. MIKSCH, was born (a twin) in Ohio but grew up at Visalia, CA. After CPT in 1940, he enlisted as an Army Air Corps Flying Cadet in March 1941. At completion of pilot training, he joined the 55th Pursuit Gp. In 1942 he was assigned to the 9th Photographic Recon. Sdn., which went to the Tenth Air Force in the China-Burma-India Theater. During 1942 and 1943 he flew 60 photo missions in F-4s, F-5s and B-25s. In 1944 he was an instructor in Photographic Reconnaissance Operational Training Units in Oklahoma and Kansas. In 1945 he went to Italy where he was assigned to 5th Photographic Recon. Gp. and flew six more photo missions in F-5s, for a combat flight total of 303 hours in two Theaters. Military retirement rank, Lieutenant Colonel.

During his first combat mission, Miksch experienced an electrical failure in the F-4 over China. The remotely-driven magnetic compass and all engine instruments for both engines were disabled. He was alone and unescorted, but he employed pilotage to return nearly 700 miles over unfriendly and unfamiliar mountains and jungles without radio aids to Calcutta, India.

After WW II, Miksch returned to the University of California at Berkeley. He worked as a registered Mechanical Engineer in aerospace until retirement in 1983.

DAVID MILES, was born in Indiana on June 11, 1920. He enlisted n the Air Corps on Oct. 16, 1941 and went through Radio Mechanics School and Radar Mechanics School before taking Pilot training. He got his wings at Napier Field, AL. Class 44-A and was an Instructor Pilot before transition training in P-39s at Ephrata, WA and P-38 at Vay Nuys, CA.

He shipped out to Naples, Italy and was assigned to the 48th F.S., 14th F.G. at Triola, Italy. He was shot down on his 17th mission which was to Odertal, Germany on Dec. 17, 1944. He bailed out over Czechoslovakia and was captured by the Germans and interned in Stalag-Luft 1. He was liberated by Russian Troops.

He received the Purple Heart and Air Medal/Clusters and several Unit citations. He was separated from the military in 1946 and attended Purdue University and Hanover College in Indiana.

He was working in Iran when evacuated in January 1979 because of the turmoil there. He and his wife Mary, have seven grown children and have lived in Jacksonville Beach, FL for over 30 years.

EVERETT MILLER, (Bud), was born in Nashua, IA and raised in Chino, CA. He joined the Royal Canadian Air Force in 1941, but was discharged shortly after when the U.S. became involved in the war and he instead joined the U.S. Air Force. He graduated at Williams Field, AZ in Class 43K.

He was assigned to the 1st Fighter Gp., 94th Fighter Sdn. (Hat in the Ring) of the 15th Air Force which was based in Foggia, Italy at the time. He flew 273 combat hours in 53 missions with four victories to his credit. His missions included escort, strafing and dive bombing.

Bud thinks he has the distinction of having flown more missions in a shorter period of time than most combat pilots. He flew his first mission Apr. 16, 1944 and his last July 26, 1944, averaging better than five hours per mission. During all these flights Bud was forced to abort only once and credits this good record to "one of the best crew chiefs in the Air Force," Everett Jacobsen of Oakland, NE.

Bud recalled one of the missions still vivid in his mind and probably in the minds of those who returned with him. It was the dive bombing raid on the cracking plant in the Ploesti Oil fields on June 10, 1944. "I never realized they had so many planes! We did the job but we paid for it dearly."

When he had completed his missions, Bud returned to the States and became an instructor at Redmond, OR and then an operations officer at Portland, OR His final duty was as a pilot in the first jet group in the U.S., the 412-31st Fighter Sdn., Bud says though he flew every fighter plane with the exception of the P-47, in his opinion there never has been a "bird"like that beautiful P-38 Lightning!

P.S. Martha J. became Martha M. and has remained so for 43 years.

ROBERT A. MILLER, was born in Chillicothe, MO and lived in Mexico, MO until graduating from high school. Entered military service with Btry. B., Mo. National Guard and transferred to the Army Air Corps in 1942. Graduating from Williams Field, AZ in July 1943, he was sent to North Africa to transition in P-38s. After training at Telergma, North Africa he was reassigned to the 37th Fighter Sdn., 14th Fighter Gp. at Foggia, Italy. During his combat tour he served as Flight leader and also led the squadron on sorties. He was part of the group of P-38 pilots sent to England to ferry replacement fighters for aircraft lost in the invasion in southern France.

After completing 56 missions and a tour of combat he returned to the U.S and completed 30

years in the Air Force. During this time he was stationed at Steward Field, NY; Vance A.F.B. Enid OK; Hanscom Field, MA; Narsarssuak, Greenland; Tyndall Field, FL; Rhein, ME, Germany and Little Rock, AR. Except of the primary flying assignments (airlift, etc.) most of the time his primary duty was in the aircraft maintenance field.

He retired from the Air Force in 1965 with the rank of Lieutenant Colonel.

GREGOR W. MILNE, was born and raised in Seattle, WA and was 20 years old when Pearl Harbor was attacked. He immediately applied for Aviation Cadet Training, was accepted, trained as a single engine pilot and graduated from Luke Field, AZ in the Class of 43-E. His first assignment found him instructing Basic flying at Marana, AZ where he graduated some 30 young student pilots before leaving to enter fighter upgrading. In this pipeline Milne flew P-39s and finally, his dream machine, the P-38. Overseas he was assigned to 15th Air Force Headquarters in the 154th Weather Recon. Sdn., Bari, Italy under the command of Gen. Nathan Twining. Flying a P-38L-5 with full armament his mission was to position himself over the projected enemy target of the day. At sunrise he began radioing information about the primary and alternate targets back to headquarters. With the aid of this 'real time' information the Commanders made the decision of where the 15th Air Force would strike, or perhaps, the decision to 'stand down'. At 23 years of age Milne was performing these missions over a wide area of Europe with such dispatch that he earned five Bronze Battle Stars on his E.T.O. medal and, further, was awarded the Air Medal with two Oak Leaf Clusters and was decorated by General Twining with the Distinguished Flying Cross.

Today Milne proudly holds Charter Membership number 1 in the National P-38 Association and lives on Catalina Island off the coast of Southern California where he and his wife, Norma, own the unique little Seacrest Inn.

ROBERT B. MINUTE, was born June 6, 1922. Attended North Ridgeville, OH schools. In high school would spend a week's wages for a flight in a Waco bi-plane at a nearby field.

Enlisted in Army Air Corps September 1942; sent to Texas Tech, Lubbock, TX for schooling.

After pre-flight at Santa Ana, CA, he started flight training at Rankin, Tulare, CA in Stearman P-17. Basic at Gardner Field, Taft, CA. He graduated from Williams Field, AZ, Class 44-D.

Assigned to Victorville for P-39 transition. In fall of 1944 sent to Santa Maria for P-38 training. He remembers demonstration and check out by Milo Burcham and Tony LeVier from Lockheed before first flight in P-38s.

Left for overseas Jan. 1, 1945 assigned to 18th Fighter Sdn., 343rd Gp., Attu, Aleutian Islands—served 12 months with that unit. Then flew five months with 344th Fighter at Shemya. His last flight in P-38 June 6, 1946 (his birthday). Discharged summer 1946.

Married Neva Manning his high school sweetheart. They have two children. Founded a wholesale floral greenhouse business fall of 1946 which the family still operates.

FRANCIS E. MIRON, was born Apr. 10, 1922 in Columbus, MT. Went to work at Vega Aircraft CO August 1940 to February 1943 when he went into the Army Air Corp. Graduated from Williams Field, AZ Class 44B. Went overseas, ETO, 9th A.F. 370th Ftr. Gp. 485th Ftr. Sdn. August 1944 till end of war.

Flew all kinds of missions, mostly dive-bombing and strafing in close support ground troops (by far the most hazardous of missions). Was shot at by everything the Germans had including an ME-262. Consider myself a lucky survivor. Ended with 54 missions, 14 in P-51D. Last 20 missions flew as flight leader. Back to work at Lockheed in 1946. Married one and only wife Marjorie in 1942 to present, two daughters, Pamela and Janis, fourgrandchildren retired 1986. Present 2nd V.P. of P-38 National Association.

FRED L. MONTGOMERY, born Dec. 24, 1919 in Morenci, AZ. Enlisted U.S. Army Air Corps Nov. 12, 1941, San Francisco, CA, at age 21. Basic Training: Keesler Field, Biloxi, MS.

Acft Mech. Training: California Flyers, Inc., Inglewood, CA. Assigned to 1st Pursuit Group, Glendale, CA, May 1942. Shortly after that, the 1st was reformed as the 82nd Fighter Gp., and Montgomery was assigned to the 96th Fighter Sdn., temporarily stationed at Muroc Dry Lake, CA (now Edwards AFB). Back to Glendale until September.

Shipped to N. Ireland September 1942 aboard the Queen Mary. (That was the voyage on which the Q.M. sliced through one of her two British submarine escort corvettes off the coast of Scotland. Montgomery was on deck and witnessed the tragedy in which more than 300 Britons lost their lives.

Combat duty: No. Africa, December 1942-September 1943. Italy, October 1943-June 1945 (approximately 18 months at Foggia).

Honorable Discharge-Ft. Bliss, TX, June 29, 1945, Staff Sergeant.

Current data: Co-editor of three annual issues of 96th Sdn. history, "The Sluggin' Jackrabbit", so named for squadron insignia adopted while forming at Muroc in the Mojave Desert in 1942.

JAMES A. MOREING, was born in Stockton, CA, Aug. 9, 1916 where he was raised.

At the time Pearl Harbor was attacked, he started working at Lockheed Aircraft in Burbank, CA. Even though he had a diferment, he applied for the Air Force, where he graduated and was commissioned at Williams Field, AZ.

He trained in P-38s at Salinas and Santa Maria in California. In April 1944 he was sent to New Guinea, where he joined the 431st Sdn. 475 Fighter Gp. On his first mission, he was flying Maj.Tommy McGuire's wing, the day Dick Bong shot down 25-26 Japs. They send Dick home for a short time after that.

After 14 months in New Guinea and the Philippines, he had 105 combat missions, shooting down three Jap airplanes. Having enough points, he was released and sent home before the war ended.

Looking back at his combat experiences, on Christmas day in 1944, in combat over Clark Field, in the Philippines, was the most exciting and disappointing experience. Three of his friends, were shot down. They were all suppose to go on leave to Sidney the next day. So Jim Moreing had to go alone.

EMORY C. MORGAN, was born Dec. 7, 1920 in Kerns, TX. Graduating from Weatherford (Texas) High School in 1939 he enlisted in the Army Air Corps and was assigned to Randolph Field (AFB) TX. He completed Armorers School at Lowry Field and was assigned to the 49th Bomb Gp. at McDill AFB where he became "Classified" as a Bombardier and Aerial Gunner. He accepted the opportunity to enter pilot training class 42C, the first class of pilots to be graduated as Staff Sgt. Pilots.

In May 1942 he was assigned to the 82nd Fighter Gp., a P-38 group being formed at Muroc Dry Lake with 86 Staff Fgt. Pilots and 12 Officers. In September 1942 he left with the 82nd via troup train to New Jersey, thence by Queen Mary to England. Commissioned 2nd Lieutenant in December 1942 he flew P-38s with the Group to North Africa where he flew 50 combat missions totaling 468 combat hours including missions to Sicily, Sardinia and Italy.

He is credited with one confirmed and two probable aerial victories.

In 1946 he entered the University of Southern California graduating and returning to the Air Force in 1948. He retired to 1966 as a Colonel having spent the last 13 years in the Strategic Air Command.

His decorations include DFC 2/two Oak Leaf Clusters, Air Medal w/nine Oak Leaf Clusters and five Campaign Ribbons. He makes his home in Carmichael, CA.

ARLETT G. MOSIER, (Art), born 1924 in Modesto, CA, entered USAAF in 1942 taking primary at Ontario, CA, basic at Merced, CA, and advanced at William Field AZ, graduated Class 44-A and was assigned to the 479th Ftr. Gp., Wattisham, England. Flew half of his tour in P-38s and last half in P-51s from D-Day through VE-Day. He was one of the first group of 12 pilots to deliver P-51 Mustangs to Stockholm Sweden. He exited service on return to Conus but was recalled for Korean Conflict.

He instructed advanced student pilots in B-25 aircraft before attending technical training for a second Aero rating as Electronic Counter-Measures Observer. He organized and set into operation the first ECM Reconnaissance Operation ever used in TAC. He completed his combat tour and returned to Conus as Pilot Instructor at the instrument Pilot Instructor School flying both jet and conventional aircraft.

In 1956 he entered Nuclear Weapons Training followed by duty in 6th Air Div., SAC, as Chief of Munitions. He returned to Combat Crew Duty in 1959 as a B-52 Aircraft Commander to fly missions on a worldwide basis including SAC Airborne Alert until his retirement in 1968. Retired with rating of command pilot, rank of Lieutenant Colonel, and a total of 20 and one half years active service. He holds a commercial pilot license with instrument rating.

After retirement from USAF he served with the California Highway Patrol in communications. After 14 years with CHP, he again retired in 1983. He presently resides in Fresno, CA and flies his Piper Cherokee 180 for pleasure.

JACK W. MULLAN, was born Sept. 17, 1924 in Ft. Dodge, IA. He grew up and went to school in the Los Angeles area. Graduated class 1944 H from La Junta, CO flying B-25s as an advanced

trainer. Graduated top of class in Photo Reconnaissance School flying P-38 F5s Photo Recon. at Coffeyville, KS.

Joined 26th Photo Sdn., 308 Bomb Wing in the Philippines. Flew 19 combat missions, including flights over Formosa and China. After end of war he occupied Korea, flew out of Kimpo Airbase, Photo Mapping South Korea. After dissolution of 26th Photo Sdn., he joined 475th Fighter Gp. in Korea. Flew training missions and patrolled 38th Parallel, flying P-38s. After war, completed education at U.S.C. Flew for two years with United Airlines. Managed aviation business in Frankfurt, Germany with a Reserve assignment as Assistant Base Flying Safety Officer at Rhein/Main, Germany. Currently a Real Estate Developer in Newport Beach, CA. Founding Director of P-38 National Association. He says "Will keep flying as long as can pass physical."

JOHN MULLINS, was born in Avon, MA. He enlisted in the Aviation Cadet program directly from high school in 1942. He graduated in Class 43-K from Luke Field, AZ. Following both P-39/P-63 training (Santa Rosa) then P-38 training (Van Nuys) he joined the 94th (Hat-in-the-Ring). Fighter Sdn., 1st Fighter Gp. in Salsola (Foggia #3) Italy in June 1944.

Mullins completed 53 missions including beachhead patrols for the Southern France invasion, divebombing, strafing and heavy bomber escort. He was awarded the Distinguished Flying Cross and Air Medal (four Oak Leaf Clusters). He was discharged from the Air Corps in 1945 at age 21 with rank of Captain.

Mullins, together with wife Phyllis went on to earn the Ph.D. degree in pharmaceutical chemistry. He recenlty retired from Alcon Laboratries R&D Division as Director, Analytical Chemistry. He and his wife Phyllis now reside in Kerrville, TX where he is compiling an history of the 94th Sdn. during the war years.

PAUL C. MURPHEY JR., was born Nov. 11, 1920 in Waco, TX. He attended Kemper Military School Booneville, MO and then the University of Texas until the summer of 1941.

He entered the service in July 1941 and was sent to aircraft armament school, the assignment to aviation cadets came in the spring of 1942. He graduated with Class 43D at Williams Field Chandler, AZ. He went immediately to P-

38 training at Muroc, CA. Then to Santa Ana, CA for final training.

He left for overseas duty in June 1943 arriving at Brisbane, Aus. with 11 other P-38 pilots. Six were assigned to the 80th and four of this group were destined to become aces. The 80th (Headhunters) were at Kila Airstrips (three miles) near Port Moresby. They made all the missions to Wewak and Rabaul from there (all points between). The 80th was moved often Dobodura, Cape Gloucester, Nadzab, Wadke and Owi Islands, Morotai, Leyte and finally Mindoro, P.I. His tour ended there after nearly two years. At Owi Island the 80th met and flew with Col. Lindberg.

In September 1944 he was part of a flight of four from the 80th joined by flights from the other P-38 Sdn. to fly a fighter sweep to Davao, P.I. This marked the first appearance of our fighter planes since the Philippines were lost to the Japanese in 1942.

Capt. Murphey became an official ace on Dec. 20, 1944 downing two Zeros and damaging two others while leading a patrol over Mindoro. He was credited with six Zero's and five probables.

He went into ranching, raising reg. Hereford cattle. He and his wife have two children and five grandchildren and still live on their ranch near Meridian, TX.

DFC with Cluster Air Medal with six clusters, Presidential Unit Citation, with Cluster, Asiatic Pacific Medal with four Battle Stars Philippine Liberation with one Battle Star.

KEITH E. MURPHY, was born in Shenandoah, IA, Sept. 26, 1923. He enlisted in the USAAF in December, 1942 as an Aviation Cadet, but was sent to Coe College, Cedar Rapids, IA for Cadet preparatory schooling. While there he met his future, and still present, wife Lois. After CTD he was sent to West Coast Training Command for flight training and graduated with Class 44-C at Williams Field, Chandler, AZ.

He flew AT-9s and P-322s in Advanced Training and had an emergency bail-out due to engine fire. He then went to Will Rogers Field, Oklahoma City for Overseas Training and then on to the CBI Theater and was assigned to General Chenault's 14th AF "Flying Tigers" in China. He was awarded the DFC and Air Medal and after 75 missions, was returned to the states and discharged with the rank of Captain in 1946. He is currently the manager of a Credit Union in Cedar Rapids, IA.

ALBERT J. MURRAY, was born in Minnesota and raised in Florida. He joined the Army Air Corps in 1941 and was discharged in 1945. He served in Australia and New Guinea with the 49th Fighter Gp. and the 5th Fighter Command as a mechanic and crew chief. He worked on P-38s, P-39s, P-40s and P-47s, but says the P-38 was his favorite. He says he can still heard those two Allison engines purring. It was a beautiful airplane to pre flight.

After the war he took up flying under the G.I. Bill; he is a licensed A-P mechanic and flies about twice a month. He and his wife, Martha, live on a lake in Central Florida. Al retired from the Post Office in 1972.

Al remembers that sometime in 1944 Maj. Borg and a Col. Lynch had their P-38s based temporarily at Nadzab in New Guinea with the 5th Fighter Command. They flew together on a mission to Hollandia, but only Maj. Borg returned. As he was getting out of the cockpit he had tears in his eyes. He said, "My friend got shot down."

CLAUDE C. MURRAY, JR., P-38 Pilot and Evader, born Dec. 30, 1921 Spokane, WA. Husband of Shirley, father of Jeffrey, Steven and Thomas. Enlisted USAAC November 1942, graduated Class 1943-J, Williams Field, AZ.

His combat assignment was to the 8th Air Force, 7th Photo Recon. Gp., Mount Farm, England August 1944 as a replacement pilot. After three successful missions, he was "hit" by an ME-262 on Oct. 6, 1944 while on a lone trip to Hamm, Germany. He learned the hard way how to bail out of a Lockheed Lightning via the "split-ess" maneuver and falling out. His good luck began with a 13-hours experience in a dinghy in the cold Zuider Zee, The Netherlands. His luck continued when he was picked out of the sea by young fisherman, Jan Dobber, and two companions, who proceeded to take him ashore and ultimately to contact with the Dutch Resistance Leader in the village of Muiden, Joh. Rozendaal. The Resistance received orders that Murray was to remain there and not to attempt escape. Although unknown to him at the time, hundreds of other airmen were also evading the Nazi while posing as deaf and dumb Hollanders. Liberated on May 5, 1945, he returned to his base in England via Camp Lucky Strike and a C-47.

Post-war he was a director of the 8th Air Force Historical Society, director of the Air Forces Escape & Evasion Society, director and secretary of the P-38 National Association. In 1979 he founded the 7th Photo Recon. Group Association and for eight years he served the 7PRG as Treasurer, Corresponding Secretary, Computer Manager, Newsletter Editor, PX Mangaer and advisor to the association presidents. He became widely known among members of 8AFHS, AFEES and the 7PRG as a "gadfly" and for his many innovations.

H.C. MUSGROVE (MOOSE), was born in Homer, LA. He joined the Army Air Corp. in January of 1942. He received his training in the Southeast Training Command and graduated from Spence Field located in Moultrie, GA, Class of 43-D. He received training in P-51s and P-40s at Tampa, FL. He was assigned to the 15th Air Force in North Africa and received ten hours training in P-38s before being assigned to the 49th Fighter Sdn., 14th Fighter Gp. After flying 52 combat missions, he was sent back to the states for reassignment. He was assigned to Muroc Air Base which is now known as Edward Air Force Base.

He left the service from Santa Rosa Air Base to civilian life in 1945.

ROBERT E. NAISMITH, was born May 19, 1918 in Los Angeles, CA. He enlisted as a Flying Cadet in November, 1940, graduating from Brooks Field, TX in the Class of 41-E. He was immediately assigned to the 71st Sdn., 1st Pursuit Gp.

On Dec. 6, 1942, when Naismith (still with the 71st) parted company with P-38F #417497 she was settling toward the bottom of the Mediterranean and her driver was paddling toward shore. She had served him well; he had served her badly.

He, like many of his colleagues, has a genuine appreciation for the heroic qualities of stability, dependability and willingness to absorb punishment in bringing their pilots home which the 38s demonstrated so consistently. His toast to old 417497: May she and her loyal sisters forever rest in peace — and their memories never die.

ALVIN ROBERT NEAL, born Dec. 4, 1921, Buffalo, NY; died Dec. 12, 1975. Many young men were eager to become aviation cadets during WWII, and the waiting list was long. Impatiently, Al enlisted in the Army Air Corps in July 1942 and was assigned for pilot training at the University of Rochester (NY). In September of 1942 this group was re-assigned to the glider program in Roswell, NM. This program was soon discontinued and, as the fighting increased, so did the need for pilot replacement. So, it was on to Eagle, Lemoore & Luke Fields-graduating 43-K. Then waiting on the west coast—McChord, Salinas, Santa Rosa, Hamilton, Ontario & Chico—with specialized P-38 familiarizing. Finally, to the Pacific in July 1944.

Al flew over 100 combat missions with the 475th Fighter Gp., 431st Sdn. All but two of these missions were in P-38s. CAPT. NEAL was commended for his leadership as both flight commander and operations officer of the 431st. Many times he flew the wing of Maj. Thomas McQuire. NEAL had four enemy aircraft to his credit.

Campaigns were: China Defensive, New Guinea, Bismarck, Archipelago, Luzon, West-

ern Pacific, Southern Philippines, China Offensive, Philippine Islands.

Decorations and Citations: Air Medal with Three Oak Leaf Clusters, American Campaign Medal, Asiatic Pacific Campaign Medal, World War II Victory Medal, Philippine Liberation with Bronze Star, Distinguished Flying Cross.

Left the Pacific for home November 1945 (Discharged Feb. 21, 1946).

Al grew up on a farm in Cattaraugus County, graduated in 1940, and attended University of Buffalo Post WWII. For 26 years he was in retail management (also a "trouble-shooter") at several points in Western and Central New York and Northern Pennsylvania, for Agway, Inc. (the worlds largest Agricultural Co-op).

In his last years, he flew gliders as a member of both Dansville and Rochester Soaring Clubs. (NY)

Neal's wife Eileen, a retired R.N., resides in Hilton, NY. His daughter Diane (Mrs. Richard LaValley) and granddaughter, Rachel, live in Rochester, NY.

MARY LOU COLBERT NEALE, was born in Juneau, AK and lived in New York City, Washington, D.C. and Manila, P.I. during her early years, as her father was an officer with the Coast and Geodetic Survey. She graduated from Wellesley College, worked in the Library of Congress, and took the CPT course at George Washington University. She continued flying to obtain her Commercial license and was accepted in the first class of the Army Air Corps Women's Flying Training Detachment, graduating in April, 1943.

She was stationed at Love Field, Dallas, TX from April to August 1943, Palm Springs Air Base, CA from August to October, Long Beach Air Base from October to December. Then it was back to Love Field until March and a return to Palm Springs for permanent duty until deactivation of the WASP program on Dec. 20, 1944. During this time she ferried many types from primary trainers to four-engine bombers, the majority of the time delivering pursuits. Of these, the P-38 was the superstar in her opinion.

ROBERT O. NEWELL, attended Eden High School in Greenfield, IN, Ball State University in Muncie, IN, Purdue University, in W. Lafayette, IN, the University of California, Berkley, CA, the University of California at Los Angeles, CA, University of Liverpool Faculty of Medicine, England - M.D., Spartan School of Aeronautics, OK and Randolph Field, TX.

He enlisted at Ft. Benjamin Harrison, Indianapolis, IN on Nov. 21, 1940 in the regular Army of the United States. Flying cadet, Air Corpus Spartan School Aero, Muskogee, OK (three years).

Elementary training Air Corps Cadets, Muskogee, OK Nov. 21, 1940 to Feb. 8, 1941. Basic training, Randolph Field, TX Feb. 13, 1941 to Apr. 28, 1941. Advanced training, Brooks Field, TX Apr. 28, 1941 to July 11, 1941, Class of 41-E.

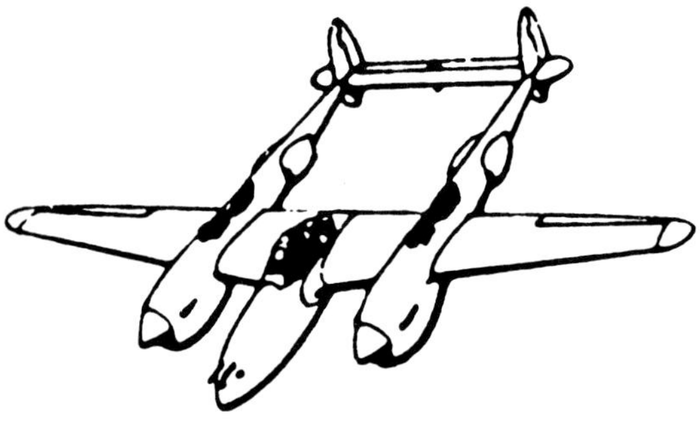

Active duty, Air Corp July 12, 1941, separated Nov. 18, 1945. Enlisted Nov. 21, 1940 Air Corps. Discharged July 11, 1941; Comm Sv EAD July 12, 1941; Reld fr AD Nov. 18, 1945; notified of final discharge Sept. 28, 1957 as a Major.

In the European Middle Eastern Campaign made 52 combat missions accruing 158 combat hours and a total of 1,400 flying hours. Received the following decorations and medals: American Defense Service Medal, American Campaign Medal, European-African Middle Eastern Campaign Medal, World War II Victory Medal, Armed Forces Reserve Medal, Air Medal, GO 46 Hq. 8th AF, Air Medal 9th oak Leaf Cluster and Silver Leaf Cluster GO 17, Distinguished Flying Cross #1722, GO 28 Hd 12th AF, Three Bronze Stars to European-African Ribbon, GO 75, WD, and recommended for Distinguished Service Cross Mar. 19, 1943 and one Cluster to Distinguished Flying Cross.

Have seven confirmed victories against German aircraft.

He is a member of the Disabled American Veterans; National Order of Trench Rats, DAV, Colonel Confederate Air Force; and The American Fighter Aces Association Air Force.

He has two children: Ronald Harrison Roberts, 29 years old, and Rosemary Carolyn Roberts (deceased Apr. 9, 1981).

ROBERT D. NEWMAN, was born July 18, 1923 in Kansas City, MO.

He enlisted in the Army Air Corps in 1942 and graduated with Class 44-F at Luke Field.

Following graduation Newman first flew P-39s at Victorville, CA then transitioned to P-38s at Santa Maria, CA.

Newman was assigned the 12th Ftr. Sdn., 18th Ftr. Gp. in the Philippines till the war ended. Along with several other squadron mates Newman spent a brief period flying for Far Eastern Air Transport out of Manila, P.I. Shortly thereafter he returned home and completed his college education.

After college Newman entered business in the paint and hardware field and currently is Chairman of several firms engaged in the manufacture of tools related to the paint/hardware and sanitary supply fields.

Newman is married and has six children and three grandchildren. He resides in the Kansas City area at Lake Winnebago, MO.

FRANKLIN A. NICHOLS, was born in Holdenville, OK, on Apr. 18, 1918. Entered the Flying Cadet program in September 1940, commissioned Second Lieutenant, and pilot in April, 1941.

First assignment was to Wheeler Field, HI, as a fighter pilot flying P-40s. In August, 1942, was transferred to the 7th Pursuit Sdn., flying P-40s in New Guinea. On June 1, 1943, he activated and commanded the 431st Pursuit Sdn. in Australia, flying P-38s, and three months later, combat ready, returned to New Guinea. By the end of the war the Squadron, with only two years of combat, had the second best combat record of pursuit squadrons within 5th Air Force, with 217 aerial victories.

He flew 185 combat missions and 420 combat hours with five confirmed, and four probable aerial victories, and one Japanese troop ship destroyed. Decorations include: Distinguished Service Medal with Cluster, Silver Star, Legion of Merit with Cluster, Distinguished Flying Cross with three clusters, and Air Medal with two clusters.

He retired with 30 years service as a Major General, United States Air Force.

A. RAY NISH, was born and educated in San Bernadino, CA. He left the University of California at Berkely following Pearl Harbor and was an Associated Press reporter in San Francisco until enlisting as a USAAC aviation cadet in October 1942.

He was assigned to the first College Training Detachment unit at Washington State College before entering preflight and flight schools. After getting his pilot's wings at Williams Field, AZ, he flew P-39s at Victorville, CA, and Moses Lake, WA, then took P-38 transition at Ontario, CA, before going to the Southwest Pacific.

He joined the 68th Lightning Lancers Fighter Sdn. of the 347th Fighter Gp. of the 13th Air Force Jungle Fighters on Middleburg Island, New Guinea. From there, the 68th moved on to Morotai in the Halmaheras, then to Puerto Princessa on Palawan Island in the Philippines. Missions were flown to targets in New Guinea, the Philippines, the Celebes, Borneo and French Indo China.

Nish, a flight leader, flew 62 missions, some of them 8.5 hours long. The most memorable was being one of five P-38 pilots on the first fighter sweep on Saigon against Japanese shipping and fighter and seaplane bases.

Nish and his wife Lois, married 46 years, have two sons, Steven and Richard; a daughter, Kathleen Jenner, and four grandchildren. He retired in 1985 after 41 years as a newspaper reporter and editor, the last 23 years as managing editor of The Modesto Bee. The Nishes live in Modesto, CA, and enjoy woodworking, gardening, golf and traveling.

CYRIL L. NOLEN, was born in Minnesota in 1922. He joined the Minnesota National Guard in 1940. In February 1941 they were activated and sent to California. He went to Kodiak, AK as a medic with the 215th C.A.A.A. in June 1941. After Pearl Harbor he volunteered for pilot duty and graduated in the Class of 43-E.

In August he went to Africa as a P-40 pilot, but due to casualties in the 1st Fighter Gp., was assigned there and flew 50 missions as a P-38 pilot. On Apr. 20, 1944 he was shot down while strafing. He claims three probables. He was a P.O.W. for 11 months. In March 1945, on the second escape, he and Dick Hart made it back to the advancing American Army.

He married Rose Fenker in 1945. They have two married children and six granddaughters.

Cyril has worked as a welder, carpenter, E.M.T. and 28 years for T.W.A. as a flight engineer and pilot. In 1967 and 1968, on leave from T.W.A., he was an Army pilot in Vietnam.

LOUIS CARL NUEHRING, was born at Lytton, IA on Nov. 23, 1920, the youngest child of Carl and Frieda Nuehring. He enlisted in the USAAC on Sept. 27, 1942 and completed single engine flight training at Williams Field Chandler, AZ on May 23, 1944. He received P-39 transition training at Victorville, CA in June, 1944 and completed tactical training in P-35s in April, 1945 at Chico, CA. He was ferried to India and was assigned to the 459th Ftr. Sdn. on May 19, 1945. He was returned to the States on Dec. 30, 1945 and separated from active service on Jan. 28, 1946. He is now retired after 35 years with the Iowa State Conservation Commission and resides in the little town of Otterville near Independence, IA.

I enlisted in the Army Air Corps and applied for Air Crew training with dreams of flying P-38s. On arriving at Preflight at Santa Ana, CA in August , 1943 and was assigned to Sdn. "13". On Friday Aug. 13, 1943 we were marched to the 13th building on 13th St., and given 13 tests. The dream was realized.

JAMES E. OBERMILLER, Lt. Col. USAF (Ret), was born June 30, 1921 in rural Allen County, near Ft. Wayne, IN. He grew up in the immediate area and graduated from High School in Columbia City, IN in 1939. The following year he joined the U.S. Army Corps was assigned to the Army's pilot training program, graduating in March 1942 from the Air Corps Gulf Coast Training Center in San Antonio, TX with the rank of Staff Sergeant Pilot. He subsequently completed transition training in P-38 aircraft and was commissioned a Second Lieutenant and ordered to the South Pacific in September 1943. He was first assigned to duty with the 67th Fighter Sdn. in New Caledonia and later with the 339th Fighter Sdn., 347th Fighter Gp. on Guadalcanal.

Colonel Obermiller's military service included wartime assignments in World War II and the Korean War with post-war service in the Arctic. Sixteen months of overseas service in the World War II included duty as a P-38 "Lightning" fighter pilot with the 339th Fighter Sdn. noted for the intercept mission in which Japanese Admiral Isoroku Yamamato was shot down over Bougainville Island. Earlier, in November 1942, on a long range escort mission to Bougainville, Obermiller was credited with one of the first victories in aerial combat involving the P-38 and the Japanese Zero Aircraft.

Prior to retirement from the U.S. Air Force in 1961, he was associated with early implementation of the military ballistic missile program and was the Air Force Project Officer for the Titan I underground launch facilities. Military decorations include the Air Medal with two Oak Leaf Clusters; the Air Force Commendation Medal with Oak Leaf Cluster; the Army Commendation Medal and the Navy Presidential Citation with star.

Since 1961 Col. Obermiller has been associated with a Defense Contractor headquartered in Lexington, KY serving as Assistant to the President and where he now resides with his wife Evelyn

GEORGE OLSEN, was born Feb. 17, 1913 in rural North Dakota the son of John C. and Lena M. (Norby) Olson.

The 48th Pursuit Sdn., March Field, CA was initial assignment May 1941 as selectee from Ft. Snelling, MN then Atcham Field, England August 1942 as Operations Sergeant prior to North Africa invasion November 1942. A C-47 ditching delayed his arrival at Maison Blanche, Algiers, several days. The 48th flew P-38 combat missions off bases from Casablanca to Tobruk. Moved to Foggia area Italy for the

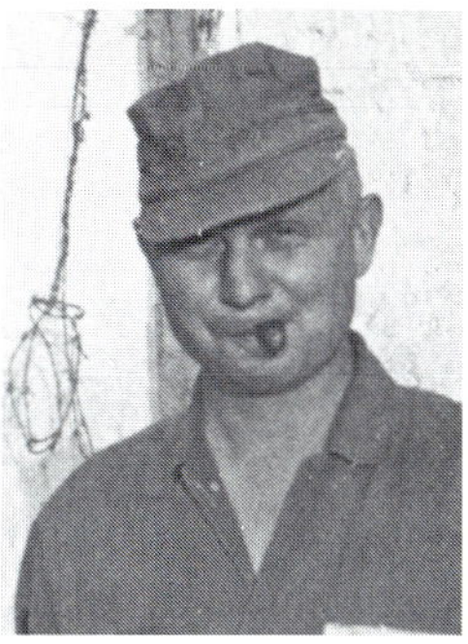

duration of WWII flying missions throughout ETO earning 14 Battle Stars and DUB.

Discharged Camp McCoy, WI September 1945 as Tech Sergeant. Resumed civilian employment until recalled EAD October 1950. Duty as 1st Sergeant 769th AC&W Sdn., Continental Divide, NM followed by tours in USAirA Office, Cairo, Egypt; Maxwell AFB, AL; Travis AFB, CA and RTAFB, Korat, Thailand retiring November 1966 as SMSergeant.

MAURICE OLSON, (Ole), born and educated in South Dakota, enlisted USAF at Ft. Snelling, MI in July of 1942. He received pilot training in the Southeast Training Command flight schools of Bennetsville, SC, Bush Field, Augusta, GA, in the graduating class of 4AA Moody Field, Valdosta, GA. Transitioned in B-24s at Charleston Air Force Base.

Olson completed 14 missions with the 376th Bomb Gp. losing his crew to P.O.W. in Sofia, Bulgaria. He transferred to the 82nd Fighter Gp. in September of 1944 completing a combat tour of 36 missions in P-38s. Twenty-six were escorting bombers to Germany, Hungary, Austria, and Czechoslovakia. Ten were strafing, dive bombing targets in Yugoslavia, Austria and Italy.

After completing combat duty, First Lt. Olson served as R.T.U. instructor in P-38s at Chico, CA Air Force Base until discharge in August 1945.

Olson resides in Everett, WA where he has raised three sons and two daughters. The entire family is involved in the retail food business.

RICHARD OSTRONIK, enlisted in the Army Air Force in late 1942 with the hopes of being an airplane mechanic. The Air Force, true to form, placed him in a teletype maintenance school at Chanute Field, IL. A friend, now forgotten, talked him into applying for cadet training. Sadly, the forgotten friend flunked out and Richard went on to Nashville, TN for placement; Montgomery, AL for pre-flight; Albany, GA for primary; Macon, GA for basic; and Spence Field, Moultry, GA for advance. He was made an instructor, but soon, with some luck, was assigned to the West Coast for P-39s as introduction into P-38s. Back across the USA for a slow boat to Naples, Italy. He arrived at the 82nd Gp. 96th Sdn. in the hot summer of 1944 at Foggia, Italy.

Until the close of the war, he flew 52 missions. He flew squadron and group leads, earning the DFC for a group lead over Yugoslavia. Has Air Medal with Clusters as well as some Battle Stars.

Returned home in June 1945, leaving the service. He then served 28 years with the Alhambra Fire Dept. He presently is retired living in San Gabriel, CA with his wife, Sylvia.

MARION P. OWENS, was born on Apr. 30, 1921 in East Point, GA. Graduate of Russell High School 1938. Attended Georgia Tech (night school) studying electrical engineering for nine months. (This cut short by Pearl Harbor).

On Apr. 10, 1942 became an aviation cadet, Avon Park, FL in PT-17 Stearman. Jan. 10, 1942 aviation cadet, Army Air Corps, Maxwell Field, AL; Ground School and P.E. Gunnery, Avon Park, FL, PT-17; Basic, Greenwood, MS, BT-13; Advanced, Craig Field, Selma, AL AT-6, P-39, P-40. Graduated Oct. 2, 1942 and instructed there from Oct. 15, 1942 thru Feb. 10, 1944 in AT-6, P-39, P-40 and P-36. March 1944, Moses Lake Air Base, P-39 Fighter Training, gunnery, dive bombing, strafing, formation and night flying. May 1944, Ontario, CA Air Base, P-38 training (same as above).

Aug. 15, 1944, 370th Fighter Gp., 471st Fighter Sdn., 95th AF, E.T.O.

Armed Recce, dive bombing, strafing, low level Napalm; 56 missions P-38s and P-51. Three missions same as above and nine Med. Alt. Escort B-26. Total missions 68.

In P-38: four confirmed kills; three ME-109, one ME-110. Also one probable and one damaged confirmed, both ME-109s. In P-51: one confirmed kill, ME-262 JRT.

Awards: E.T.O. Campaign w/three Stars, Distinguished Service Cross, Air Medal w/11 Clusters, British Distinguished Flying Cross, Presidential Unit Citation, American Campaign Medal and Army of Occupation (Germany) Medal.

Dec. 23, 1946, USAAF, Major.

January 1947 to August 1949, Western Electric Co., Atlanta, GA. Opened Owens Sporting Goods October 1949 in Albany, GA. Also Dougherty Realty Co. in Albany , GA September 1963. Have served as President of both since opening. Still active in these businesses.

Married Marjorie Block Jan. 3, 1946; celebrated our 41st Anniversary Jan. 3, 1987. Both the P-38 and P-51 (named "My Bama Baby" for her. Four children—two boys, Ronnie and Larry; two girls, Charlotte and Linda. Ten grandchildren—nine boys, one girl.

Nickname was "Snuffy" while with the 401st.

KENNETH PAARLBERG, was born in Kenneth, MN, May 18, 1916. The family moved to a farm near Chicago Heights, IL in 1920, where he was raised and farmed in partnership with his father until WWII called him to active duty.

He was assigned to the Air Force, and joined the 475th Fighter Gp. January 1944 in Nadzab, New Guinea. He was a statistician for the 433rd Fighter Sdn. keeping records of all the planes in the squadron - plane hours, engine hours, missions, pilot hours, fuel consumption - whatever required the recording of data.

He completed the New Guinea campaign, the Philippine campaign, and went to Korea with the organization as occupation troops spending two years with the group overseas.

He vividly remembers the Leyte invasion and all the air activity involved in re-taking the country from the Japanese. The p-38s were very effective in the success of this campaign, and they out-flew the Zeros as they flew cover missions for the bombers, and did bombing and napalm bombing in support of the ground troops. The P-38 was superior to the Zero in air combat and over 500 Zeroes were destroyed by the 475th Fighter Gp. alone during the war in the Pacific.

CHET PATTERSON, had never been to an airport or walked around an airplane. Yet, thinking a war was in the making he joined the Aviation Cadet Program while attending University of California at Berkeley in mid 1941. By the time he had completed his primary training at Tex Rankins Aviation School he loved flying and had the makings of a born fighter pilot, (he almost got expelled for stunt flying way under school height limits before his girlfriend for the graduation ball).

WorldWar II began while he was in Aviation Cadet School, and upon graduation he was assigned to the 55th Fighter Gp. The group was just being equipped with P-38s and Patterson felt he was in "fighter heaven". The Group was sent to England as the first long-range fighters operational against the Germans.

Patterson shot down four enemy aircraft, was the 338th Sdn's Flight Commander when the 55th became the first American aircraft to fly over Berlin, and completed his first tour before D-Day. Feeling D-Day was just around the corner he volunteered for a 25-hour extension of his tour. He completed this in three days during the Invasion. Because of the P-38s unique configuration the 55th Fighter Gp. was assigned the beach heads for simple identification by the Allies. No German aircraft came close to strafing the Invasion Forces.

Patterson started his second tour after a 30 day R&R in the States. The 55th F.G. had been switched to P-51s by this time, and after eight more missions Patterson was assigned to Headquarters 8th Air Force working directly under Gen. Doolittle.

Patterson's awards include DFC with three clusters, Air Medal with seven clusters, Bronze Star, French Croix de Guerre and Belgian Award of Merit.

MERLE D. PEARSON, was born Jan. 4, 1920 in Oregon City, OR, he grew up in Sidney, NE. February 1942 he enlisted in Parachute Inf., completed jump training at Ft. Benning, GA. Assigned to 82nd Airborne Div. Ft. Bragg, NC. November 1942, applied for Aviation Cadet training was accepted. Requested twin engine fighters. Graduated from Williams Air Force Base (Chandler, AZ) Class of 43I. Was sent to South Pacific to join the 5th Air Force and 475th Fighter Gp. to fly P-38s. First stop was Port Moresby, New Guinea. My squadron commander was Thomas B. McGuire. While in New Guinea was privileged to meet and talk to Charles A. Lindberg. Our group moved up the Island Chain to Clark Field. Flew 115 combat missions with over 400 combat hours. Tour of duty was completed in April 1945.

Decorations include Air Medal with three Oak Leaf Clusters. Attained the rank of 1st Lieutenant. He retired from Sears Roebuck in January 1983 after 35 years with that Co. Merle and his wife Pauline live in Salinas, CA, they have two daughters, Dayna Barron and Kris Pearson, their son Merle Jr. died in 1965.

ARTHUR CAMPBELL PERKS, was born in Port Arthur, Ontario, Canada on May 4, 1922. He moved to Rockford, IL on or about the fall of 1929. He graduated from Rockford High School in 1940. Completed two years at Colorado School of Mining, before enlisting in U.S. Army Air Corp in July 1942. He received basic training at Santa Anna, CA, primary flight training at Rankin Field, Tulare, CA, primary flight training at Rankin Field, Tulare, CA, basic flight training at Chico, CA, graduated from advanced training at Luke Field, Phoenix, AZ and was commissioned 2nd Lieutenant in U.S. Army Air Force in September 1943. He then went to basic instructors school at Randolph Field, San Antonio, TX and upon completion was assigned as instructor at Chico, CA. He was transferred in April 1944 to Moses Lake, WA for P-39 transition, then to 443rd AAF Ontario, CA for P-38 transition. Upon completion of transition training, Cam was assigned to the 459th Fighter Sdn. in November 1944 which was then based at Chittagong, India. Flew combat missions until September 1945 and then he returned to the United States in November 1945. He remained in the Air Corp Reserve until December 1946 when he received a honorable discharge as a 1st Lieutenant. While in foreign service he received the Air Medal with Oak Leaf Cluster and the Distinguished Flying Cross.

He married Joanne Lundholm on Aug. 2, 1947. They had four children, two girls and two boys. He has been employed by Eclipse Inc., Rockford, IL since 1947 and presently holds the position of Chairman of the Board of that company. His present home address in 6939 Severson Road, Rockford, IL 61102.

I. NEWTON PERRY, (Newt), born Chicago, enlisted 1942 Cadet Training while in college, New Haven, CT. Earned Wings at Luke Field, - then to Randolph Field for Instrument Instructing, Chico, CA instructor - to P-39s Ephrata, WA. Then to Van Nuys, CA and P-38 Training near Santa Barbara, CA where he has lived since 1946 .

Shipped to Italy - (God - 30 days and night on a Liberty ship, four bunks high and it was "high" otherwise. But, it was worth it because "we" - (that includes my life-long buddy Greg Milne, who is the No. 1 Charter Member of the P-38 National Assoc.) ended up at 15th Air Force Headquarters, Bari-Italy flying 38s in the 154 Weather Recon. Sdn. This was the most enviable job of the war - flying ALONE - in a "no-torque Cadillac", arriving over Germany Target areas arriving at dawn to report weather conditions to Bomber Command, thus saving aborted missions. Now—on the way home we could go sight-seeing the Austrian Alps - Venice-wherever....so, alot of missions, the D.F.C. for both of us, (alot of buddies didn't come home but they earned them too). We returned to the U.S.-signed for another tour in the Pacific but the War ended, Thank God. What an experience.

CLIFF PETERSON, was born and lived in Oakland, CA. After high school he entered the University of California at Berkeley and spent two years there before joining the cadet train-

ing of the Air Force in the Spring of 1944. He trained in and around California, receiving his wings as a twin engine fighter pilot.

Cliff was assigned to the 9th Fighter Sdn., 49th Fighter Gp., in the Philippines. In April 1945, Cliff went on a long (seven hour) but routine mission north of his base. On the return trip, about 50 miles from his base, Cliff developed a fuel problem in both engines, turning his fighter into a glider. Still ten miles at sea, he was able to glide from 9000 feet to about 2500 feet over an island beach and bailed out. He landed hard but safely, and was immediately surrounded by farmers who were working with the guerrillas. That night he stayed at the guerilla headquarters just back of the line where the guerrillas had a Japanese General surrounded on a nearby hill. Americans were fairly new in that part of the island; however, he was able to radio his base. The next day they sent a plane up to a small strip nearby and his adventure was over.

After the war, Cliff finished his education at UCLA, and later was president of this own firm in the LA area.

ROBERT R. PHIPPS, JR., LTC USAF (RET). Born Nov. 7, 1920 Annapolis, MD. Spent his youth in Forty-Fort, PA where he graduated from high school in 1939.

Enlisted in the Army Air Corps Mar. 4, 1942. After basic training, assigned to Hickam Field, HA as first replacements after Pearl Harbor. Obtained the rank of Corporal.

Returned stateside in 1943 as an Aviation Cadet. Graduated Class 44-A Williams Field, AZ.

Tour of duty in ETO flying P-38s and P-51s, out of Kings Cliffe, England. Flew shuttle run to Piryatin, Russia, Sept. 19, 1944. Completed 56 missions — 270 combat hours. Awarded DFC and Air Medal — four clusters.

Stateside December 1944. Married Kathleen Wintermute, Forty-Fort, PA, Sept. 8, 1945 — one daughter. Separated from service Oct. 31, 1946 with rank of Major. Active member of the Air Force Reserve until retirement in 1970 with the rank of Lieutenant Colonel.

Retired from Chrysler Corp., Production Control Department, Apr. 30, 1980.

JAMES M. POOLE, JR., born July 1916 Statesville, NC took Preflight at Santa Ana, Primary, Ryan Field, AZ. Basic at Marana, AZ. Graduated Williams Field, AZ 43D. To Lockheed Engr. Burbank, CA. Assigned Photo Reconnaissance, Peterson Field, CO August 1943. Group formed, overseas in March 1944 with 10th Photo Gp., 31st P.R. Sdn. in England. He completed 70 combat missions in 13 months. Flew many Invasion related missions. Killed August 1948, B17- crash. Buried Arlington.

His awards included the Distinguished Flying Cross w/cluster Air Medal w/12 clusters Northern France Star, Ardennes Star Southern France, Germany Star, WWII Victory Medal, American Campaign Medal, Theater Ribbon w/ Air Offensive Europe Star 31st Sdn. issued The Presidential Citation.

FRANCIS JOSEPH POPE, born Oct. 17, 1916 in San Francisco, CA. His military career began in 1935 when he enlisted in the Naval Reserve Sdn. at Oakland Airport, CA.

He graduated from San Francisco State College in June 1940 with a Bachelor of Arts Degree. His major was economics, with a math and science minor. He enlisted as a flying cadet, receiving primary training at Lindbergh Field, San Diego; basic training at Randolph AFB, TX; and advance training at Kelly AFB, TX. He graduated as a 2nd Lieutenant Feb. 8, 1941.

He was assigned to the 35th Pursuit Gp., 34th Pursuit Sdn., at Hamilton Field, CA. He commanded the 54th Fighter Sdn. in the Aleutians, where he flew 75 combat missions. He was transferred to Italy in 1944 and commanded the 27th Fighter Sdn., 1st Fighter Gp., flying out of Foggia, Italy, where he amassed a total of 105 missions. 400 Combat Hours.

After the war he served in Santa Rosa, CA, and closed the base as Wing Commander. He was the Operations Officer for Hamilton Field. In 1948 he attended Command and Staff School at Montgomery, AL. He was then assigned to March AFB as Operations Officer of the 1st Fighter Wing.

He attended the Academy of Instruction at Selma, AL, in July 1949. He was then assigned as PAST, Professor of Air Science and Tactics, at the University of Nevada for one year and Loyola of Los Angeles for three years.

In September 1953 he was assigned to the 374th Troop Carrier Wing in Tachikawa, Japan, as Deputy Wing Commander. He returned to the States in July 1956 and attended the Naval War College at Newport, Rhode Island. In June 1957 he was assigned to the Secretary of the Air Force Office of Information as Director of Community Relations. In March 1959 he was assigned as Special Assistant to the Deputy Chief of Staff of Operations, United States Air Force. In August 1960 Colonel Pope departed for WPAFB, where he was assigned as Director of Recruiting Operations.

The Colonel is a command pilot and has logged more than 4,500 hours military flying in all types of aircraft.

Col. Pope has been awarded the Distinguished Flying Cross, Air Medal, American Defense Service Medal, American Campaign Medal, Asiatic-Pacific Campaign Medal w/ three Bronze Stars, European-African-Middle Eastern Campaign Medal w/one Silver Star and four Bronze Stars, World War II Victory Medal, National Defense Service Medal, Korean Service Medal, Air Force Longevity Service and Award Ribbon w/four Oak Leaf Clusters, French Croix De Guerre/with Palm, Thailand's Most Exalted Order of the White Elephant (Third Class), and the United Nations Service Medal.

He is married to the former Frances Ann Bush of Burlingame, CA. The Popes have four children: Andrea Marie, (Mrs. (Capt.) Lino F. DaMichieli, who is now in Saigon), Jan Carol, 21, Francis Joseph, Jr., 13, and William Francis, 9.

ALBERT STANLEY PRAZAK, was born in Caldwell, TX May 7, 1915 to Marie and Rudolph Prazak. he was educated in the Caldwell Public School. After he graduated from high school, he went to work for the Ford dealership in Caldwell.

Prazak was married to Elizabeth Dickerson of Giddings, TX on Nov. 8, 1941.

In January 1942, he enlisted in the Air Force. His basic training was done at Sheppard Field in Wichita Falls, TX. He was transferred to Hamilton Field, CA. There he was assigned to the 49th Fighter Sdn. Prazak served as Crew Chief, and later a Flight Chief, on the P-38 Fighter. He served with the 49th in England, Africa, Sicily, Corsica and Italy. Prazak returned to the States in May 1945, and was discharged July 8, 1945 with the rank of T/ sergeant.

After discharge, he went back to work for the Ford dealership in Bryan, TX as their Service Manager. Prazak worked for the dealership forty-five (45) years and retired from there in 1979.

He and Elizabeth have one daughter, Cherry Lynn, and three grandchildren. The Prazaks have attended 49th reunions since 1976.

WILLIAM R PREECE, was born in Salt Lake City, UT on June 24, 1924. While attending Sacramento, Jr. College he applied for and was accepted into the aviation cadet program. Called to active duty in March 1943 he graduated from Williams Field, AZ in June 1944. After training in photo recon. and F-5s, he was sent to the 32nd Photo Recon. Sdn. at San Severo, Italy via liberty ship. He flew 14 missions over Italy and deep into Germany until the wars end in May 1945.

He was recalled in April 1951 to serve two years flying B-29s for the 56th Weather Recon. Sdn. over Korea, after which, he returned to his home in Woodland, CA. He is now retired and living and enjoying an adult community in Sun Lakes, AZ.

His decorations include the Air Medal, ETO medal with five bronze Battle Stars, Korean Service Medal with two Bronze Battle Stars.

JOE PRICE, grew up and still resides in Johnson City, TN. While attending East Tennessee State University there in 1940, he enrolled in the Government Civilian Pilot Training Program and went on to earn his Commercial License and Instructions Rating.

His dream was to fly the P-38 in combat though. He joined the Air Corps in the Fall of 1942, went through Flying Schools on the West Coast and graduated with the Class of 43-I at Williams Field, AZ, Oct. 31, 1943.

After approximately 100 hours combat training in the P-38, Lt. Price was assigned to the 475th Fighter Gp., 433rd Sdn. in the Southwest Pacific. He flew 82 missions from bases on the Northern Coast of New Guinea and Biak Island.

He crash-landed -out of fuel - in a spectacular race with an A-20 to the strip at Saidor, New Guinea on "Black Sunday", the day the 5th Air Force suffered its worst loss of the War. The 433rd alone, lost three pilots, and six P-38s to the weather that day.

Flying wing man for Col. Charles A. Lindbergh on three missions ranks among his most memorable experiences.

Price is officially credited with one victory.

GERALD C. PRIMM, born Jan. 31, 1922, Sanford, NC. Military service April 1942 - January, 1946. Rank: Captain P-38 pilot June 1943 - October 1945. Certified Flight Commander with Instrument Pilot rating.

He flew P-38s from bases in California, North Africa, Sardinia, Italy, Corsica, England, Gibraltar, Alaska and Aleutian Islands. He flew 56 combat missions, including flying Lt. Gen. Jacob. L. Devers, Deputy Supreme Allied Commander of the Mediterranean Theater of Operations on a "piggy-back " P-38 mission to observe troop landings at the Southern France Beachhead on D-Day, Aug. 15, 1944, for which he received a written personal commendation from the General.

He was awarded the Distinguished Flying Cross and the Air Medal with nine Oak Leaf Clusters. His group, First Fighter, was awarded a Unit Citation by Maj. Gen. Nathan Twining for "outstanding performance of duty in armed conflict with the enemy" on a Ploesti, Rumania bomber escort mission May 18, 1944. He flew nine combat missions to Ploesti.

Family: wife, Ethel; sons, John and Mark. Pastor of five Baptist Churches from 1949, the last being Eller Memorial Baptist Church, Greensboro, NC (1959-1987). Home address: 1307 Summit Avenue, Greensboro, NC 27405.

THOMAS F. PUGH, born at Vallejo, CA June 11, 1920. Lived and went to school in Napa, CA. Employed by Mare Island Navy Yard until enlisting in the Army Air Corps in 1942.

As a Fighter Pilot, flew P-38s in combat in North Africa and Sicily with the 82nd Fighter Gp., an original member of the 449th Ftr. Sdn., that flew their planes from North Africa to China arriving at Kunming in July 1943. Missing in Action Sept. 20, 1943. Found wounded by Chinese peasants and returned to combat when recovered.

Stationed at Kunming, Lingling, Suichwan and Chengkung China. Returned to United States September 1944 and assigned as combat instructor P-38s and P-63s on West Coast.

Attended college at the Redwoods and Humboldt State University in California and was employed as a Gas Engineer and Supervisor of Operations for a major West Coast Utility.

Now retired and lives in Trinity Village, CA with wife, Glenna. Has two grown sons Michael and Fred. Has been active in various civic endeavors and was a founding member of the Board of Trustees of the College of Redwoods.

GUY M. PURDY, was born in Monroe, WI on May 10, 1924; he attended school in Racine, WI, volunteered for the draft from Racine and was inducted into the USAF at Ft. Sheridan, IL in March 1943.

He received basic training at St. Petersburg and Miami Beach, FL April 1983; was sent to New Orleans, LA for airplane mechanic training in June 1943 and received P-39 maintenance training in November 1943 at the Bell Maintenance Training School, Niagara Falls, New York. In January 1944 he was assigned to the 479th Fighter Gp., 434th Fighter Sdn., 5th Air Force at Glendale and Lomita, CA where they began their P-38 combat training under the leadership of Lt. Col. Kyle L. Riddle, becoming known as "Riddle's Raiders." He became Crew Chief for P-38s under the training of T/Sergeant John Gilchrist. They were shipped overseas and docked at Gourock, Scotland May 14, 1944, and traveled to the airfield at Wattisham, England where they became a part of the 8th Air Force and distinguished themselves in combat over Europe. In May 1945 he was reassigned to the 9th Air Force at Augsburg, Germany. In November 1945 he left from Camp Twenty Grand, France to return home by Liberty ship to New York. He received an honorable discharge from Ft. Sheridan, IL on Dec. 7, 1945.

A vivid memory he has from the 434th Fighter Gp. is when 1st Lt. H. (Bud) Grenning of Glenview, IL for whom he was Crew Chief, gave him a "piggy-back" ride in a P-38 over the English countryside one day.

After discharge, he attended Spartan School of Aeronautics in Tulsa, OK and obtained his A&E License. He retired on Oct. 1, 1986 from the airlines where he had been Crew Chief for 39 years, first in Chicago, IL and then Denver, CO.

He is married with three children, a daughter and two sons, and has five grandchildren. He resides in Aurora, CO.

JOHN E. PURDY, (Jack), Capt., was born in Wyandotte, MI June 17, 1919 and has resided in Dayton, OH since the end of World War II.

In June of 1941, he was drafted into the army, and served his basic training in the cavalry at Ft. Riley, KS. After completing his basic training he was sent to Ft. Snelling, MN, where he remained until the outbreak of World War II in December of 1941, at which time he applied for, and was accepted, into the Army Air Corps Flying Cadet Program. He graduated from Luke Air Force Base at Phoenix, AZ, in May of 1943.

After extensive stateside training in P-40s, P-39s, and P-38s he was sent to the Southwest Pacific Theater for Combat duty. He was assigned to the 475th Fighter Gp., 433rd Sdn., 5th Air Force, and was stationed in New Guinea, flying the Lockheed Lightning P-38.

Capt. Purdy, during the next 18 months, flew 184 combat missions from New Guinea bases at Port Moresby, Dobo-Dura, Finchaven, Nadzab, Hollandia, Biak, and Philippine bases at Leyte, Mindoro, and Luzon. While stationed at Min-

doro he was appointed Operations Officer of the 433rd Sdn. and subsequently led many of the Squadrons combat missions.

Tally Record: seven kills, two probables, all air.

Awards: Distinguished Flying Cross with Two Oak Leaf Clusters, Air Medal with six Oak Leaf Clusters, Purple Heart.

Capt. Purdy's dubious claim to fame was the fact that he survived four crash landings: one in the San Francisco Bay while flying a P-39. The other three were in P-38s: one off the end of the Dobo-Dura air strip in New Guinea, one in the Leyte Gulf, Philippine Islands and the last in Cavite Province, Southeast of Manila, Philippine Islands.

In the Cavite crash Capt. Purdy was behind enemy lines, and spent 16 days with Philippine guerillas, recovering from injuries, before being transported to the coast for eventual pick up by an air-sea rescue PBY.

HAROLD EDGAR RAMBECK, was born on a farm in Parshall, ND April 1923. Moved to Tacoma, WA when 13.

February 1943 joined the Army Air Corps; reported to Lincoln, NE for Basic Training; Graduated at Williams Field, Chandler, AZ class of 44-D, as a Pilot Officer. He was introduced to the P-38 at Santa Rosa, CA. On his first flight the port engine failed just after take-off. Following emergency, single engine, procedures he landed safely on the power of one engine.

Sailed from Boston to LeHavre, France January 1945. He was assigned to the 485th Fighter Sdn., 370th Fighter Gp., 9th Tactical Air Force, strafing, dive-bombing and escort.

Returned home in the summer of 1946 and was honorably discharged. Married Eva Johnson September 1947 and the father of one daughter and two sons. Joined Tacoma Fire Department in 1946 and retired 27 years later as Captain.

CHARLES J. RATAJSKI, was born in Shawand, WI. He joined the Army Air Corps in July 1940 and retired from the USAF in August 1962 as a Lieutenant Colonel, completing 22 years of military service.

His state side assignments were: Chanute Field, IL; Mitchell Field, NY; McGuire AFB, NJ; Hamilton AFB, CA; and Klamath Falls, OR. His overseas assignments were: Elemdorf Field, AK, Kisarazu AB, Japan and Tule, Greenland.

He spent 22 months in the Pacific during WW-II flying a P-38 with the 432nd Fighter Sdn., 475th Fighter Gp., the famous "Satan's Angels." He flew 142combat missions logging 442 combat hours. He destroyed four enemy aircraft in air-to-air combat.

He joined the 475th Fighter Gp. at Amberly Field, Australia in July 1943 and flew his last mission in May 1945 from Leyte Island in the Philippines. This last mission in May 1945 from Leyte Island in the Philippines. This last mission took the life of a very good friend of his, Capt. Paul W. Lucas. This happened while strafing a group of parked aircraft on Negros Island, Philippines. Capt. Lucas was hit by ground fire and as a result, crashed. His P-38 broke in two, the tail section severed and the P-38 was on fire. The flight of P-38s circled a few times but because of enemy aircraft in the area, they had to leave. His aircraft was later found to be completely burned.

Subsequent to retirement from the USAF Col. Ratajski taught school for 17 years in Mesa, AZ. He is an alumnus of Arizona State University in Tempe, AZ.

JOHN G. REEMSNYDER, was born in Canton, OH. Drafted in April 1942. First Army Air Force Unit was at Mather Field, CA where I made Corporal. Trained there as an Aircraft Mechanic. Then sent to Lowry Air Base at Denver, CO and trained for Aircraft Armorer.

First base was at Moses Lake Air Base as an Armorer on the Flying Fortress B-17s. Went to the CBI Theater in November 1944 on the USS General Mann. Landed in Bombay and joined the 459th Fighter Sdn. in January 1945. C.O. was Col. Verl Luehring. We had the Twin Dragons — the P-38s. The (our) pilots were fond of these planes. The Japanese were afraid of them!! Reason—the P-38s could out-maneuver their planes.

After war our outfit came back to New York on the USS General Brooke—November 1945. All in all, the CBI Theater has been a great influence on my life. I t was very educational.

RICHARD READE, (Dick), transferred from the Iowa State University Army R.O.T.C. Horse Drawn Artillery Program for pilot training in 1943. After earning his wings in B25s at LaJunta, CO, he transitioned to F5 Recon. Training at Coffeeville, KS. He was then assigned to the 28th Photo Sdn. located in Okinawa. The Squadron, attached directly to the 7th Air Force, provided solo aircraft photo missions to China, Korea, and Japan. Of note, Dick took the first photo of Japanese carriers in the Sasebo Harbor and in addition to pre and post bomb damage assessment missions he spotted a downed B29 in South China waters calling a Dumbo to the scene. The Nagasaki "A" bomb mushroom cloud was witnessed as a amazing and unusual phenomenon. He received the Air Medal and several Battle Stars and following hostilities became Personnel Assignment Officer for the 8th Air Force and 1st Air Div. flying B25s, P47s, and P61 Black Widows attached to Capt. Bob Tyldesley's 418th Fighter Sdn.

After the war and college graduation, Reade chose an aviation career in Agricultural Aviation and is still in all phases of the industry, and was the first president of the National Agricultural Aviation Association and currently Chairman of Mid-Continent Aircraft Corp, Hayti, MO.

JAMES N. REICHARD, was born in Hominy, OK in 1919 and grew up in Tulsa, OK. Learned to fly in CPT Program 1941-42 at University of Oklahoma in Piper Cubs and Waco UPF-7s (for aerobatics). Entered Aviation Cadet Program at Santa Ana, CA, March 1943; Primary at Hemet, CA in Ryan PT-22s and Basic at Marana Field, Tucson, AZ in Vultee BT-13As Flew AT-6s, AT-9s RP-322s (P-38) and graduated Williams AFB, AZ, Class 44-A. OTU, P-38s, Van Nuys, CA.

First mission June 18, 1944 with 79th Fighter Sdn., 20th Fighter Gp., 8th Air Force, Kings Cliffe, England in P-38s. Flew only 14 missions in P-38s before converting to P-51s mid-July 1944. Completed Tour December 1944, but continued on extensions for total 69 missions 337 Combat Hours. Missions were mostly bomber escort with some strafing, dive bombing and train busting after "D" Day and strafing "targets of opportunity" after breaking escort on missions to Germany. Credited with 1/2 FW-190 shot down. Last Mission, Apr., 13, 1945, delivered new P-51 to Swedish Air Force, Stockholm.

Rotated to U.S. with Group October 1945 as Captain. Engineering Technician, Wright Field, OH with TDY Ladd Field, AK Winter 1946-47. Separated May, 1947, joined Reserves as Major and later Oklahoma Air National Guard flying P-51s while obtaining Mechanical and Petroleum Engineering Degrees, at University of Oklahoma. Married Rachelle Organ in 1948.

Joined Magnolia Petroleum Company and resigned from the Air National Guard (due to relocation), June, 1950. Retired from same Company, now Mobil Oil Corporation in 1984 after 34 year career which in last years supervised Project Engineers constructing offshore oil/gas processing facilities.

Jim and his wife continue to live in New Orleans, LA and presently own a Piper Twin Comanche Airplane. They have three children and four grandchildren.

RICHARD H. REINKE, was born in San Jose, CA, July 27, 1919. On Sept. 19, 1941 he enlisted in the USAAF. He received training in aerial photography at Lowry Field, Denver, CO.

He was assigned to the 29th Photo Recon. Sdn. based at Peterson Field, Colorado Springs, CO. Later the 29th was transferred to Will Rogers Field, Oklahoma City, OK. Key personnel were transferred into the 31st Photographic Recon. Sdn.

His duties, aerial photographer (940) documentary and public relations photographer and photo copy specialist. Maneuvers were held at Cox Field, Paris, TX.

On to England, Chalgrove (Oxford) the 31st Photo Recon. Sdn. came under the 10th Gp. directly under the 9th Air Force. They worked for Gen. Patton's 3rd Army and Gen. Simpson's 9th Army mapping the coast lines for enemy anti-aircraft installations and gun emplacements of France, Belgium and Holland.

Preparing for the forthcoming invasion they filmed Normandy Beach at 10-50' above ground. The 31st Sdn. was awarded the Presidential Unit Citation for their part in the invasion. He was discharged as a Staff Sergeant.

He is married to Deloris Bottin and father of two girls, Denise and Donna. Sixteen years in Audio Visual training field. Twenty-two years with EDL/McGraw Hill Book Co.

CHARLES EDWARD RHODEBECK, born Nov. 16, 1924 at Mount Gilead, OH. Joined U.S. Army Air Corps December, 1942. Pre-flight San Antonio, TX.

Primary - Ballinger, TX. Basic - San Angelo, TX. Advanced - Moore Field, Mission, TX, Class of 44B. P-39 training at Concord, CA. Arrived Chittagong, India, July 1946. Assigned 459th Fighter Sdn. Later moved to Rumpkapalong, Burma. Was youngest pilot of 459th. Airplane was named, "Chasley's Chariot". Separated Randolph Field, TX, 1946.

Graduated BSAeE, Indiana Institute of Technology, Ft. Wayne, IN, 1950. Employed by

North American Aviation, Inc. and Rockwell International, Inc., Columbus, OH for 33 years as Test Operations Manager. Retired March, 1984. Married, reared four children. Favorite pastimes are golf, sailing, and model building. Have attended all squadron reunions to date. Welcome any and all correspondence from classmates of 44-B, especially those from Moore Field, Mission, TX, February, 1944.

HAROLD D. RIGNEY, was born in Glasgow, KY and raised in Detroit, MI. He joined the RCAF in 1941 and soloed on skies at Chatham, New Brunswick.

In 1942 he transferred to the USAAF and received his Wings at Spence Field, GA. He was assigned to the North African Campaign as a P-39 pilot—but soon transitioned to P-38s. He joined the famous "Hat-in-the-Ring" Sdn. of the First Fighter Gp., then based at Mateur Tunisia.

On August 30 while on a bomber escort mission to Naples, Rigney was observed being fired on by a P-38 with black crosses. With his engine on fire and being aggressively attacked by five ME-109s he was forced to ditch. He was rescued and taken prisoner by the Italian. He subsequently escaped and evaded for nine months.

After combat in the Pacific and Korea, Lt. Colonel Rigney retired in San Diego with 16 decorations. *(See photos, page 140)*

ROBERT B. RILEY, was born near Auburn, CA on Nov. 11, 1920. He entered Cadets in April 1941 and graduated at Luke Field, AZ on Dec. 12, 1941. After checkouts in B-18s, P-40s, and P-66s, he was assigned to the 27th Pursuit Sdn. of the 1st Pursuit Gp. at Mines Field, CA (now L.A. International Airport) where he checked out in P-38s in January 1942. This unit participated in Project Balero ferrying its aircraft to the U.K. — starting in May 1942. He flew his own P-38 to England by way of Labrador, Greenland and Iceland. After several sorties escorting B-17s over France he again flew with the group to North Africa in November 1942 (eight hours, ten minutes enroute).

Missions there were primarily B-17 escort and later dive bombing. His combat record included one shared victory, one ME-109 probable, one damaged, many railroad marshalling yards and airfields damaged and his own P-38 belly landed on the way home all shot up.

He served a full career in the USAF on flying status and retired in 1968. He served with two atomic tests in the Pacific, the Office of Atomic Energy in Washington, DC, a NATO Hq. in Norway, and the United Nations Command Hq. in Korea.

He married Charlotte Rachford of Xenia, OH in 1943. They currently live in Auburn, CA, as does their daughter Krisanne Burge. Bob golfs and fishes in his spare time. He served for six years on the Executive Board of the 1st Fighter Gp. Association and as its President from 1984-86.

JIM ROACH, first saw the P-38 Lockheed Fighter at Selfridge Field as new member of the 49th Fighter before we moved to Palm Beach, FL before Pearl Harbor was struck in December of 1941. I was a weatherman and gained the rank of Staff Sergeant.

After the defense of Darwin in 1942, and the 49th Fighter moved on into New Guinea I was placed on detached service with the Royal Australian Air Force weather branch. From service with the RAAF I was picked up by the 15th USAAF weather Sdn. and worked the balance of overseas duty at Port Moresby, New Guinea.

On the wall of the officers club at Wurtsmith AFB you will see newspaper articles about Gen. Wurtsmith the first C.O. of the 49th Fighter, and Gen. Ennis Whitehead with whom I served Port Moresby - New Guinea.

JAY T. ROBBINS, Lt. Gen., is Vice Commander of Military Airlift Command (MAC) with headquarters at Scott Air Force Base, IL. MAC operated the Nation's global military airlift system - -plus air rescue, air weather, photo audio-visual, and aeromedical evaluation systems throughout the world.

Gen. Robbins was born on Sept. 16, 1919, in Coolidge, TX, where he graduated from Coolidge High School in 1936. He received a bachelor of science degree from Texas A&M University, in 1940, and a commission as second lieutenant through the Reserve Officers Training Corps (ROTC).

In July 1941 Gen. Robbins joined Army Air Corps at Randolph Field, TX, and began flying training as an officer at Corsicana, Randolph and Victoria, TX. In July 1942 he received his pilot wings and began training in fighter-type aircraft as a member of the 55th Fighter Sdn., 20th Pursuit Gp., Morris Field, NC, and Drew Field, FL.

In September 1942 during World War II, Gen. Robbins was assigned to the 80th Fighter Sdn., 8th Fighter Gp. of the Fifth Air Force in the Southwest Pacific area. In January 1944 he became commander of the squadron in September 1944 became deputy commander of the group. He flew 607 hours on 181 combat missions in P-39 and P-38 aircraft and scored 22 aerial victories, all of which were Japanese fighter-type aircraft. This represents the fourth highest number of enemy aircraft destroyed by an Army Air Corps pilot in the Pacific Theater of Operations. He twice destroyed four enemy fighters during single missions and was awarded the Distinguished Service Cross for each of these battles.

Assigned as Commander of the 434th Army Air Forces Base Unit (AAFBU) at Santa Rosa Field, CA, from February to November 1945. Served as squadron operations officer with the 412th Fighter Gp. (later redesignated 1st Fighter Gp.), and March Field, CA, the first organization in the Air Force to be equipped with F-80 jet fighter aircraft, and later held several operational staff positions.

In June 1947 assigned to Headquarters Tactical Air Command (TAC), Langley Air Force Base, VA, where he held key staff positions in plans and operations. In 1949 he served in operational posts in Headquarters Twelfth Air Force, Brooks Air Force Base, TX. He attended the Air Command and Staff School at Maxwell Air Force Base, AL, in early 1950 and, after graduation in June, returned to Twelfth Air Force.

In August 1950 assigned as Assistant Chief, Tactical Air Operations Branch, Headquarters Continental Air Command (CONAC), Mitchel Air Force Base, NY, and in January 1951, joined the operational staff of the newly established Air Defense Command (ADC) at Ent Air Force Base, CO.

From June 1953 to September 1955, served as Plans and Programs Officer of the War Plans Division, Directorate of Plans, at Headquarters U.S.Air Force. He then became a member of the Joint Strategic Plans Group of the Joint Chiefs of Staff. While in this position, he represented the Joint Chiefs of Staff on a special continental defense subcommittee of the National Security Council.

Deputy Commander and later Commander of the 20th Tactical Fighter Wing in England during the period of July 1957 to August 1961.

In July 1962 he was graduated from the National War College, WA, D.C., and then was named Director of Flight Safety. In January 1963 he became the Director of Aerospace Safety at Norton Air Force Base, CA. During his tenure as Director of Aerospace Safety, the U.S. Air Force established two consecutive record years in low worldwide aircraft accident rates.

Became Commander of the 313th Air Div. in July 1965. In March 1967 he was assigned to the position of Chief of Staff, Pacific Air Forces (PACAF), with headquarters at Hickam Air Force Base, HA.

From July 1968 to February 1970 was Commander of the Twelfth Air Force with headquarters at Bergstrom Air Force Base, TX. In February 1970 he was named Vice Commander, Tactical Air Command at Langley Air Force Base, VA, where he remained until August 1972 when he was appointed as Vice Commander of Military Airlift Command. He retired on Sept. 1, 1974.

Is a jet-qualified command pilot with nearly 5,000 flying hours, most of which is in fighter type aircraft. His military decorations and awards include the Distinguished Service Cross with one Oak Leaf Cluster, Distinguished Service Medal with two Oak Leaf Clusters, Silver Star with one Oak Leaf Cluster, Legion of Merit with one Oak Leaf Cluster, Distinguished Flying Cross with three Oak Leaf Clusters, Air Medal with six Oak Leaf Clusters, and the Air Force Commendation Medal with one Oak Leaf Cluster.

Gen. Robbins met and married the former Ina Louise Priest of Winchendon, MA, during WWII while she was serving in Australia as an Army nurse. They have two sons, Ronald and Robert, who are both Air Force pilots.

L.D. ROBERTSON

NEWELL O. ROBERTS, M.D., educated at Eden High School, Greenfield, IN: Ball State University, Muncie, IN; Purdue University, W. Lafayette, IN; University of California, Berkley, CA; University of California at Los Angeles, CA; University of Liverpool Faculty of Medicine, England; Sparton School of Aeronautics, OK; Randolph Field, TX.

Childrens names and ages: Ronald Harrison Roberts, 29 years, Rosemary Carolyn, Deceased Apr. 9, 1981.

Appointments Held: Atomic Energy Research Commission University of California at Los Angeles, CA. September 1948 to August 1949.

Wadsworth General and Medical and Surgical Hospital Internship, November 1955 to November 1956.

Veterans Administration Mental Hygiene Clinic - Chief (15-10) Rate at present time. Oct. 1, 1973 - Mar. 9, 1987.

Enlisted Ft. Benjamin Harrison, Indianapolis, IN. Nov. 21, 1940 Regular Army of the United States of America. Flying Cadet, Air Corpus Spartan School aero, Muskogee, OK (three years).

Elementary training air Corps Cadets, Muskogee, OK Nov. 21, 1940 to Feb. 8, 1941 Basic Training, Randolph Field, TX Feb. 13, 1941 to Apr. 28, 1941. Advanced Training, Brooks Field, TX Apr. 28, 1941 to July 11, 1941, Class of 41-E.

Active Duty, Air Corps July 12, 1941. Separated Nov. 18, 1945.

Enlisted Nov. 21, 1940 Air Corps, discharged July 11, 1941; Comm Sv. EAD July 12, 1941; Reld fr AD Nov. 18, 1945; notified of final discharge Sept. 28, 1957 - Major.

Pilot license #60662 - 41 Pleasure Lunscomb, Taylor Cub, Aeronica, Ryant Sta, Beachcraft, 100 hrs. Sept. 29, 1941.

In the European-African Middle Eastern Campaign made 52 combat missions accruing 158 combat hours and a total of 1,400 flying hours.

Awards and decorations received:

American Defense Service Medal, American Campaign Medal, European-African Middle Eastern Campaign Medal, World War II Victory Medal, Armed Forces Reserve Medal, Air Medal, GO 46 Hq. 8th AF Oct. 23, 1942, Air Medal 9th Oak Leaf Cluster and Silver Leaf Cluster GO 17 June 27, 1943, Distinguished Flying Cross #1722, GO 28 Hd 12th AF Feb. 17, 1943, (Awarded March 1943 by Gen. J.A. Atkinson, Overseas), Three (3) Bronze Stars to European-African Ribbon, GO 75, WD, 1943, and Recommended for Distinguished Service Cross Mar. 19, 1943 and One (1) Cluster to Distinguished Flying Cross. Member of the American Fighter Aces Association.

Confirmed Victories seven qualifying as an Ace.

Publications

Collier's Magazine article titled, "Three Seconds to Fight in Combat with a Yank Fighter Pilot, by Capt. Newell O. Roberts", Sept. 4, 1943.

PHILIP OWEN ROBERTSON, born Sept. 1, 1920 near Munising, MI.

Attended school in Masonville, Rapid River, and Mainstique, MI graduated from ManistiqueHigh School in 1938. Northern State College, Marquette, MI 1938-39.

CMTC program Ft. Brady, Sault Ste Marie, MI, 1937, 8, 9, 40. Enlisted in Army Reserve, 339th Inf. Reg. August 1939. Enlisted in regular Army, Co. D., 1st Bn., 2nd Inf. Reg. 5th Inf. Div., Ft. Sheridan, IL, October 1940. Detached to Hq. 5th Corps Area in Chicago November 1940 to May 1941.

Appointed Flying Cadet. Primary flying training in PT-17s at Sikeston, MO, June 1 to Aug. 15, 1941. Basic in BT-13s and BT-15s at Brady, TX, August 15, to November 1. Advanced in BC-1s and AT-6s at Brooks Field, San Antonio, November 1 to Jan. 8, 1942. Commissioned 2nd Lieutenant and rated pilot Jan. 9, 1942.

Assigned to 1st Mapping Sdn., Bradley Field, Windsor Locks, CN, January 10, to Apr. 11, 1942 flying F-2s (photo C-45) and A-29s (Hudson). Pilot course at the Photo School, Lowry Field, April 12, through June flying F-2s and F-4s (P-38E). Checked out in the F-4 on Apr. 23, 1942.

Assigned to the 9th Photo Rec. Sdn. July 2, 1942 to mid-February 1944. Arrived in CBI August 1942. Flew 72 successful photo missions plus eight or ten which were unsuccessful due to weather or mechanical problems. Most missions flown in F-4s and F-5s (P-38G), with quite a number flown in B-25s which had been stripped of all but two guns and most armorplate and equipped with cameras. Awarded DFC with two Oak Leaf Clusters and Air Medal.

On return to ZI was assigned briefly to Will Rogers Field, Oklahoma City, as IP. Requested overseas duty and assigned to 39th Photo Recon. Sdn. as Ops Officer. Trained at Muskogee, OK, and deployed to ETO December 1944. Assigned to 9th Air Force and based near Metz in France, Beek in Holland, and, when Weisbaden was taken, operated from there until VE-Day. Flew 27 successful missions in F-5Es (P-38L).

Final flight in a P-38 model was at Weisbaden on June 7, 1945. Total flying time in all P-38 models was 460 hours and 40 minutes.

When WWII ended I had the temporary (AUS) rank of major, reserve rank of 1st Lieutenant and the exalted Regular Army rank of private.

1946-7 Commander, 24th Combat Mapping Sdn. (re-named 5th Recon. Sdn. (VLR), Clark Field, PI. Commissioned in Regular Army in 1947.

1947-63 assigned to the Strategic Air Command. 1947-8 MacDill AFB, 55th Recon. Gp. - pilot, 1948-9 1st Motion Picture Sdn. -TDY locations entire period. 1949-50 9th Strat Recon. Wg, Fairfield-Suisan AFB, (Travis AFB). 1950-51 28th Strat Recon. Wg, Ellsworth AFB, 1951-54 Hq. SAC, Deputy then Chief, Recon. Div, D/OPS. 1954-55 DCO, 506th Strat Ftr. Wg, Dow AFB & Tinker AFB, F-84Fs. 1955-57 DCO, 4070th Support Wg, SAC (5th pilot to fly the U-2).Command pilot rating Jan. 9, 1957. 1957-60 Commander, Det. 1, 3rd Air Div., SAC, Yokota AB, Japan. 1960-63 DCO, DCM, and Vice Commander, 98th Bomb Wing, Lincoln AFB, NB, B-47s.

1963-66 Hq. USAF as branch and division chief for reconnaissance requirements. Retired Apr. 1, 1966 with rank of Colonel.

Worked for North American Aviation for two years and for Hycon Corporation for three years in market research. Took up the pursuit of trout and golf balls in earnest in 1971.

Married to Susan Taylor of Great Britain in August, 1946. Three sons, three daughters, and one grandchild.

JAMES O. ROEBUCK, was born on Sept. 13, 1915, near Seymour, TX and was raised on a farm near Olton, TX. He was an avid model airplane builder while in Olton High School. He majored in electrical engineering at Texas Technological College. On Nov. 27, 1937, he soloed a Piper Cub after only two hours and 50 minutes of dual time. After completing 3-1/2 years of college, he entered the U.S. Army Air Corps for pilot training at Ontario, CA on June 9, 1941. (Capt. Robert L. Scott of the "China Flying Tigers" fame, was the C.O. at the time.) Mr. Roebuck graduated pilot training at Stockton, CA, class of 1942-A. He took a Photographic Officer's Course at Lowry Field in Colorado and was checked out in the P-38 (F-4) photographic aircraft. He was assigned to the 9th Photo Recon. Sdn. and arrived in Karachi, India in July, 1942. On Dec. 25, 1942, he made the first aerial photographs along a jungle trail in northern Burma (from Mogaung to Shingbwiyang) along which the Ledo Road to China was later built (from Asam, India to Bhamo, Burma). Most Photo Recon missions were flown to gather information on military targets and/or to assess bomb damage to targets. He received the Air Medal on Aug. 30, 1943, returned to the U.S. on September 1943, and was assigned to the Photo Recon. Pilot Replacement Training School at Coffeyville, KS, as a supervisor of instrument flight training, flying safety officer, and base operations officer until discharged from Extended Active Duty in October, 1945.

He was married in 1945, received a B.S. Degree in Electrical Engineering in 1947, and worked for 20 years as an instrumentation engineer for the Pratt and Whitney Aircraft Div. of United Technologies Corporation. He retired from the USAF Reserve in 1975 as a Lieutenant Colonel, and retired from Pratt and Whitney Aircraft in 1979. James Roebuck has four children, six grandchildren, and presently lives in retirement at Leisure World, Mesa, AZ.

FRANK D. ROGERO, was born in St. Augustine, FL, Aug. 9, 1920. He enlisted in the United States Army Air Corps as an Aviation Cadet on Feb. 13, 1942, and went through single engine flight school, graduating from Spence Field, GA, Class of 43-B. He won the top aerial gunnery award of the class. After six weeks as an assistant instructor at Spence Field, he was transferred to the Panama Canal Zone and flew P-40s and P-39s on intercept missions. In August 1943 he and several others were ferried to Constantine, Algeria, and were checked out in P-38s. He was transferred, in November 1943, to the 82nd Fighter Gp., 96th Sdn. at Lecce, Italy. His group moved to Foggia in 1944. He flew 51 missions and was reassigned to Santa Rosa, CA, AAB as an instructor in fighter tactics. Separated from the service in June 1945. He remained in the Air Force Reserve and retired as Lieutenant Colonel in August 1980.

Married Marian L. Watson on Feb. 16, 1943 (graduation day) and is the father of three children. He was in private business in St. Augustine for 20 years before joining the Barnett Bank, from which he retired as Vice President in 1985.

PLATT C. ROGERS, was an airplane mechanic having graduated from the Air Corps Technical School at Chanute Field, IL in November 1940. After graduation I joined the 8th Fighter Gp., at Mitchell Field, NY. At the time we only had P-40s or Warhawks. The 8th Gp. went to Australia in January 1942 and they must have gotten the P-38s after they went to Australia and New Guinea. I wasn't with them then in Australia, as I was transferred to Cadre at Baer Field, Ft. Wayne, IN. I finally arrived in New Guinea in July 1944 with the 348th Fighter Gp. We had P-47s then and by the time we got to Leyte, PI we had P-51s.

I must tell you that Maj. Richard I. Bong, ace with 40 Jap planes, ate in our mess hall in Leyte (Tannan Strip). He was on his way back to the U.S. Shortly afterward he was killed at Lockheed in Burbank testing a P-80 jet.

I remember two things about the P-40 and the P-38. They both had Allison engines made in Indianapolis, IN.

I am now 69 years old and retired.

LEROY ROSS, was born in Chalybeate, MS and raised in Blytheville, AR. He joined the Army Air Corp in 1942 and received his preflight training in Santa Ana, CA, his primary training in Stearmans at Tulane, CA, his basic training in BT-13s at Chico, CA and his advance training at Chandler Air Force Base near Phoenix, AZ where he flew AT-10s, AT-6s and the training version of the P-38 called the P-322. After graduation he trained in P-38s at Salinas, CA and San Diego, CA. He was then assigned to the 433rd Sdn. of the famous 475th Fighter Gp. in the Southwest Pacific.

He flew 144 combat missions, had three victories, and was decorated with the Air Medal

and the Distinguished Flying Cross. He is now retired, living in Blythesville, AK with his wife Billie. The couple has two children, Patsy Lee Sellers and Joseph Ross, and two grandchildren.

GERALD LYNN ROUNDS, was born Apr. 26, 1921 in Imlay City, MI third child of four to Thelma Fern (Bailey) and Ora Andrew (Commings) Rounds.

He attended the Hillman one room school S. of Fenton, MI through 8th grade and graduated Hartland City High School June 1939.

Enlisted in Army Air Corps July 23, 1940. After graduating from Aircraft Mechanics and Instructor's Schools of December 1940 Chanute Field, IL stayed on as Instructor of Electricity until receiving orders to flight training August 1941.

Flight schools were Muskogee, OK, Brady, TX and Kelly Field San Antonio, TX, receiving his wings Mar. 7, 1942 Class 42-C as Staff Sergeant Pilot.

Assigned to the 82nd Ftr. Gp., 97th Ftr. Sdn. at Muroc, CA for check-out in the P-38. After further training and sea patrol out of Long Beach, CA and promotion to 2nd Lieutenant shipped to Scotland on the Queen Mary, arriving Oct. 6, 1942. More training and check-out of his combat P-38 took place in Ireland. He left Lands End, England for his longest flight in the P-38, nine hours, to Gibraltar on Dec. 23, 1942, then to Oran, N. Africa.

His Group entered combat against the German and Italian pilots within days. Winter fields were grass and mud which required slewing the plane sideways to stop on slick tires before taking out the fence or an Arab hut. This was interesting when returning with a 1,000 lb. bomb on one rack. Summer fields were hot and dusty.

Most combat involved escorting B-25, B-26 and B-17 bombers with some dive skip bombing and strafing. Air to air combat usually involved engaging enemy fighters who were, or about to attack the bombers.

Combat was flown over N. Africa, Med. Sea, Sicily and Italy. During 82 combat missions shot down five ME-109s. Awarded 16 Air Medals, two Distinguished Flying Crosses and one Silver Star. Upon completion of 50 missions Flight Surgeon decided he was burned out and sent him to the States, arriving Oct. 4, 1943.

Round trained other P-38 pilots in Glendale and Ontario, CA until honorably discharged as Captain Nov. 22, 1945.

He immediately enrolled in college and received the Bachelor of Science, major in Chemistry and minor in Math from University of Michigan June 1949. Worked as Chemical Engineer for one year and Research Chemist for 27 years to retire as Director Environmental Quality Control Dept. Kaiser Steel Corp 1980. During school years married and is father of two children, Richard and Laurinds and has one grandchild, Michelle.

Gerry now lives as single person outside Cedar City, UT, enjoying fishing, hunting, gardening and Amateur Radio (KC7ZH).

ELWOOD DELBERT RYLAND, born Nov. 15, 1921 in DeSmet, SD. Enlisted in the Air Force Reserves in Minneapolis, MN. Entered active duty in January, 1943. He received preflight training in Santa Ana, CA, basic in Marana, AZ, and Advanced at Williams Field, Chandler, AZ. His P-39 transition was at Moses Lake, WA, and P-38 in Ontario, CA. In December 1944 he arrived in Foggia, Italy with the 82nd Fighter Gp., 95th Sdn. He flew 24 combat missions. Separation from service took place in December 1945.

He received a Bachelor's Degree in Economics from St. Olaf College, Northfield, MN, in 1947. He was employed by Studebaker Corporation and Ford Motor Company in New York, New Jersey, California, and Massachusetts. From 1962 until 1988 he has served as owner and president of a Ford/Volkswagen dealership in Duluth, MN.

He has a wife, Mary, and two sons, David and Stephen.

LESTER NORMAN SANVIK, born Dec. 11, 1921, Starbuck, MN. Enlisted in the National Guard of Minnesota, Sept. 16, 1940, then entered military service Feb. 10, 1941. June 22, 1943 he was commissioned First Lieutenant in the Army Air Force, Williams Field, AZ. Aerial Photo Training completed October, 1943 Colorado Springs. He left New York port of embarkation on Jan. 17, 1944 in a convoy of 44 ships; destination, E.T.O. He arrived in Chalgrove, England Feb. 2, 1944, where he was with the 9th Air Force, 30th Tactical Recn. Sdn., 67th Tactical Recon. Gp. After completing 24 highly successful photo missions, he was shot down on July 19, 1944 spending the next six months in England hospitalized, and on fracture boards for a back injury. Many details of this last mission earned him the Distinguished Flying Cross. He also was awarded the Air Medal with five Clusters, and the Purple Heart. In Dec. 17, 1945 he was discharged at S.F.A.A.F., Sioux Falls, SD.

April 2, 1949, Les wed Anita Helmrich. Home was the San Fernando Valley, CA, where they raised their daughter, Clarice. Les owned and operated an upholstering business both in

Westwood and the Valley areas, where his expertise was widely sought. Five days before their 38th Wedding Anniversary, Lester's soul winged its way to heaven, on Mar. 27, 1987

JOHN EDWARD SAUM, (Jack), was born in Hagerstown, MD, Feb. 4, 1918. He started flying in 1937, enlisted in the Army Air Corp in December of 1941, and earned his wings at Luke Field, AZ Class of 42 - K.

He received his P-38 training at Muroc Dry Lake, CA, then on to the 55th F.S., 20th F.G., March Field, CA before joining the 48th F.S. 14th F.G. in the Spring of 1943 in the Mediterranean Area.

He flew 50 combat missions, 235 combat hours, awarded the Distinguished Flying Cross, Air Medal with ten Oak Leaf Clusters and three Bronze Service Stars.

He was transferred to the States where he instructed pilot trainees in P-40s and P-47s.

After completing Fixed Aerial gunning course at the 3rd Air Force Gunnery School and the Air Force Gunner School, TX. He became a Charter Member of the newly formed "First Fighter Command Gunnery Instructors School", Suffolk, A.A.F. Long Island, NY, instructing instructors in Aerial Gunnery.

At the War's end, Capt. Saum reverted to inactive status, but continued in general aviation. His participation in plyon Air racing includes unlimited events (P-38s), Formula Ones (he built his own F-1 Racer); long range races (Oshkosh 500) and record attempts.

The two world speed records he holds for Class C-I-A aircraft are the 500 Kl. and the 1000Kl.

For the 1000 Kl. record he was awarded in 1986 the Louis Bleriot Medal by the Federation Aeronatique Internationale for the best world record turned in for light aircraft.

JOSEPH F. SCHAFFHOUSER, place of birth Franklin, NJ. Inducted in the Army Dec. 31, 1942. Assignments bases stations Ft. Dix, NJ. Issued military gear and assigned to the Army Air Force. Basic Training at Miami Beach, FL. Training and duty at Scott Field, IL. Jefferson Barrack Advance Combat Training for Overseas. Hammer Field, CA reassignment for overseas. Assigned to the 474th Fighter Gp., 429th Fighter Sdn., to the Armament section November 1943. Our group went overseas and assigned to the 9th Air Force (February 27 to March 10 on the high seas on the troop ships Exceisior and dropped anchor at Glasgow, Scotland.) Our new home Warmwell March 12.

Our stations overseas were Warmwell, England (Normandy Beach, St. Marceau September 2, Peronne September 12, France) Florennes Belgium October 2, (Strassfeld March 30, Langansalza April 22, Schweinfurt June 16, Germany) I left my squadron for home, September. Camp Kilmer and discharged at Ft. Dix October 1945 and went back to work in November for the Federal Government. I was a Ammunition Inspector. I retired after 34 and half years. I inspected lot of Air Force Ammunition, bombs, missiles.

JACK SCHETLER, born June 14, 1923 in Riverside, NJ. After enlisting as an aviation cadet, he completed training and received pilot wings in Marianna, FL, Class 43-C. He was assigned to the ETO as a P-39 pilot, transitioned to P-38s at the Fighter Training Center near Constantine, North Africa. Schetler joined the 14th Fighter Gp., 49th Sdn., at St. Marie Du Zit, Tunisia. He flew 50 combat missions from North Africa and Foggia, Italy. His missions were primarily bomber escort, including targets in the Ploesti oil fields, aircraft factories in Austria and the sub pens in southern France.

After returning to the States in 1944, he served as an instructor at Portland and Redmond, OR and flew in a "pin-ball" outfit in Boise, ID.

Separated from the service in September 1945, Schetler worked in the building trades and had a career in Air Traffic Control (CAA-FAA). He retired in 1979.

WILLIAM J. SCHILDT, (Bill), was born in Kendall, NY, on Nov. 18, 1918. After graduating from high school in Kendall, he attended Cornell University. On Sept. 14, 1940, he enlisted in the Army Air Corps. Shortly thereafter he was selected to be in the first class of "Flying Sergeants" and was sent to Cuero, TX. When the attack on Pearl Harbor occurred he was in flight training at Brady, TX and went on to receive his wings at Kelly Field in San Antonio.

His entire flying school class of 42C was assigned to form the 82nd Pursuit Gp. which was flying P-38 Lightning Aircraft. The 82nd

Pursuit Gp. was sent overseas to Northern Ireland to receive additional training and shortly before Christmas of 1942, the Group flew from England to assist in the battle for North Africa. A combat tour was completed and he was returned to the USA. He then went on to the Air Transport Command and flew C-54 Transport Aircraft across the Pacific Ocean routes.

Maj. Schildt checked out in jet aircraft and in 1954 he was sent to Korea as the Material Officer of the 18th Fighter Bomber Gp. where he completed a non-shooting tour in F-86 aircraft.

Schildt retired from the Air Force in 1960 as a Lieutenant Colonel.

Lt. Colonel Schildt is credited with destroying one ME-110, one ME-109, three JU-52s, and one unidentified Italian aircraft.

Aerial Combat: Early on Palm Sunday, 1943, Lt. Col. Schildt was with a 15 aircraft P-38 fighter sweep over the Sicilian Straits between the islands of Sicily and Tunisia when there appeared a flight of 21 German JU-52 transport aircraft escorted by numerous fighter aircraft (ME-109). The German fighter cover was engaged by eight of the P-38 lightning fighters and the remaining P-38s jumped the transport aircraft. A short but furious hassle ensued with transports being shot down, enemy fighters crashing into the blue Mediterranean and even one transport putting flaps down and ditching, power off, into the water to "escape."

The sea was covered in area by burning gasoline and smoke from the jerry can cargo load of the transports and when the last transport was downed and the score tallied at home base, the 95th had destroyed all 21 of the JU-52 transports and six enemy fighters while losing three of the P-38s.

CHARLES E. SCHREFFLER, (Schreff), was born in Coshocton, OH and raised in Tampa, FL and Anderson, IN. He joined the Army Air Force on Mar. 13, 1942 and progressed from pre-flight at Santa Ana, CA through primary at Oxnard, CA, to basic at Gardner Field, Taft, CA and on to Luke Field, AZ, where he graduated on Dec. 3, 1942 as a member of Class 42-K.

Charlie checked out in the P-38 during the latter part of December 1942 at Muroc Army Air Field (now Edwards AFB) and flew some operational flights out of Van Nuys Municipal Airport, CA, before shipment to North Africa and the First Fighter Gp., (71st Ftr. Sdn.) at Chateaudun-du-Rhummel, Algeria. From there the group moved to Mateur, Tunisian and temporarily Gerbini, Sicily where he completed his combat tour during the invasion at Salerno, Italy. During his combat tour "Schreff" received credit for one confirmed victory, two probables and one damaged. He also spent a combat tour in Korea with the 18th Fighter Gp. and later a tour with the 20th in England. He retired in 1970 from Holloman AFB, NM.

His hairiest mission was a B-26 escort mission to Aversa, Italy on Aug. 30, 1943. The mission was attacked on the in from the IP by about 70 ME-109s with belly tanks retained while the P-38s attempted to get the bombers to the target and back to the coast on their back to home base. In this they were successful as the B-26s accurately dropped their bombs without a loss of a single aircraft. In the meantime the 38s were catching it from all sides. Charlie saw about ten P-38s, and about the same number jumped from about 5 o'clock. When the flight broke into the enemy, four more attacked from the other side and picked off the number four man. During the ensuing dogfight his left engine was hit by a 20mm explosive shell and the right aileron was damaged and the right fuel selection valve was rendered inoperative. Fortunately, the enemy pilot was a lousy shot and he missed the rest of the aircraft. The 38s were able to fight their way out and some of the worst of the war. The First Fighter Gp. was awarded the Presidential Unit Citation for the efforts that day.

HARRY H. SEALY, was born Sept. 12, 1918. Attended schools in Oregon and Hawaii. Commissioned in Hawaii 2nd Lieutenant August 1941, was assigned to Ft. Benning, GA then to flight training in Arizona at Thunderbird, (PT-17) Marana (BT-13) and Williams. At the latter he flew the RP322. Following graduation flew P-38 F&G at Muroc Army Air Base and Lomita Flight Strip, CA on submarine patrol and gunnery training, then transferred overseas to 459th Fighter Sdn., 80th FG, CBI. Flew combat in P-38 G&H from 1943 thru 1945. His G model was retired from combat after a mission to Rangoon during which Tojo scored hits on the right tail boom, right engine and left prop. That second engine was just like a ticket home. He made 550 miles on the left with its badly damaged prop. He credits Crew Chief Gus Kogel of New York with the aircraft's excellent performance.

Following his tour he returned to Hawaii for R&R, then assigned to Redmond, AAB, OR to fly P-38 L on alert for Japanese fire balloons riding the north Pacific jet streams. Later assigned to Chico AAB, CA then to Paine AAF, WA where he wrapped up his P-38 flying regretfully. Other fighters he has flown include the P-47, P-59, F-80, F-84, F-86, F-100, F-102 and the F-106 in which he was current when retired as a Colonel in 1970. Now resides in Southern California.

FRED A. SELLE, was born on a ranch near St. Anthony, ID on Apr. 3, 1921. He enlisted in the Army in September, 1940, and entered aviation cadet training in 1942. He graduated with class 43C receiving his wings and commission at Luke Field, AZ

After checking out in P-38s at Muroc, CA, and flying coastal patrol from Orange County Airport at Santa Ana, he shipped out to North Africa in May and joined the 96th FS, 82nd FG, 12th AF. He flew missions that included bomber escort, dive bombing strafing, beach patrol and weather recon. These missions included the first bomber escort missions to Italy, the Balkans, and Austria and provided cover for the Salerno invasion.

On Aug. 26, 1943, he participated in the long range, low level strafing mission to neutralize the 13 Foggia, Italy, airfields prior to the Salerno invasion. This mission was so successful that the 82nd FG received its first DUC. Then on Sept. 2, 1943, while escorting bombers to the Naples area, he was credited with the destruction of three ME-109s and damaging two in the biggest air battle of that theatre. The 82nd FG received its second DUC for this mission for destroying or damaging 37 enemy aircraft and bringing all the bombers home safely.

After completing his combat tour in November, 1943, he was assigned as an instructor at 12th AF FTC in Algeria and returned to the states in March, 1944. During a brief stateside stay he instructed in P-39s and then joined the newly formed 508th FG, 7th AF, flying P-47s on coastal patrol in the Hawaiian Islands. In March 1945, he was assigned to the 73rd FS, 318 FG, on Saipan and as the war was coming to an end was flying the long range P-47N on missions from the Ryukus to Japan, China, Formosa, and Korea. On Aug. 9, 1945, while on a diversionary fighter sweep, he witnessed the atomic bombing of Nagasaki.

Released from the service in 1946, he later returned to active duty in the Artillery branch of the U.S. Army. After commanding a combat unit with the 7th Inf. Div. in Korea, he served with various tactical units in the US and Europe, retiring with the rank of Lieutenant Colonel in July, 1963.

During these periods of service he participated in 12 campaigns in the ETO, AP, and Korean theatres. His awards include the DFC, AM w/11 Oak Leaf Clusters, Bronze Star, and DUB w/one Oak Leaf Cluster.

Since retiring from the service he has been active in ranching and real estate and served a term as mayor of Garden City, UT. He presently resides in Salt Lake City, UT, with his wife, Shirley.

ANTHONY SGROI, (Tony), Capt., born in Pittsburgh, PA, Feb. 27, 1920. Attended University of Louisville, KY. Entered the service Dec. 12, 1941. Married Ethel Knapstein of New London, WI, in 1944.

Flew 80 missions in P-38s in North Africa in 1942 - 152 hrs. of combat flying.

He was stationed at Muroc AAB in the 441st AAFBU at Metropolitan Airport, Van Nuys, and at Santa Maria, CA. He was at Portland AAB, Oregon in 1945. From 1946 to 1947 he was in Germany.

He was awarded the Distinguished Flying Cross, and the Air Medal with three Oak Leaf Clusters.

He had two daughters, Mary Jo and Teresa.

He died in a P-80 Shooting Star crash near

March Field, CA, in 1947. He was buried in New London, WI.

ROBERT W. SHARE, was born in Columbus, OH on May 8, 1920, but has spent most of his life in and around Cortland, NY. On Dec. 16, 1941 he enlisted in the Army Air Corps and completed his basic training at Keesler Field, MS. He was subsequently assigned to the 27th Pursuit Sdn. of the First Pursuit Gp. Along with the other two squadrons of the group, the 71st and the 94th locations were established at various facilities along the west coast, with the 27th operating from Mines Field which is now part of the Los Angeles International Airport. It was here that Share developed a deep respect for the pilots and a fond admiration for the fascinating P-38, especially fostered by the fact that the 1st was the first unit to be equipped with this sleek and unique new fighter plane.

Following a period of intense training on the west coast the 1st. ferried their combat ready P-38s across the continent and the North Atlantic to take up training and orientation duties with the R.A.F. at several locations throughout England, and became one of the first U.S. Air Force units to enter combat in Europe. The later became part of the 8th Air Force flying fighter sweeps and bomber escort missions over the continent. Meanwhile, Share had been assigned to the 27th's communications section where he performed maintenance on airborn radio equipment. He also operated homing stations and airfield traffic control equipment.

The 1st landed in N. Africa on "D" Day and was eventually assigned to the 12th Air Force. During those early weeks following the invasion Share acted as an interpreter for deposed French Air Force pilots who were learning to fly the P-38.

The year that followed saw the group flying dive bombing, strafing and bomber escort missions over southern and eastern Europe, including the disastrous low level attack on the Ploesti oil fields in Romania, from bases near Biskra and Constantine in Algeria, and the Bizerte area of Tunisia.

Share's last action with the 27th was in Italy where the 1st operated for over a year as part of the 15th Air Force. It was here that he became the Squadron Photographer, an assignment he had long sought, and which proved to be his most satisfying. He produced many photographic records of squadron and group action in the air and on the ground, including some outstanding art work painted on the planes by his close friend and co-worker Warren Perrymore.

The most memorable and impressive experience which Share had during his tenure with the 27th through their seven campaigns in the E.A.M.E was his photo coverage of ceremonies at Foggia III in Italy when the 1st was awarded one of its three "Distinguished Unit Citations". Present for that event were high ranking Air Force officials, including Generals Jimmie Doolittle and Carl Spaatz of S.H.E.A.F. and Gen. Nathan Twining, Commander of the 15th Air Force.

Share was separated from the 27th by rotation in March of 1945 and discharged from the service in July with the rank of Sergeant. He is presently serving his fifth year in politics as a County Legislator following his retirement as a manufacturing executive with Smith-Corona Corp. after 43 years with the typewriter maker.

ALLAN W. SHAW JR., was born May 3, 1919, at Daytona Beach, FL, oldest child of Allan White and Virginia Maley Shaw. He attended two years of college in West Virginia and married Evelyn Wilson Jan. 29, 1941. He enlisted as Aviation Cadet, Army Air Corps Mar. 18, 1942. Preflight training was at Santa Ana, Primary at Tulare, Basic at Lemoore, CA, Advanced at Luke Field, AZ in AT-6s and AT-9s, and rated Pilot and Commissioned 2nd Lieutenant Dec. 3, 1942. December 5, to Feb. 3, 1943, assigned to P-38 Operational Training Unit at Muroc AAF,CA thence to NATOUSA Fighter School and Replacement Training Center, Berrechid, Morocco, for four months as P-38 instructor and test pilot.

Sept. 1, 1943 was assigned to 71st Sdn., First Fighter Gp. Flew 53 combat missions in P-38 G, H, J, K and L Models. Fighter escort missions for B-17, B-24, B-25, B-26 and Wellington Bombers from Africa, Sicily, Sardinia, Corsica and Italy to targets as far as Ploesti, Poland, Greece, Northern France, Italy, Austria, Germany, Czechoslovakia, Fighter sweeps and bombing - all in P-38s. Awarded the DFC, Air Medal with 12 Oak Leaf Clusters, EAME Campaign ribbon w/12 battles stars. Shot down 2-FW 190s confirmed and three ME-109s unconfirmed in Air Combat.

Retired from USAF as Regular Major rated Command Pilot Nov. 1, 1962.

JACK MAURICE SHELTON, (Jack M.) was born Pittsburg, KS 1922. U.S. Army A.F. September 1942, Aviation Cadet: Phoenix, AZ, Bakersfield, CA and Phoenix again. 2nd Lieutenant January 1944. Assigned Salinas, CA, Madras, OR, Santa Maria, CA - Flying P-39s and P-38s. Assigned Pacific Theatre June 1944 - New Guinea, Biak Island, Philippines 1944-45 - P-38 Fighter Pilot in 475 Fighter Gp. - 432 Fighter Sdn. Flew early and late model P-38s through P-38-L5 on bomber escort, dive bomber, strafing, and air-to-air combat for 106 combat missions.

Released from active duty December 1945. Married Mary Jane Thomas October 1946 - still married to Mary Jane with two daughters and two grandchildren. Joined A.F. Reserve then mustered out and had activity as Reserve Pilot and Research and Development Officer until 1972 - transferred to Retired Reserve as a Lieutenant Colonel. Graduated U.C. Berkeley, CA, Wildlife Conservation in 1951 - Retired National Marine Fisheries Service 1978 - Fishery Biologist.

GENE W. SIDWELL, was born May 15, 1921 and raised in Los Angeles, CA. Graduated from Luke Field, Army Air Force Fighter School, Class of 44-C. Flew P-39s, P-40s, P-63s in preparation for P-38 training at Santa Rosa, CA. Joined the 95th Fighter Sdn., 82nd Fighter Gp., 15th Air Force at Foggia, Italy in November 1944. Completed 27 combat missions in P-38s when the war ended in Europe. The mission ran the gamut, high level bombing with a "Droop Snoot," dive bombing, strafing, escort for B-17s and B-24s and chaff drops.

He now resides in the Hollywood Hills, and is president of A-1 Roofing Service in Los Angeles. Married and has two daughters.

RICHARD SMITH SIMPSON, was born Sept. 7, 1917 on a farm near Corning, IA where he lived until he and his family moved to Fresno, CA in 1930.

He enlisted in the Army in 1936 and served in the Coast Artillery in Hawaii until 1939. Dick was working for the Civil Service as a storekeeper at Ft. Ord in February 1942 when he applied for the Aviation Cadet Program. He took his ground school at Santa Ana, primary at Dos Palos, basic at Merced, CA and advanced at Williams Field, AZ receiving his P-38 transition at Santa Ana and Murco, CA.

2nd Lt. Simpson arrived Nuthampstead England in October 1943, from here flew 73 missions over Germany including, escorting the first bombing raid over Berlin and cover for the Normandy Invasion, all with the 38th F.S. of the 55th F.G. Returning to the US in November 1944 Capt. Simpson was based at Stockton, CA as a test pilot, Lukefield, AZ Transitioning Chinese pilots into P-40s and Barksdale AFB, LA as an Instrument Instructor until his release from active duty in 1946 as a Major.

Dick is still pursuing his post war career as a physical therapist in Carmel, CA spending his weekends at his home in St. Helena,CA.

Personal Interest Stories:

Returning from a mission over Germany one day Dick ran out of gas and belly landed the airplane in an Englishman's potato patch!

On take off for a mission a leak in a fuel line pouring gas into the super charger caused a fire. Dick belly landed the airplane, jumped out of the airplane but was stopped short by a strap still secured inside of the airplane—he was able to break the strap and escape before the P-38 exploded.

EDWIN E. SIMS, born June 5, 1923, was accepted for flight training in May 1942 and called to active duty in January 1943.

He went through single engine training being stationed at Santa Ana, CA for preflight, primary at Ontario, CA, basic at Taft, CA, and advanced Luke Field, AZ where he graduated with the Class of 43K on Dec. 5, 1943.

After flying P-39s at Santa Rosa, CA, he was transferred to Santa Maria, CA where he was introduced to the fabulous P-38.

He was shipped to the ETO on June 5, 1944 and was assigned to the 82nd Fighter Gp. 97th Fighter Sdn. at Foggia, Italy.

In addition to escort, dive bombing and strafing missions he participated in several level bombing raids with the "Droop Snoot". He made a "shuttle mission" to Poltava, Russia and while there the group went on a strafing mission to Milec Aerodrome in Poland. After the strafing run, he the Group Command and another pilot were separated from the rest of the formation and on returning to Poltava were bounced by three Russian Yaks. Upon turning into the attack, the Russians recognized the distinct configuration of the P-38 and broke off.

After his combat tour he went to instructors school at Waco, TX and completed his active duty as an OTU instructor at Richmond, VA flying P-47s.

Upon separation, he returned to his home in Happy, TX where he still resides as a farmer-rancher. He married Ora Mary Gurley in March 1946 and they have two daughters and five grandchildren.

SAMUEL E. SIMUNIC, was born in Chicago, IL on Apr. 29, 1912. He worked in the Steel Mills and also as a warehouse man. Sam passed away suddenly on Oct. 9, 1986.

Sam enter active service on Jan. 23, 1942 in Los Angeles, CA. He served as an Airplane Armorer with Sdn. S 233rd AAF Base Unit in the following battles and campaigns: Air Combat Balkans, Northern France, Naples-Foggia, Tunisia, Sicily, Air Offensive Europe, Rome-Arno, Italian. Arrival in the EAME Theater was Oct. 4, 1942 and he departed the Theater on July 18, 1944.

Decorations and Citations include: Bronze Star Medal, WW-II Victory Medal, European African Middle Eastern Campaign Medal, one Silver and three Bronze Service Stars, Distinguished Unit Badge w/two Oak Leaf Clusters, Good Conduct Medal and American Campaign Medal.

S/Sergeant Simunic was separated from the service on Sept. 28, 1945 at Davis-Monthan Field, Tucson, AZ. *Submitted by his sister Margaret C. Simunic*

ARTHUR F. SMITH, 1st Lt., pilot in the 37th Fighter Sdn., 14th Fighter Gp. in the 15th Air Force in Italy from July 1944 through March 1945. Completed 50 combat missions, and is the holder of the Air Medal with four Oak Leaf Clusters and the Distinguished Flying Cross.

Worked for the New England Telephone Company from 1947 to 1983. Was an Outside Plant Engineer and later a Staff Manager responsible for the Outside Plant Construction Budget on a Company Basis.

Married in 1947, and has a daughter, Lynda. With his wife, Jeanne, moved to Chatham, MA on Cape Cod after retirement in 1983.

CORNELIUS M. SMITH JR., (Corky), was born on Nov. 11, 1918 in Baltimore, MD. Graduate of Roanoke College, Salem, VA in 1940. Began flying in the U.S. Government "Civilian Pilot Training Program" (Roanoke College Senior year), Woodrum Field, Roanoke, VA, Fall 1939 to May 1940. Received private pilot license.

Entered aviation cadets on Dec. 15, 1941 at Maxwell Field, Montgomery, AL. Pre-flight at Maxwell Field, thru Jan. 5, 1942; Primary at Avon Park, FL Jan. 10, 1942 thru Mar. 30, 1942 — flew Stearman PT-17 bi-plane in Primary; Basic at Shaw Field, Sumter, SC April thru July 1, 1942 — flew Vultee BT-13, mid-wing monoplane; Advanced at Spence Field, Moutrie, GA July 5, 1942 thru Sept. 6, 1942—flew North American AT-6, low wing monoplane—received commission as 2nd Lieutenant and rated Pilot on Sept. 6, 1942.

Assigned to 69th Fighter Sdn., 48th Fighter Gp. (P-39 training) at Sarasota, FL Sept. 10, 1942. Received 50 hours training in P-39. Departed Oct. 15, 1942 for Southwest Pacific where assigned to 80th Fighter Sdn. (Hq.), 8th Fighter Gp., 5th AF, Milne Bay, New Guinea Nov. 23, 1942. This was a P-39, P-40 unit at this time. On Feb. 6, 1943 unit sent back to Mareeba, Australia for re-equipping with P-38 aircraft. Returned to New Guinea Apr. 10, 1943 with P-38s. Stationed at Kila Drome, Port Moresby. During the next 13 months the Squadron served at the following locations (PCS): Dobadura, New Guinea; Cape Gloucester, New Britian, Nadzab, New Guinea. I came home as a Captain in late May 1944.

Flew total of 169 combat missions from Nov. 24, 1942 thru May 15, 1944 — 25 in P-39 and P-40 aircraft and 144 in P-38s. Retired as Colonel

from USAF on Dec. 1, 1968 (medical retirement).

Since that time, golf, fishing, fun!! Married Dorothy Louise Harman on May 31, 1944 (college sweetheart); two daughters; four grandsons.

Combat decorations earned in the P-38: Silver Star, DFC w/three Oak Leaf Clusters, Air Medal w/five Oak Leaf Clusters and Presidential Unit Citation w/two Oak Leaf Clusters.

Types of missions flown in P-38s: From review of my Form 5s (flight record obtained from USAF) primarily as bomber escorts. Approximately 83 bomber escort missions averaging three hours 40 minutes (longest mission 6:15). Also flew five missions escorting C-47 and C-46 transports averaging three hours and ten minutes per mission.

Fighter sweep missions: precede bombers into target area, engage enemy a/c, attempt to draw enemy a/c to altitudes below bomber attack altitudes (clear area of enemy fighters). Average 4:20 per mission.

Missions providing cover for amphibious landing operations and patrol of beach areas following landings. Average mission of four hours duration.

Missions devoted to providing air cover for Allied shipping on the high sea. Average duration three hours.

Missions involved with patrol of vital areas and/or local air defense. Average duration 1:30.

Missions of ground attack (machine gun and 20 mm cannon) against enemy ground installations. Average duration of mission 2:40.

The P-38 was the foremost long range aircraft in our theater. We had no P-51 Mustangs in New Guinea during my tour. During my months in New Guinea our outfit never carried bombs or rockets. Our four 50 calibre and one 20 mm cannon comprised our armament.

WAYNE O. SNEDDON, Lockheed Tech Rep., born in Salida, CO, Mar. 20, 1917. Graduated from Colorado State University (ME) 1941. Began flying as a Tech Rep during the war, but did not get private pilot licenses until after the war ended.

Became a Lockheed Tech Rep in 1942. Assigned to the US in New Jersey, Long Beach etc. My first foreign assignment was to Iceland to cover the 50th Fighter Sdn. and the North Atlantic Ferry route (November 1942).

In September 1943 I was given orders to proceed to my next assignment in India. On arrival in Kurmitola I was assigned with the 459th Fighter Sdn. This was a new squadron (formed in combat—later disbanded in combat) its member coming from the 80th Group of the 10th Air Force. Most of its members had never even seen a P-38.

In the first 53 days of combat the squadron shot down an average of 2-1/2 Jap planes a day and destroyed an equal number (or more) on the ground.

In January of 1945 a Capt. Hill and I were ordered to take complete control of the Hindus-

tani Aircraft Factory and modify ten P-38s to droop snoots in 90 days or less. Capt. Hill was a stress engineer in his pre-military days and he handled the paperwork while I, having some manufacturing knowledge took charge of manufacturing the droop snoots. The excellent work of the Hindustani personnel allowed us to build ten droop snoots in 53 days, and nine of these were delivered to the combat dispersal area in 61 days. (One aircraft was damaged beyond immediate repair in a ground accident.)

On completion of this assignment I returned to Burbank. The war had ended, but I stayed on with Lockheed as a Tech Rep until my retirement in 1974. My first commercial assignment took me to Ireland working with Aer Lingus and the Constellations. Then on Holland to work a short time with KLM. I was privileged to work with Quantas in Australia and to see them go from near extinction as an airline to one of the world's top carriers. I spent some months at the factory working with the Engineering Dept. on the Electra. An assignment with Air France (and other lines needing assistance on stop-overs in Paris), and a return to the US to work with the Western Airlines at LAX. By now the P-3 was in the hands of the military, and it was time for me to work with the Navy—first at Moffatt Field, and later going back to Patuxent River, MD and Johnsville, PA to assist in the work at NADC.

I was married July 3, 1947 to Ellen E. Cary (who had spent the war years as a secretary in the Field Service Dept.) We have two children (a daughter born in Australia and a son born in California). Our daughter is married and lives in Iowa with her husband and two sons.

STEVE SOBAK, was born Mar. 8, 1925 in Detroit, MI. Received BSME from Lawrence Institute of Technology. Aviation Cadet training Feb. 13, 1943 at Nashville, TN, Newport, AR, Jackson, TN, Marianna, FL, Eglin Field, FL, Camp Springs, D.C., and Millville, NJ. Aviation Cadet Class 44-A.

Fifth Air Force, 8th Fighter Gp., 36th Fighter Sdn., Fighter sweeps, dive bombing, convoy and troop landing cover, strafing, bomber escort cover, troop support. Eighty-three missions in Pacific (Philippines, Formosa, China, Japan).

1st Lt. Sobak separated on Feb. 26, 1946, all before he was 21 years old.

He is a retired engineer. He and wife Ramona have two children, Steven and Michael; four grandchildren.

ROLAND SPERRY, received a Private Pilot Certificate on his 17th birthday and the Commercial Pilot Certificate and Instructor's Certificate at age 18. He joined the Army Air Corps on Dec. 18, 1941 at Los Angeles, CA, and completed Air Corps Civilian Pilot Training at Hill Field, UT in May of 1942, and was commissioned a 2nd Lieutenant. After graduation as a fighter pilot, he was sent to China to join Gen. Claire L. Chennault and the Flying Tigers. He served under Col. Robert L. Scott, author of "God Is My Co-Pilot", and flew wingman to Col. Scott with the famous 23rd Fighter Gp. As a flying Tiger, Roland Sperry flew 237 combat missions, totaling 642 combat hours. He flew Curtiss P-40s and the North American P-51 Mustang. He was shot down twice by ground fire, injured, and evaded capture twice on both occasions. His records include four Japanese fighters (confirmed) and seven Japanese fighters and bombers (probables) which he shot down as a Flying Tiger.

Maj. Sperry resigned his commission in 1951 after returning to the United States in 1945 and flying Lockheed P-38 Photo Recon. at Will Rogers Field in Oklahoma City, and Martin B-26 Bombers at Long Beach, CA. He then received an Aeronautical Engineering degree from South Western Aeronautical Institute in Dallas, TX.

Mr. Sperry has remained active in aviation as a company and demonstration pilot, flight instructor, narrating and performing in air shows, air racing, aircraft sales, and with his own aviation radio program: "Aviation Showcase", on KFOX, 93.5 FM.

Military Decorations: Distinguished Service Cross, Distinguished Flying Cross with six clusters, Air Medal with 11 clusters, Purple Heart with one Cluster, China Freedom Medal — China's most distinguished award given to Americans, Chennault Memorial Award for distinguished service.

Civilian Awards: 1948 Young Man of the Year, U.S. Chamber of Commerce, 1950 Inglewood Chamber of Commerce National Recruiting Award, 1950 Thompson Trophy for Speed run in a P-51 Mustang, Van Nuys, CA to Cleveland, OH. Speed of 427 MPH, 1954 Toast Master International Speakers Award, 1956-1957 Dale Carnegie National Speakers Award, 1961 Gavel Club U.S. Speakers Award.

Active in various community affairs, Chamber of Commerce, Kiwanis Club, Gavel Club, Boosters Club, and Air Show Commentator, Member of Combat Pilots Association of America, Flying Tigers of 14th Air Force Association, P-40 Club, Walkers Club.

His total military and civilian flying time is 20,000 + hours.

A. DEWEY SPIKER, of Scotts Valley, CA, who produced the first space shuttle handbook for the U.S. Air Force in 1982, died Thursday November 3. He was 66.

Mr. Spiker moved with his family to Scotts Valley from Santa Cruz in 1987. The family had moved from San Jose to Santa Cruz in 1984.

Mr. Spiker, a communicant of San Agustin Catholic Church in Scotts Valley, was a district governor of the Toastmasters in 1961, the same year he was hired by Lockheed in Sunnyvale. He retired as Lockheed publication engineer.

Before that he was employed at FMC Corp. in San Jose as a technical writer.

He was a U.S. Army Air Corps veteran of World War II and flew a twin-fuselage P-38 Lightning fighter aircraft in the Pacific Theater.

Mr. Spiker was born in Lucas County, IA. He attended Iowa State College after graduating from Chariton High School in Chariton, IA.

Mr. Spiker leaves his wife of almost 44 years, Bettie Spiker of Scotts Valley; two sons, Robert Spiker of Scotts Valley and William Spiker of San Jose; four brothers, Gene Spiker of Mesa, AZ, E.D. Spiker of Allentown, PA, W.R. Spiker of Baton Rouge, LA and Ray Spiker of Chicago; and four grandchildren.

ROBERT W. SPITLER, was born in Santa Paula, CA on Jan. 18, 1923. He was educated at Fillmore Elementary, Junior High and High Schools in Fillmore, CA and at Ventura College in Ventura, CA, earning an Associate of Arts Degree in June 1942.

He joined the US Army Air Corps Reserve in June 1942. He was called to preflight school at Santa Ana, CA in September 1942, taking primary training at Morton Air Academy at Blythe, CA. Basic at Minter Field, CA and Advanced training at Williams Field, AZ, graduating as a Second Lieutenant and Pilot on May 20, 1943.

After receiving Tactical P-38 training at Muroc Air Base and Van Nuys, CA. He was sent to North Africa in September 1943 joining the First Fighter Gp., 71st Sdn., flying P-38s later moving to Sardinia and Italy ending at Foggia.

He flew 50 combat missions on strafing, dive bombing, medium bomber escort and heavy bomber escort. He held the rank of Captain and had three planes confirmed in aerial combat. Received DFC and Air Medal with ten clusters and group was awarded two Presidential Unit Citations during his tour.

After returning to the States in September 1944 he instructed new pilots in P-38s in transition and combat training at Chico, CA and Paine Field at Everett, WA. He was separated from service in October 1945.

He worked as a Public Accountant for 26 years, owned and operated a retail liquor store for 14 years and is currently half owner of a highway striping company in Fillmore, CA.

HARRY SPRING

EARL SPRENGEL, was born in New Berlin, WI on May 9, 1918 and passed away Sept. 17, 1982. He enlisted in the Army on Mar. 10, 1942. He was stationed at Selfridge Field, MI. Pilots were needed; he passed the required test. He was sent to Anaheim, CA. Preflight training December 1942 - February 1943. Basic training for three months Ryan Field in Tucson, AZ, 11, AAFFTD. Got his Wings on Aug. 13, 1943, Class 43-H at Williams Field, Chandler, AZ. In November 1943 sent to Santa Ana Army Air Base, Santa Ana, CA for practice flying.

He was stationed in Foggia, Italy as a fighter pilot. Seven of his missions were over Ploesti, Rumania, the major source of German oil. This refinery was the third most heavily defended German Point after Berlin and Vienna. His squadron, 27th Fighter Sdn., 15th Air Force, 1st Fighter Gp., received the Presidential Unit Citation for outstanding performance of duty in an armed conflict with the enemy on May 18, 1944—raid over Ploesti.

Excerpts of his personal diary detail some of his missions. He received the Distinguished Flying Cross for his strafing mission of Apr. 18, 1944 over Udene, Italy where he shot down a 109 Messerschmitt and damaged it. Shot down five enemy planes, bellied in his P-38. He also received the Air Medal and eight Oak Leaf Clusters for meritorious achievements in aerial flight. Following his 50th mission he was sent to Walla Walla, WA in August 1944. He flew P-63 King Cobra pursuit planes simulating attacks on Liberator bombers to train bomb crews. January 1945 stationed in Chico, CA, C.A.A.F. 4B 3BW Section 11, flight instructor. October 1945 he was stationed in Portland, OR. He went to class for P-38 fighter planes with rockets. In 1946 he was sent to Landsburg, Germany, Army of Occupation. In January 1947 he was sent to Oberphaffenhafen; he had an accounting class there. Returned to the US May 1947 and was discharged from the service in July 1947.

Earl married Aug. 1, 1942 to Ethel Sharpeng. A daughter, Cheryl, was born in Chico, CA on May 14, 1945.

WILLIAM H. STARKE, was born Apr. 4, 1919, in Chicago, IL. He left Illinois Wesleyan University to enlist in the Army Air Corps, Dec. 26, 1941. After graduating from Clerical School in Ft. Logan, CO he applied for flight training and was sent to San Antonio Aviation Cadet Center and then was transferred to Santa Ana, CA Cadet Center.

Starke's primary training was at Cal-Aero Academy, Ontario, CA, basic training at Minter Field, Bakersfield, CA; and advanced training in P-38s, Class 43-G, at Williams Field, Phoenix, AZ, graduating July, 1943. Combat training at Edwards Air Base, Muroc, CA, and at North Island Naval Air Station, San Diego, CA. He joined the 44th Fighter Sdn., (Vampires) at Guadalcanal, Solomon Islands, November 1943.

Flying P-38s, he served first as pilot then element leader, flight leader, and finally operations officer during 121 combat missions. Many missions were 8-11 hours long and included bomber escort, dive bombing, Napalm bombing, skip bombing, and strafing, spread over 18 months. He is credited with shooting down one Jap bomber.

Capt. Starke was awarded the Air Medal with seven Oak Leaf Clusters, was in eight major battles, flying out of Guadalcanal, Bougainville, New Guinea, and the Philippines, returning to the U.S.A. in May, 1945. A graduate of the University of Wisconsin, he has been a Ford-Mercury dealer in Ft. Atkinson, WI for 39 years.

He is also the author of a book about the 44th Fighter Sdn. entitled "Vampire Squadron".

JOHN A. STEGE, was born in Los Angeles, CA Oct. 26, 1920. Moved to Lancaster, CA in 1923 and attended Antelope Valley High, graduating in 1938. Attended Antelope Valley Junior College, also attended Compton Junior College, CA.

He joined US Air Corps as a Cadet in 1941, graduated with the class of 41-I on Dec. 12, 1941, Luke Field, Phoenix, AZ. Joined the 48th Ftr. Sdn., 14th Ftr. Gp. at March Field, CA. In January 1942 he checked out in P-40s, P-43s, P-66s and later on that year checked out in a P-38 at North Island, CA. Flew training missions and patrol on the west coast. In the month of June the squadron received 25 P-38s Fs and all pilots were assigned a new P-38 aircraft. Flying a squadron across the States. Amarillio, St. Louis, Dayton, and into Bradley Field, CT. Trained with B-17s crossed the Atlantic, stopping at Goose Bay, BW#1 Greenland, Reykjavik, and in to Scotland, and then to England. First combat mission was Oct. 13, 1942 out of Tangmere, England. Moving from Atcham, England to Gibraltar and on to Tafarour, Africa. Flew 30 missions out of Youks Les Bains, aerial combat with 109s and 190s, bomber escort strafing and local patrol. Crashed landed one P-38 in the desert. Returned to the United Sates in February 1943. As an instructor in P-38s trained pilots at Muroc Field, Salinas, and Santa Maria, returned to combat in the Pacific. Joined the 13th Air Force 13th Ftr. command 347th Ftr. Gp. 68th Ftr. Sdn.; flew 70 missions in this area later Squadron Co.; received the DFC, Air Medal six Oak Leaf Clusters; Rank of Major, Honorable Discharge in 1945; went in retail business, married Charlene M. Mertes in Illinois, five children. Semi-retired lives in Newport Beach, CA. His experience in P-38s, approx 1,000 hours in P-38s, models E thru L.

KURT ERIK STERNER, was born in New York City in 1924. Accepted for pilot training, he went through the Southwestern Training Command and graduated in Class 44-A, 2nd Lieutenant, from Williams Field, AZ, Jan. 7, 1944. He received P-38 training in 440th Base Unit AAF Santa Maria, CA, and was sent to Europe (15th Air Force, Foggia, Italy) 14th Fighter Gp. July, 1944. He flew 17 bomber-escort missions (P-38-J) during July.

Reassigned to the 154th Wea Recon. Sdn. (P-38-J) in August, 1944, he completed his missions, returning to New York Dec. 24, 1944.

He flew 302 hours in the P-38-J, and received the DFC and Air Medal with five clusters. The group was awarded the Presidential Unit Citation.

Recalled to active duty in August, 1950, as 1st Lieutenant all-weather interceptor pilot, he was integrated into the regular Air Force. He held various positions with the Air Force Systems Command and retired Mar. 31, 1969, as Lieutenant Colonel, command pilot. He joined a major aerospace company as design and development engineer, and is currently involved in DoD contracts concerning electronic warfare hardware.

LAWRENCE I. STOLAND, Lt., from Lawrence, KS. Father was a professor at University of Kansas. Graduated from Luke.

Died July 22, 1942, in the Aleutians - lost in weather. 54th FS

BENNIE HIGH STONE, was born in McGregor, TX in July, 1921. He was flying P-38s in the Aleutian Islands in 1943-1944 where he survived a ditching of a P-38 having been picked up by a surface vessel after dual engine failure.

After the war he was assigned to the 94th Ftr. Sdn. of the 1st Ftr. Gp. in Santa Maria, CA flying P-59s and at March Field, CA flying P-80s. He was killed in a P-80 near Greenville, SC just a few days before his 25th birthday. *Submitted by Colin D. MacDougall*

GLEN R. STOVER, (Smokie), born and currently living in Findlay, OH. Enlisted in Flying Cadet Program, October, 1940.

Trained at Parks Air College, Randolph Field and Brooks. Graduated in Class of 41E.

Assigned to 71st Pursuit Sdn. - 1st Pursuit Gp., Selfridge Field.

Participated in training maneuvers in Louisiana, the Carolinas and night maneuvers in Savannah, GA until Dec. 1, 1941.

Dec. 9, 1941 North Island Naval Station flying patrols and interception missions.

January, 1942 group transferred to Los Angeles area.

April, 1942 group transferred to Bangor, ME flying new P-38s and joined by 97th Bombardment Gp.

May, 1942 left Bangor, ME with 17s navigating for England via Labrador, Greenland, Iceland, Scotland, and finally Bournmouth, England (was the first time Fighters had flown the North Atlantic).

June, 1942 group transferred to Goxhill, England for combat operations until invasion of Africa.

Flew P-38s to Africa from England. Missions in Africa—everything from tree top to high level escorting Bombers, Fighter Sweeps, skip and dive bombing etc. with flights over Mediterranean, Sardinia, Pantelleria, Sicily and Italy.

Received DFC Air Medal with nine Oak Leaf Clusters during tour.

Returned to States, promoted to Major and assigned to Pilot Replacement School, Ontario, CA as section commander and instructor.

IRVINE M. STYER, (Irv), was born in Los Angeles, CA. Graduated Class 42-J, Luke Field as a Staff Sergeant Pilot, October 1942.

Flew P-39s at Hamilton Field until January 1943. Transferred to Metropolitan Airport, Van Nuys, flying P-38s.

Sent to North Africa in late May 1943, joining the 27th Fighter Sdn., 1st Fighter Gp.

Logged his first combat mission June 15, 1943. His "mission to remember" occurred Aug. 21, 1943. He assumed leadership of the group when the commander aborted with mechanical failure. Near the target 40 to 50 ME-109s, FW-190s and Machi-202s attempted to break through the bombers. They were repeatedly driven off by the First under Styer's command. The B-26s completed their bombing run unmolested, but coming off the target divided into two waves. The first wave was quickly escorted to lesser danger. The 1st Fighter Gp. then swung back to pick up the second wave.

A fresh, numerically superior force of enemy fighters had been scrambled and were immediately engaged by the group. The attacks continued 40 miles out to sea.

During the flight home, Styer turned back nine times to protect damaged planes. He shot down one E.A. and probably destroyed two others. The tactics of the 1st Fighter Gp. in

protecting the bombers and themselves resulted in all American planes returning safely from the mission.

For his responsibility in leading the mission, Styer received an Oak Leaf Cluster to his Distinguished Flying Cross.

Three weeks later he flew his 51st and final mission, over the Salerno Beachhead. Two months later on the Island of Sardinia, Styer was a passenger in a car that was struck head on by a truck. Thereafter he spent almost three years in Army hospitals recovering. His flying career was over at the ripe old age of 21-1/2 years. This convinced him the only safe means of transportation was flying.

RICHARD T. SYKES, was born Mar. 24, 1924 at Evanston, IL. At the age of 12 Dick and his family migrated to California. He learned to fly before the war, taking his first instruction at the age of 15. He resided in Beverly Hills at the time of Pearl Harbor. As soon as he was 18 he applied for and was accepted for Aviation Cadet training. He graduated from Luke Field, AZ in class 43K as a single-engine Fighter Pilot. He flew P-39s and P-63s in various RTUs on the West Coast and finally was lucky enough to find his way to Van Nuys where he checked out in the P-38 and was immediately shipped overseas.

He was assigned to the 94th Fighter Sdn., First Fighter Gp., 15th Air Force. He flew his first of 50 combat missions in June of 1944. He would have flown more than 50 missions (the quota for a Tour at the time) but for the fact that he was shot down on his 50th mission on Dec. 6, 1944 by a flock of Messerschmitts. He found himself behind enemy lines with a broken ankle and other injuries about 60 miles South of Budapest, Hungary. He evaded capture and found his way to the Russians who were advancing in the Balkans at the time.

He was returned to his squadron in January of 1945 and because of his project "R" status was sent back to the States. He immediately applied for overseas duty in the Far East. He was rewarded by being made an AT-6 instructor at Luke Field for one year. He stayed in the Air Force until May of 1948 flying Fighters with the 82nd Fighter Gp. at Grenier Field, NH the 57th Fighter Gp. at Anchorage, AK, and finally flying C-54s with the 54th Troop Carrier Sdn. in Alaska.

After his release from active duty he joined the California Air National Guard and went to school to become a lawyer. He flew with the California Air Guard regularly until they made him stop flying because of his Legal Officer status. He thereupon bought himself two AT-6s and organized the Condor Sdn. The Condor Sdn. of the Van Nuys Airport operates 16 AT-6s and participates in air shows all up and down the West Coast. Dick also is an air racing fan. He won the Reno National Championship AT-6 Air Race in 1983 setting a new world record. He is still active flying with the Condor Sdn.

He was called up for active duty during the Korean War and spent time in the Pentagon as a Liaison Officer between the Secretary of the Air Force and the Congress.

He holds the Distinguished Flying Cross, the Air Medal, and the Purple Heart. He has been credited with one enemy aircraft destroyed. During his 50 missions he had the opportunity to go to Ploesti six times and was selected to participate in dive bombing and strafing missions on ten occasions.

ROBERT SYMONS, born Apr. 2, 1910 at Bishop, CA. Died Apr. 19, 1958 at El Mirage, CA in a B6-12A glider crash when the wing of the glider fell off while he was being towed for a recreational flight. He developed his flying abilities when he was at Riverside Junior College. Robert married Mildred Wells in 1935 and had two daughters, Emily and Patricia.

Bob got his pilot's license in 1941. Bob logged 18,000 hours of flying before his death in 1958.

Scientist gathered at Bishop, CA in 1950-1952 to study the Jetstream for the Air Force. Gliders were used to gather statistical data on temperature and magnetic declinations from North, because of the minerals in the Sierra. Bob was a most experienced pilot with over 300 glider flights in the wave and hosted the group.

Bob served as a water commissioner, volunteer fireman, and search and rescue pilot to many stranded desert travelers and mountaineers,. He was also a Chamber of Commerce Member. He operated Symons Flying Service, Soaring and Stratospheric Research, Cloud seeding using Cal Electric's P-38, snow survey for Southern California Edison Co., invented hydraulic operated split-apart airplane skis. He received the Warren E. Eaton Memorial Trophy, 1949 for soaring, flew record distance flight (192 miles) in a glider, from Bishop, CA to Las Vegas, NV, 1952.

ROBERT F. TAYLOR, JR., was born on Nov. 4, 1919, in Ft. Worth, TX. He was educated in Virginia and graduated with B.S. Degree and commissioned at Virginia Polytechnic Institute May 1942. Transferred to Army Air Corps and graduated in the Class of 43F. He was assigned to P-38s in 94th Fighter Sdn. 1st Fighter Gp. located in Foggia, Italy. In 1944-45, he became Squadron Operations Officer. He flew 40 combat missions, logging 225 combat hours. He was

one of the pilots chosen to fly escort for President Roosevelt to the Yalta Conference.

After WWII he flew P-80s with 4th Fighter Gp. In 1950-52, he served as an Exchange Officer with the RAF at Odiham, England, flying Vampires and Meteors. In 1959, assigned as Squadron C.O. in the 5th AF, Fuchu, Japan. Held various operational staff assignments at TAC Headquarters, 9th AF and 7th AF. He has been awarded the DFC, Bronze Star, A.M. (three Oak Leaf Clusters), Meritorious Service Medal, and seven Campaign Ribbons. Retired September 1969 as Lieutenant Colonel.

Since retirement, he has been a Life Insurance Agent with New York Life Insurance Company for over 18 years. He is married to Janice Carruth. They have four children, Lynn, Cynthia, Douglas, and Mary Kay.

RALPH THIESSEN, 1st Lt., 15th AF, P-38 Pilot from Hamburg, IA, the son of Mr. and Mrs. Harold Thiessen. He received his wings from Luke Field, AZ and joined the P-38 Sdn. in May 1943.

He received a Distinguished Flying Cross and a Silver Star.

DALE O. THISTED, was born May 10, 1924 in Freeborn Co., MN. Attended high school at Alden, MD and then St. Olaf College. C.T.D. at McAllaster College, MN, February 1944. Entered Aviation Cadet training at Santa Ana, CA on March 1944. Training at: Dos Palos, CA (Primary); Marana, AZ (Basic); Williams Field, AZ (Advanced); P-39s Moses Lake, WA, and P-38s Ontario, CA.

Assigned to 475th Fighter Gp., 432nd Fighter Sdn. and served in the Pacific. Flew escort, patrol, dive bomb, Napalm bomb and fighter sweeps, total number 49. Was a 1st Lieutenant at time of discharge.

Civilian occupations have been flight instructor, fixed base operator and construction. Married Barbara Walker in 1947; four children, Ronald, Blair, Ellen and Dana; two grandchildren.

ARTHUR THORSEN, born Chicago, IL, July 11, 1918.

School attended: Chicago Academy of Fine Arts.

Began Flying: Feb. 8, 1943. Cal-Aero Academy, Ontario, CA. Stearman PT 13B.

Entered Aviation Cadet Training, June 1942, Santa Ana, CA.

Training Bases: Cal Aero Academy, Ontario, CA, War Eagle Field, Lancaster, CA, Williams Field, Chandler, AZ. Muroc, CA, Salinas, CA, Lomita, CA, Grand Central Air Terminal, Glendale, CA.

Assigned to: 55th Fighter Gp., 38th Sdn., Nuthamstead and Wormingford, England.

Theatre: European.

Missions: Mostly escort, but many strafing sweeps.

No. of Missions: 67.

Kills: 1-1/2 ME 109s.

Retired USAFR: July 11, 1978

Rank: Lieutenant Colonel.

Occupation after retirement; Script writer, CBS

Married 1965: Barbara Hess. Three children from former marriage. Her three children also from a former marriage.

Grandchildren: three: Christopher, Naomi, Julie.

GUNVALD THORSEN, (Bert), was born in Norway on Mar. 1, 1924 and grew up in Brooklyn, NY. He enlisted in the Army Air Corps in August, 1942 and took the cadet exam at Chanute Field, IL in December of that year.

Flight training took place at Santa Ana Army Air Base, CA, Thunderbird Field in Phoenix for primary in PT-17s, Marana at Tucson for basic training in BT-13s, and Luke Field, Phoenix for advanced in AT-6s. He graduated in November 1943 in Class 43-J.

After graduation, Bert went to Redding, CA to check out in P-39s and to Santa Ana for P-38s. He joined the 49th Fighter Sdn., 14th Gp., in Triolo, Italy in March 1944. He flew 52 missions before returning stateside in September, 1944.

Bert instructed in P-38s at Santa Rosa Army Air Base, CA until he was discharged in August, 1945 at the age of 21. He is currently living in Whittier, CA and flying ultralights for fun.

JOHN T. TIPPS, was born Jan. 25, 1923 at Wilson, OK. He entered the A/C program in 1942 at Santa Ana, CA. He graduated at Luke Field in the Class of 43I where he flew P-40. He was stationed at Santa Rosa, CA with the 367th F.G. 192 F.S. where he flew P-39.

He checked out in P-38 at San Diego, CA and went overseas as a replacement pilot.

He was assigned to the 479th F.G.434th F.S in England where he flew 64 combat missions and received the Distinguished Flying Cross Air Medal with clusters, and the Purple Heart.

He shot the first ME-109 down in the 479th F.G. while flying tail end charlie. (Hot for one Day). Bailed out Oct. 18, 1944. Was retired for disability Oct. 8, 1945.

He and his wife Irene now live in Ardmore, OK.

RAYMOND F. TOLIVER, Col., USAF, (ret), was born and raised in Ft. Collins, CO. He joined the Army Air Corps in 1937, graduating from Randolph and Kelly Fields, TX, in October 1938, Class 38C, as a Pursuit Pilot, Second Lieutenant.

He was stationed at Mitchell Field, NY and in early 1940, due to the so-called "Phony War" and getting only four or five flying hours per month, joined TWA Airlines as a pilot. Needing more money than a co-pilot drew at that time, and having accumulated much flying time in twin and four-engine aircraft (DC-3s and Boeing 307B Stratoliners) he took a leave of absence from TWA in March 1941 and joined the RAF Ferry Command and flew Hudsons and Liberators across the North Atlantic to Prestwick Scotland.

In March 1942, he came back into the USAAF as a test pilot stationed at Wright Patterson Fields in Dayton, OH here he tested P-38s. He went to the Pacific Theater of Operations in January 1945 but saw no combat. Since WWII he has served as the commander of the USAF

20th Tactical Fighter Wing, two tours in the Pentagon (very reluctantly!) and retired as a Colonel in October 1965. He has over 9000 hours as a pilot in more than 225 different types of aircraft, including about 4000 hours in single-engine jet aircraft.

Toliver is the author and co-author of several books about fighter aces including BLOND KNIGHT OF GERMANY, HORRIDO!, FIGHTER ACES OF THE LUFTWAFFE, FIGHTER ACES OF USA and THE INTERROGATOR. He is presently working on the official biography of Luftwaffe fighter Gen. Adolf Galland, two books on the Japanese aces and a novel. BLOND KNIGHT OF GERMANY has been published in Germany under the title HOLT HARTMANN VOM HIMMEL and is currently in its 52nd hardback printing.

He was the Historian of the American Fighter Aces Assn. from its inception in 1960 until 1984. He is a member of the Society of Experimental Test Pilots, Quiet Birdmen, Combat Pilots Assn. and the Authors Guild. He is also a member of the Order of Daedalians and Honorary Member of the German Fighter Pilots Assn, and the Japanese Zero Fighter Pilots Assn.

GEORGE I. TOUMANOFF, was born in Russia while his parents were escaping from the Bolsheviks. His family arrived in the USA via Turkey, just in time for his fourth birthday. From then on it was growing up in America, graduating from M.I.T. just after Pearl Harbor and going to work for Lockheed as an aeronautical engineer. Shortly he found himself in the Experimental Flight Test Department and a piggy back engineer-observer in numerous P-38 test flights.

By far the most interesting program was converting an experimental, extended fuselage, genuinely two place P-38 into a flying wind tunnel. Experimental laminar flow airfoils were built as "gloves" over the wings between the nacelle and aileron on each side and extensive instrumentation was installed to measure lift, drag, boundary layer profile and eventually a boundary layer control suction system installed on the left side.

An in-flight photo of this airplane, with test pilot Ernie Claypool up front and George in the rear cockpit, appears on page 22 of the 23rd issue of Lockheed Horizons which was distributed at the 50th Convention.

GEORGE I. TRIPP, JR., was born in Bergen, a small rural farming community in New York. Enlisted in the Air Corps in October 1940 graduating from Chanute Field as an airplane mechanic. When eligible, transferred to cadet training and graduated with class of 43-D. After considerable time in the P-39, was transferred to Santa Ana for P-38 transition.

Graduated with the Class of 43-D at Luke Field, Phoenix, AZ. Was assigned to the 444th FS at Hamilton Field for transistion in P-39s. Later helped train new pilots until October 1943 when transferred to Orange County Airport for checking out in P-38s. After 20 hours was sent to England as a replacement pilot in the 370th FB Gp. - 485th FB Sdn. - 9th Air Force. This group had trained in US in P-47s but when they arrived in Andover, England were given P-38s. Being new to the Squadron, had to wait for others to train in the P-38.

After 20 hours was sent to England to the 9th Air Force. On June 25 a mission to Rouen was hit by flak causing fire behind cockpit and bailout. After free falling to 1200' practically fell into friendly resistance people near Lisieux. With the aid of French helpers evaded capture for two months before returning through front lines to British forces. Returned to the US and trained new P-38 pilots at Santa Rosa, CA. Released from active duty in August 1945 and remained in reserves until December 1965. *(See photos, page 140)*

NICK UDORVICH, Madrid, IA. Entered Air Force in 1941. Basic Training at Jefferson Barracks, St. Louis, MO. He joined the 459th when it was formed. Nick left the 459th, July 1, 1945. He was discharged from the Air Force Nov. 16, 1945 as T/Sergeant.

Nick has been in sales work almost entirely since leaving the service. He has been selling speciality advertising for the past 18 years and is still selling part-time.

Being a crew chief on P-38s for a couple of years was a very rewarding and interesting experience. The plane was quiet easy to maintain, except for changing the hydraulic pumps when the aileron booster control was added.

I am proud and honored to boast of my pilot, the late Capt. Walter F. Duke. To me, he is the most unsung hero, having 20-1/2 Japanese planes to his credit. Who else in CBI could top this?

Capt. Duke was a rather quiet individual, soft spoken and all business. Once he got in the cockpit for a mission, this flight leader was very intense and ready to get his job done.

We of the 459th Twin Dragon Sdn. have a convention every two years which we enjoy. My regret is not having Capt. Duke to renew memories with as so many crew chiefs and pilots do of our past experiences.

WALTER J. URBANK, was born on Jan. 18, 1922 in Los Angeles, CA. He grew up in this area, attended school, and received a scholarship to the University of Southern California. It was at this time that the bombing of Pearl Harbor took place in 1941. Walter, having always wanted to fly, convinced his parents that it was best he enlist in the aviation cadet program. At that time, most cadets were being sent to Santa Ana, CA for the beginning of their training. But not Walter. He wound up in Nashville, TN and entered the Southeast Air Force Training Command. Primary was at Carlstrom Field, FL, basic flying at Bainbridge, GA, and he graduated from advanced single engine flying school, Marianna, FL, in April of 1943 as a Second Lieutenant. Although flying schools continually had a symbol depicting "prepare for combat," Walt wound up as a basic flying instructor in Newport, AR. He stayed there from April 1943 until about October of 1944, at which time he arranged for a transfer to P-38 Photo Recon. Replacement school in Coffeeville, KS. He finished the flying training in Photo Recon. and joined the 8th Photo Recon Sdn. in the South Pacific, in the Philippines, in 1945. He flew missions throughout 1945 from bases in the Philippines and Okinawa and as part of the occupational Air Force in a little town called Chofu, Japan.

Photo Recon. missions were flown either singly or in twos and covered a very large area in the South Pacific from Borneo up through Formosa to Indo China. They were generally flown at altitudes around 20,000 feet, unless there was reason to fly low oblique photo missions over areas such as Mindanao or other areas where troop operations were deemed to be probable. There was not much evidence of Japanese fighter influence in the South Pacific by this late in the war. Most of the activity putting photo recon pilots in peril included long-distance missions, bad weather, and occasional Japanese anti-aircraft.

The closest Walter ever came to being shot down occurred actually two days after the war was over when Gen. MacArthur had indicated that photo missions would continue to be flown, at least over the southern part of the Japanese island of Kyushu. At that point, two airplanes, Walt's being one of them, had to fly at a relatively low altitude over Kyushu to take photos of airdromes and certain cities. The Japanese apparently took issue with this, and chased he and his buddy all the way across Kyushu back to the island of Okinawa. He escaped unscathed throughout the war. He blames it on pilot skill, but there might have been someone else watching over him. He received the Air Medal and several Battle Ribbons for the various battles evident in the South Pacific. He stayed in the reserves for awhile, subsequently resigned his commission, and has been employed as a personnel manager for Rockwell International until this time.

Walt was able to document his P-38 flying history during the war in an article published in the July 1984 issue of *Air Classics Magazine,* by Challenge Publications, entitled, "Photo Joe."

ROBERT ARTHUR WADE, was born on June 8, 1920, in Mattawamkeag, ME, youngest child of Flora (Call) and William Wade. At the age of eight they moved to Hartland, ME where he was educated and spent the rest of his life. At the age of 21 he married and was called in the draft to help defend the country. He was placed in the USAAF. He received his basic training in Windsor Locks, CT. He was transferred to Peterson Field in Colorado Springs, CO where they lived in tents until the airfield construction was completed. Robert then went to Will Rodgers Field in Oklahoma City, OK. Maneuvers were at Cox Field, Paris, TX in preparation for overseas duty. He was assigned to the 31st Photo Recon. Sdn. and worked in maintenance, camera repair, small motors, etc. The point of embarcation was Camp Shanks, NY.

On Mar. 13, 1944, the Isle de'France left New York Harbor transporting this unit as well as 12,000 replacement troops to the E.T.O. They arrived in Chalgrove, England Mar. 23, 1944. The 31st came under the 10th Gp. which came directly under the 9th Air Force. They worked for Gen. Patton's Third Army and Gen. Simpson's Ninth Army mapping the coastlines for enemy anti-aircraft installations and gun emplacements of France, Belgium and Holland preparing for the forthcoming invasion. Road and rail movements were photographed also. The planes were unarmed. Speed, skill and surprise their only defense. They filmed Normandy Beach at 50' and the Squadron was awarded the Presidential Citation for their part in the invasion. They went into France from Omaha Beach, Aug. 7, 1944. Their first stop was Rennes, but as Patton kept advancing it was a short stay. They lived in tents, trailers, leantos, anything available, trying to keep the photos ahead for the infantry. It was work, work and more work. After crossing France they ended up in Nurenberg, Germany on Apr. 30, 1945. V.E. Day was May 7, 1945. After the end of the war the 31st was to be among the Occupational Air Forces. With 93 points Robert was to be discharged under the point system as were many others.

The planes that flew on the 2,872 photo missions were F-5 Lightnings, F-6 (P-5), C-78, AT-6 and L-4 models. Robert was a laboratory equipment repairman. He inspected, repaired and maintained physical equipment of photo labs and darkrooms. Repaired small electric motors, printing machines, enlarging machines and generators. He also worked on the line fueling planes, doing light mechanical work, etc.

Robert was discharged Oct. 28, 1945 with the rank of Sergeant at Ft. Devens, MA. The battles and campaigns were air offensives in Europe, Northern France, Normandy, Rhineland, Ardennes and Central Europe. He received the Good Conduct Medal, Campaign Ribbon for European African Middle Eastern Theatre and the Distinguished Unit Badge as well as the Presidential Citation for the unit.

Robert married Muriel T. Brooks of Pittsfield, ME on Mar. 21, 1942. He was the father of four children—three daughters and one son. He has six grandchildren. After discharge he worked at the local Tannery, shoe shop and retired from Cooper Weymouth and Peterson, a machine shop in Clinton, ME. He died Sept. 30, 1986 at the age of 66.

JACK G. WALKER, Aviation Cadet Class 42-E, Stockton, CA. Assigned 55 Pursuit Gp., Paine and McCord Fields, WA. Reported 52nd Fighter Gp., Long Beach, CA P-38 Air Defense. Sailed to England on the "Queen Mary". Had collision at sea with H.M.S. Curacao. Sent to northern Ireland for new P-38s. Checked out in Spitfire and Magister. Started mass flight to North Africa, but had a single engine on grass field in Wales, England—ran out of runway—Class 26—one Lightning—one month hitchhiking to Algeria. Flew combat in Tunisian Campaign. Tactics instructor in Casa Blanca until rejoining 82nd Fighter at Foggia, Italy. Flew 51 missions: four destroyed, two probables, two damaged.

Stateside Ontario, CA A.A.B. P-38 and P-59 (Jet) instructor. Joined California Air National Guard P-51s and F-80s. Recalled for Korea—no combat. Instructor Air Command and Staff and R.O.T.C. (U.C.L.A.).

Returned to California Air National Guard C-47s, C-45s, B-25s and T-33s. Retired as full Colonel. Now flying with the "Condor" Sdn., a Civil Air Patrol and Reserve T-6 "Texan" unit based at Van Nuys, CA. Currently resides in San Diego, CA and has a son flying for United Airlines.

WILLIAM ALLAN WALKER, Col. USAF (Ret.), was born and raised in East Texas. He graduated from high school in Vivian, LA, and attended the University of Alabama, from which he graduated cum laude in 1940, with a degree in aeronautical engineering and a commission in the Army Corps of Engineers Reserve. He also was a graduate of the pioneer class of the Civil Aeronautics Authority Civilian Pilot Training Program.

After graduation "Tex" Walker worked as an engineer for Pan American Airways at La Guardia Field in New York City. With the war in Europe heating up, he was ordered to active duty with the Army Air Corps at Wright Field, OH, in May, 1941. He left there for flying school in December, 1941, and graduated at Foster Field, Victoria, TX, in August, 1942, in Class 42G. From flying school he went directly to the Flight Test Division at Wright Field. After a brief but intensive indoctrination in flight testing methods, which included checking out in a couple of dozen types of airplanes, he was sent to the Los Angeles area for duty as acceptance test pilot, rotating through various aircraft plants, and filing each C.O.'s in-basket with requests for transfer to combat duty. He *hated* spending Saturday nights dancing with the starlets at the Beverly Wilshire and Roosevelt Hotel Officers' Clubs.

Lt. Walker was assigned to Lockheed Air Terminal in the spring of 1943, and immediately fell in love-with the P-38. He was given a piggy-back ride in a piggy-back configured YP-38 called "Porky" by Maj. Clarence Shoop (later a major general and commander of the California Air National Guard), checked out and told to get busy. In due course Walker was put in charge of Army Flight Operations at LAT, and settled into the wonderful job of trying to fly P-38s as fast as Lockheed could build them. A part of the job was to give visiting military brass rides in Porky. This coveted privilege was also extended to certain deserving Lockheed employees. One of the most deserving just happened to be a luscious young lady who worked in the Lockheed flight office.

Poor Lt. Walter never got a chance to shoot down a Zero or a Messerschmitt, but he did manage to bag a P-38. It happened one day while he was doing an armament system test at the Muroc gunnery range. As he pulled up out of a firing run on a ground target he flew into a swarm of richochet 20mm slugs, which shattered the canopy and necesitated an emergency landing at Muroc. For a moment he thought the Japs were here for sure.

The Lieutenant never got to the wars, but he went on to a full career in the Regular Air Force, almost all of his service being in research and development, including six years at the Flight Test Center at Edwards Air Force Base. His flight records show pilot time in more than a hundred types of aircraft, many of them X-types which never made it to production.

Col. Walker retired in January 1969 at Wright Field, less than a hundred yards from where he started in 1941. He now lives, with his wife Gloria, in Santa Monica, CA.

GEORGE TAPPEN WALTER, was born in Atlantic City, NJ, Oct. 10, 1922. Joined the United States Army Air Force October 1942.

Pre-flight, Santa Ana, CA; primary flight training Santa Maria, CA; basic Lemoore, CA; advanced, Williams Field, AZ; class 43H.

P-38 Operational Training Unit at Muroc, CA; Salinas, CA: Orange Co., CA; Lomita, CA; and Berteaux, Algeria. Assigned 37th Fighter Sdn., 14th Fighter Gp., 15th Air Force, February-August 1944, Triola #7, Foggia, Italy.

Completed 50 combat missions totaling 250 hours. Received Air Medal with five Oak Leaf Clusters. 14th Fighter Gp. received Distinguished Unit Citation.

Returned stateside September 1944 at P-38 Operational Training Unit instructor at Santa Maria Army Air Force Base Unit, CA, October 1944 to July 1945. Transferred to Gowan Field, Boise, ID, until discharge in October 1945.

Married Shirley Blanche Stratton June,

1948. Graduated BSME Lafayette College 1950. Fathered three sons; one granddaughter. Retired June 1988. Resides, Buckingham, PA.

BRUCE C. WARD, Major, born Strathmore, CA July 11, 1920. Entered service Santa Ana, CA Nov. 28, 1942. Graduated Williams Field, AZ Nov. 3, 1943 Class 43-J.

Married Norma Underhill Nov. 13, 1943. Trained in P-38 Salinas, Lomita, Van Nuys, Ontario, CA.

Arrived Firth of Clyde March 1943, Stone, Goxhill, Honington 364th Fighter Gp. 384th Sdn. replacement pilot.

First mission Apr. 30, 1944. Flew 35 missions, 149 combat hours in P-38J. Flew longest mission to date. May 13, 1944. To Demmin Ger.* Flew 36 missions 167 hours in P-51 D 5NA. Both planes named, "My Norma" AM w/ seven Oak Leaf Clusters, Distinguished Flying Cross w/one Oak Leaf Cluster.

Medical Discharge resulting from back injury caused in high speed dive pull out. ME 109 went in. Pilot apparently dead from gun fire. Wrinkled wings of P-51 returned to base at 175 top speed.

Norma and I live in Strathmore, CA. We have just retired after 38 years in our own Insurance Agency. Two daughters four grand-daughters. Incidentally Norma's first flight ever was piggy-back in a P-38 December 1943 at Salinas. Have started seeing USA in fifth wheel rig purchased Jan. 8, 1987

RUSS WARD, graduated from Aviation Cadets in Class 44-C, Oklahoma City, Photo Recon. training April and May 1944. Flew 74 missions in 31st Photo Recon. Sdn., 10th Photo Gp. July 1944 to May 1945.

Graduated from Oregon State College in May 1949. Did some Civil Engineering work. Started flying for United Airlines Feb. 9, 1953, retiring from United Airlines on Mar. 1, 1985, co-pilot DC-3 to Captain B-747.

EDWIN E. WASIL, 1st Lieutenant, born Chicago, IL, May 14, 1920. Attended schools in Chicago. Enlisted aviation cadet April, 1942. Entered service at Santa Ana, CA. Graduated Nov. 23, 1943, Class of 43J at Williams Field, AZ. Flew the RP322 during advanced training. Further training at Salinas, CA, Grand Central Airport, Glendale, CA and Ontario, CA.

Departed for England from Ft. Hamilton, NY, March of 1944 on the Queen Mary. Arrived England at the Firth of Clyde March, 1944, Stone, Gox Hill, Chorley, England.

Assigned to 20th Fighter Gp., 55th Sdn. stationed at Wittering, England, A.R.A.F. Base, April, 1944. Flew Droop Snoot Mission.

Wasil proved the P-38 was virtually indestructible. While strafing, flew too low and clipped the tops of a pine tree forest. Though losing 3" off of the right wing and damaged right rudder and Aerilon, and right engine fire, was able to feather right engine and return to base at Wittering after flying over Brussels at 18,000 ft. somehow evading ground flak flying on single engine.

On another mission, he proved that you could split-s the P-38 chasing a ME109 from 25,000 ft. and recover to talk about the experience. His plane was named "Mama's Boy".

Flew escort for B-17s and B-24s. Dive, skip bombed and strafed ground target, locomotive, trains and air fields. Flew air cover for invasion of Normandy D-day -1 to D-Day +5. The 20th Gp. was assigned P-51s in July of 1944.

The 20th escorted B-17s on their Frantic VI shuttle run to Russia, Italy and return to England. Participated in 68 missions, a total of 271 combat hours. Completed tour of duty September, 1944. Decorations: D.F.C., Air Medal with six Oak Leaf Clusters, Presidential Citation.

On return, instructed at advanced twin engine school, Marfa, TX. Was stationed at McChord A.F.B. as motor pool maintenance officer unit separation September, 1945.

Married Lt. Carmen Olaiz, Army Nurse Corp. in July, 1946. Have two children and four grandchildren to date. Resides at Palos Verdes, CA.

President and charter member of 20th Fighter Gp. Association, charter member of P-38 Association.

JOHN E. WATKINS, from Galena, KS enlisted in the USAAF on Dec. 24, 1940. His basic training was at Jefferson Barracks, MO. His first duty assignment was Selfridge Field, MI. Left Selfridge in April 1942 in a cadre to form a P-38 Gp. (82nd Fighter Gp.) at Muroc, CA.

The group went overseas in late 1942 to North Ireland, North Africa, Sicily and up through Italy to Foggia. After the war, returned to the States to finish his career of 23 years.

He retired from the Air Force in December 1963 with a medical disability. He owned and operated a service station in Smyrna, TN until April 1976 when he fully retired. His jobs in aircraft maintenance were: assistant crew chief, crew chief, flight chief and quality control inspector. He married Margaret Somes of Sarnia, Ontario, Canada, has three children, seven grandchildren, and resides in Smyrna, TN.

GUY E. WATSON, was born in Los Angeles CA on Nov. 1, 1923. His favorite pastime as a child was carving propellers from used popsicle sticks and watching airplanes at the local airport. Guy's family moved around the country during his early years and WW II found him in San Francisco enrolled in Mechanical Engineering at the University of California at Berkeley. He joined the USAAF in October 1942 and entered active duty in January 1943. His basic training was at Lincoln, NE, College Training Detachment at Missoula, MT. Classification was at Santa Ana, CA and he was assigned to pilot training at the primary field at King City, CA. He won the "Hot Pilot" award for his primary class squadron and was one of six pilots selected to fly AT-6's in Basic Flying School at Marana, AZ and P-38s in Advanced Flying School at Williams Field, Chandler, AZ. The P-38 program was cancelled by the time he reached Williams Field so he had to wait until later to check out in the P-38. Graduation was at Williams in April 1944.

No leaves were being given and Guy's class left for gunnery school at Ajo, AZ the day after graduation. From there, he was assigned to P-39 transition at Victorville, CA and then to Portland, OR where he checked out in P-63s. One day in September 1944, a Lockheed representative flew in from Burbank and proceeded to beat up the airfield with loops, Immelmans, rolls and other nip-ups mostly on single engine making believers out of the crowd gathered to watch. On October 4 Guy first flew the P-38 for one hour after having a 40 minute check-out in a TP-38. The CO at Portland was Maj. George Laven, who crossed paths with Guy again in the 49th Fighter Gp.

After completing Operational Training at Redmond, OR, Guy was sent to Hamilton Field, CA awaiting assignment to overseas. He was issued a complete set of winter flying gear and put on a DC-4 for Nadzab, New Guinea. Needless to say the winter gear was traded in at Hawaii.

Guy joined the 49th Fighter Gp., 8th Sdn. in March 1945 at Lingayen Gulf, Philippines. His longest mission was six hours and 40 minutes flying cover for a Navy convoy East of Formosa on Aug. 29, 1945. In addition to being a Flight Leader, he was Group Training Aids Officer. On August 16, his squadron moved from Lingayen to Okinawa. He was on the mission over Kyushu when Lt. Hall and Col. Tice became the first Americans to land on Japanese soil. His last combat mission was on Sept. 5, 1945 making 30 total missions and 100 combat hours.

His next move was to Atsugi, Japan on September 15. In January 1946, the squadron gave up its P-38s and were given P-51s. Guy, however had one last flight with Maj. James (Duckbutt) Watkins in a TP-38. The 49th left Atsugi in February 1946 for Chitosi, Hokkaido. Guy was assigned additional duty of Post Engineer. He contracted malaria in May, was taken to the Army hospital in Tokyo and was sent home arriving in San Francisco in June 1946. He was promoted to Captain and was discharged in December 1946. Guy joined the California National Guard and flew P-51s with them until he graduated from the University of California in June 1950.

Guy retired as Technical Consultant with Lockheed Missiles and Space Company in January 1988. He currently is president of Watson Mark Corporation, making wooden propellers for antique and experiment airplanes...only now he makes them out of birch, walnut and mahogany and not used popsicle sticks.

RALPH K. WATTS, was born in Los Angeles, CA graduating from Compton Junior College in 1940. He joined the Army Air Corps Flying Cadet Program graduating from Randolph and Brooks Field in the summer of 1941 with Class 41-G. Assigned to the 4th Mapping Sdn. and later as a charter member of the 8th Photo Sdn., he flew the F-2 Beechcraft and the Lockheed Hudson Bomber.

In early 1942 he was given one and a half hours of flying time and two landings in a P-38 E at Burbank, CA. He then deployed with "A Flight" of the 8th Photo Sdn. to Australia and New Guinea aboard the USS Coolidge. His orders stated he was "Combat Ready and a Qualified P-38 and F-4 Test Pilot". On arrival, he started test flying aircraft transported disassembled aboard the USS Coolidge. The right engine ran away on the first test flight and a normal one-engine landing was effected.

Test flying and combat missions were flown in New Guinea, Guadalcanal, Solomons and adjacent areas in both the P-38 Photo Reccon and B-17 Photo Reccon aircraft. He was awarded the Silver Star, Distinguished Flying Cross and Air Medal for combat actions.

Watts pursued an Air Force Career serving in both the Strategic and Tactical Air Commends. He flew Jet bombers and fighters including the B-47 Jet Bomber and the Mach 2 F-104 Star fighter. His greatest thrill and fun flights were made in the P-38 and F-104 Star Fighter, both Super Aircraft designed by Kelly Johnson and built by Lockheed Aircraft. Watts retired as a Regular Air Force Colonel and resides in Lebanon, OR.

GEORGE S. WEART, was born in Chicago, IL, entered the U.S.M.A. (West Point, NY) in 1939. In June 1942, Air Corps aspirants went to the Flying Training Commands and joined the Class of '42 K. Donned Silver Wings in December 1942. Graduated from U.S.M.A. on Jan. 19, 1943. Reported to Guadalcanal in December 1943 assigned to 12th Fighter Sdn., 18th Fighter Gp. Transitioned to P-38. Combat from Treasure Island escorting B-24s and B-25s over Rabaul and hunting Jap ships and smaller boats.

On to New Guinea (Sansapor) and the Halamaheras whence the top mission was B-24 escort against part of the Jap Fleet in Brunei Bay-1000 miles to target and 1000 miles back to base-accomplished thru use of one 310 gallon Ferry Tank and one 165 gallon normal tank. Capt. Weart had, while still in Guadalcanal, developed the safe release technique of the large tank.

Next high point of P-38 tour was leading squadron into Lingayan Gulf Airstrip returning from B-24 escort over Formosa.

Shot down two Zeros in Battle for Leyte. Shot down Nick twin engine night fighter South of Formosa.

Commanded the 366th and the 35th Tac Fighter Wings, at Phan Rang Air Base, South Vietnam.

Col. Weart was retiree in 1972 and resides in Visalia, CA.

ROGER D. WEATHERBEE, formerly of Lincoln, ME, entered service in November, 1942, as a cadet at age 18. He completed flight training as a twin-engine fighter pilot at Williams Field, Chandler, AZ, in January, 1944. After further training in P-39s and in P-38s he was sent to Italy by boat in June, 1944. He joined the 49th Fighter Sdn. 48th Fighter Gp. located north of Foggia, Italy.

He flew 50 combat missions with the 49th completing 42 of these before turning 20 years of age. His missions included escorting bombers, strafing, and dive bombing. He made four trips over Ploesti, Romania. He was assigned to Corsica in September, 1944, to cover the invasion of Southern France. Action included strafing airfields where losses were very heavy. He assisted in destruction of many ground targets including trains, bridges, and aircraft. On eight of his missions he flew as Squadron Leader.

He returned to the States where he married Lorraine Bradstreet and was based in Florida flying P-38s and P-51s until June, 1946. He was next assigned to Kimpo Air Base in Korea flying P-51s until February, 1947, when he elected to leave active service.

He obtained a BS degree from MIT and a MS from RPI. He was employed at Hamilton Standard in June, 1951 and is now retiring after 38 years as an Engineer. He and Lorraine have four children and three grandchildren and are living in Windsor Locks, CT. He remains active in flying and will never forget the thrill of flying the P-38!

EDGAR R. WEST, was born in North Kingsville, OH, 1921. In Parma, ID, at about age six, 1927, he observed a bi-plane barnstorming out of a pasture. Now hooked on flying but never expecting to have the opportunity. He went to High School in Caldwell, ID, and a brief period at Washington State, Pullman, WA. He went to Santa Ana Pre-Flight; PT-22s at Morton Air Academy, Blythe, CA; basic at Marana, AZ; then to Williams AFB and coveted Silver Wings.

He flew P-39s at Victorville, CA, P-63s and P-38s at Chico, CA, thence to the 485th FS, 370th FG in Belgium, ETO. He was out of the AF from 1946 to 1951. He flew P-51s with the Idaho ANG 1949-1952 (1951 and 1952, the 190th FBS Idaho ANG was on active duty). Tom Lanphier, who was credited while flying P-38s with downing Adm. Yamamoto, was the Commander of the ANG at Boise, ID. Ed flew F-86Fs at Victorville, CA, 1953, and F-86As and F-84Fs at Bentwaters, England, 91st FBS, 81st FBW, 1954-1957. The 81st Wing (F-84Fs) was the winner of the USAFE and USAF Special Weapons Competition (over the Shoulder, Dive and Toss bombing) in 1956. He was in ATC at Laredo, TX 1958-1963 instructing in T-33s and T37s had first class of STUs from AF Academy. Instructed in T-37s and T-38s at Craig AFB, Selma, AL.

Then Vietnam, 2nd AD, Thon Son Nhut 1964-65 as head of the Tactical Analysis Section with one 105, 100, F4, and 104 pilot to evaluate units and tactics. He flew AIE, 0-10 and Blue Canoe (Cessna 310). He worked for Gen. George Sisler, who later was killed in a T-38 while Commanding General of ATC. Ed West retired as Lieutenant Colonel from Williams AFB in 1968. He now spends most days golfing, driving a car with license plates NO 1 GUN! Ha! Particular memories of combat in P-38 obviate activities that could be done only in a P-38.

He took off behind another heavily ordnanced 38 who aborted early. He was able by use of combat flaps to rise like an elevator and easily clear. He dropped a thousand pounder thru an already cratered upper deck of Rhine River Bridge at Cologne, thus cratering the lower deck. Dive flaps on P-38 provided a more stable platform and long dive run (provided one's guts allowed with beaucoup ACK-ACK) leading to extremely accurate delivery. He shot up a fuel truck, on a road in a canyon, headed for ME 262s on autobahn in South Germany.

This was possible only because of nose 50s and 20 MM and dive flaps. The 370th Gp. flew mostly ground support in 9th TAC though on one occasion, while escorting A-26s (top rt cover) he chased an ME 262 at 450 plus indicated down hill and a good angle but the 262 accelerated away. That was the first instance he saw another aircraft do something a P-38 could not do better. His last 20 missions were in P-51s and you can bet the "pucker" factor for he and his buddies was greater than when in the Counter-Rotating P-38s!

RICHARD L. WEST, was the leading ace of the 35th Fighter Sdn., 8th Fighter Gp., 5th Air Force. He was an ace in two types of aircraft having scored six kills in two missions with the P-40 and eight kills with the P-38 for a combined kill of 14 Japanese planes (12 fighters and two bombers). He was awarded the Distinguished Service Cross, the Silver Star, the Distinguished Flying Cross with one Cluster, the Air Medal with seven Clusters and was entitled to wear two Presidential Unit Citations.

West did not make the military a career, having returned to civilian life after the war in the Pacific ended.

He is the author of a published book of poetry and he and his wife, Norma, are the parents of one daughter, Dr. Lynda West, an assistant professor at the University of Missouri, Columbia.

He resides today in Chillicothes, MO: was born in Trenton, MO on July 12, 1921 and he grew up over the Pacific Ocean between New Guinea and New Britian on Sept. 22, 1943.

WARREN E. WHIPPLE, was born Sept. 19, 1919 in Oakland, CA. He enlisted in the Army Air Corps in November 1942 and graduated as a single engine pilot from Williams Field in the Class of 44-C. He received his initial combat training in P-39s before undergoing flight training in the P-38 at Ontario, CA.

Overseas he arrived in New Guinea and was assigned to the 44th Fighter Sdn., 13th Air Force. He flew combat missions in support of the Philippine campaign, China offensive and the Borneo invasion. From bases in the Philippines, he flew wide ranging missions in the P-38 to strike Japanese target in French-Indo China. Some missions exceeded eight hours which was pushing the absolute range of the P-38.

He served as flight leader and flew 53 combat missions totalling 344 combat hours. His decorations include the Air Medal with two Oak Leaf Clusters.

Whipple continued military flying in P-51s until the Korea conflict. He recently retired from a major aerospace company as project manager on a satellite program. He currently resides in Southern California.

JACK WILKES

RALPH P. WILLETT, was born in Holyoke, MA in 1925 and enlisted in the Army Air Corps as a Cadet in 1943. I got pneumonia in North Carolina, spent three and one half weeks in the hospital, was sent home two weeks to recuperate, and when I returned, my cadet class was gone. Reassigned as an aircraft mechanic, I was sent to Italy.

I became a crew chief on F-5s. The war ended and I became a civilian. Under the G.I. Bill I got my commercial pilots license along with an instructors permit. I am doing research on P-38s and also my old 3rd Photo Group. I build and fly radio controlled model airplanes. I am a supervisor with Northeast Utilities and hope to retire in another year. I am married to the former Marielle Godbout and we have two boys and one girl; two grandchildren.

ALFRED A. WILLIAMS, born Portland, OR, on Apr. 5, 1912. Schools: Cathedral Grade, Columbia and Lincoln Hi, all in Portland; Mt. Angel College, OR; St. Edward Seminary, Seattle, WA; St. Patrick Seminary, Menlo Park, CA. Ordained priest for Portland Archdiocese June 3, 1939.

Joined US Army Reserve, June, 1940. Called to active duty as 1st Lieutenant, and assigned as Chaplain of 55th Fighter Gp., Portland Air Base. Assigned to Alaskan Army Command, September, 1942. Sent to Adak Island, the Aleutians, October, 1942 and named Chaplain of 343rd Fighter Gp. of 11th Air Force. Transferred to Amchitka Island in July, 1943. While in the Aleutians also served in Chaplain duties to various other organizations of Army and

Navy who were without their own Chaplains. Reassigned to Stateside duty in October, 1943 and reported to 3rd Air Force station at Drew Field, FL (Aircraft Warning Training Units). Transferred to Hunter Field, GA as Base Chaplain to 3rd AF Staging Wing. In May, 1945, transferred to Topeka Air Base, KS, later to Merced Field, CA (named Castle Field later while I was there). In April, 1946, assigned as Ass't Staff Chaplain, 2nd Air Force, Colorado Springs, CO. In October, 1946, was given honorable discharge at Andrew Air Base, Washington, DC and then reported back to my Archbishop in Portland, OR for church assignment, November, 1946.

Experience with P-38 outfits came during assignment as Chaplain of the 55th and 343rd Fighters Gp. discharged as a Major.

ROBERT J. WILLIAMSON, Robert J. Wilamson was born June 10, 1924 in Chicago, IL. Attended the Northern Illinois College of Optometry. In 1943 Bob began flying and entered Aviation Cadet Training in Texas. Trained at Pine Bluff, AR, Coffeyville, KS and Moore Field, TX. He was a 2nd Air Force instructor, retiring as a Lt. Colonel in 1984. Bob is an Optometrist and he and his wife, Marguerite, have six children and five grandchildren.

RICHARD E. WILLSIE, Col. (Dick), was born in Michigan and raised in Southern California. Joined the Army Air Corps in 1942.

After graduating from flying school, Dick became one of the first American Night Fighter pilots. He trained in Florida flying P-70s specializing in night intruder tactics. Mid 1943 he was in Africa as a squadron pilot in a Beaufighter Night Fighter Sdn.

While in Africa he flew A-20s and managed an in theater check out in P-38s. Dick was then transferred to Italy where he joined the 82nd Fighter GP. in December 1943. He was assigned to the 96th Sdn., and flew combat missions for one year. During this time, he shot down two a/c, flew 82 missions was promoted to major, and became Commander of the squadron He was shot down while strafing airfields near Ploesti, but in a hair raising manner was rescued by his fellow pilots and Dick Andrews who landed his plane and in piggy back fashion flew to a safe landing in Russia, the first such rescue of the war.

After the war Willsie became an airport owner operator, until he was recalled for the Korean War. In order after being recalled, he flew F-84s in Texas and Japan during the Korean War, as Deputy director 42nd Air Div. and Squadron Commander 560th Strategic Fighter Sdn. In England he flew FI-Is as Commander 92nd. Tactical.Fighter Sdn.

Was the USAF Advisor to the Indian Air National Guard. Attended the Armed Forces Staff College. Was the Chief General War Plans Division Tactical Air Command. Then Chief War Games and Analytical Studies USAF Tactical Air Warfare Center. During the Vietnam War he flew Sky Raider a/c as Commander 602nd Air Commando Sdn. A concurrent overseas assignment to Italy as Chief Air Ops. NATO Southern Region Headquarters. Attended the US Air Force Air War College. Then back to NATO as Chief Air Operations Supreme Headquarters Allied Powers Europe. (SHAPE) his final tour of duty completing 32 years with the Air Force was the United States Air Representative to 'Live Oak' the semi secret Political/ Military organization.

Tri-National, England, France, United States, responsible for maintaining open air-ground access routes to the city of Berlin.

CALVERT W. WILSON, was born and raised near the town of Cadiz, KY on Jan. 3, 1919. On Oct. 3, 1940, he enlisted in the Army Air Corps at Chanute Field, IL. In May 1941, he was assigned to the 94th "Hat-in-the-Ring" Pursuit Sdn. at Selfridge AFB, MI, as a aircraft mechanic. Dec. 10, 1941 moved to North Island Naval Air Station. Mar. 1, 1941, promoted from Private to Corporal. March 2, 1942 assigned to first aircraft as a helper on the 322 (British P-38).

In May, 1942 he was assigned to Muroc Lake, CA to the newly formed 82nd Fighter GP. (97th Fighter Sdn.), as a P-38 Crew Chief. September 1942, was sent to Europe on the Queen Mary and had a collision in the Atlantic with a British ship and they lost over 300 people. Docked at Londerry, Ireland and stayed there till January 1943, at which time was sent to North Africa. Wilson was on P-38s all during this time as a Crew Chief and later as a Flight Chief. Most of his time was spent in Italy (Foggia #11). He had 11 Battle Stars and two Presidential Unit Citations. He came home in June of 1945 as a M/Sgt. and overseas time of 32 months. He is now retired and lives on Table Rock Lake in the heart of the Ozarks. Twenty years military and ten years Civil Service. All with the good old United States Air Force. Amen

OSCAR R. WILSON, Capt., U.S. Army, born Jan. 14, 1918, Navarro Co., TX. Oscar entered active duty Oct. 1, 1943, 23rd Photo Recon. Sdn., 3rd Photo Recon. Gp., 90th Wing ETOUSAFF, Italy. His battles and campaigns are as follows: Rome, Arno, Southern France, Rhineland, North Appenines, Balkans WDGO 105-45 (Air). His decorations and citations are: Air Medal w/six Oak Leaf Clusters GO 31 HQ 12 Air Force, Feb. 16, 1945; DFC GO 9 HQ 12 AF Jan. 5, 1945; Dist. Unit Badge GO 2 HQ AAF/MTO Jan. 16, 1945; Good Conduct Medal, Air Medal, Def. Service Medal, American Theater Service Medal, EAME Theater Serv. Medal, and World War II Victory Medal.

As a Photo Reconnaissance Pilot, he operated a two engine reconnaissance aircraft to obtain photographic and intelligence information of enemy territory and installations. Took off and accomplished mission by photographing enemy positions. Reconnaissance made by camera and eye observations. To perform these duties it was necessary to have a thorough knowledge of meteorology, navigation and hazards due to low barometrical pressures.

As a Base Operations Officer, he directed aircraft clearance and maintained liaison between the base commander and Squadron commanders.

ROBERT A. WINDOFFER, (Bob) was born in Roll, IN on Sept. 16, 1922. On Feb. 2, 1942 he enlisted in the Army Air Corps and was sent to Ft. Hayes, OH. He applied for Aviation Cadet Training, was accepted and graduated Class 43-K at Aleo Field, Victoria, TX as a 2nd Lieutenant. He went through P-39 RTU training at Thomasville, GA, then P-38 RTU training at Van Nuys, CA. After this, he was assigned to the 7th Fighter Sdn., 49th Fighter Gp. and flew combat in New Guinea and the Philippines. While at Lingayen Gulf, he flew several combat missions during the Nipo Dam Operation to save the water supply for Manila which was successful. Until the end of the war, he flew combat missions in P-38s out of Motaba Strip, then volunteered for six months' occupation duty in Japan.

After the war he attended four years of college at Indiana and Purdue Universities. In January 1951 he was recalled to active duty, flew 150 combat missions in Korea and in 1967 he spent one year in Southeast Asia working days in MACV Hq. and flying combat out of Ton San Nut airport. He worked five years in the Pentagon and retired after 30 years' service as a Lt. Colonel. His decorations include the Air Medal with Clusters, D.F.C. with Clusters, the Bronze Star, ten Bronze Battle Stars PUC with two clusters and several foreign awards.

He retired September 1975 and resides in Austin, TX with his wife, Claire.

ROSSWELL R. WINKS, (Ross) was born in Palermo, NY and entered into the service on Jan. 20, 1942. When asked what service he wished to enlist in, he said Air Corps. In those days it was known as "Army Air Corps." He reported for service at Ft. Niagara, NY. He was assigned to the 48th Fighter Sdn., 14th Fighter Gp. His unit was 449th Fighter Sdn., attached to 14th Air Force. On Jan. 24, 1942 he was transferred to Baer Field, Ft. Wayne, IN and on

Apr. 26, 1942 on to Muroc Bombing and Gunnery Range in California.

On May 8, 1942 he was sent to North Island Naval Air Station in San Diego, CA and then to March Field, Riverside, CA. He went overseas to England on Aug. 5, 1942 aboard ship "West Point". He served in England from August 1942 to November 1942, then went to North Africa from November 1942 to July 1943. They asked for volunteers to go to China and India as an air base could be established, and he volunteered. He was in the India-China area from July 1943 to Sept. 23, 1944 He was a Staff Sergeant at time of medical discharge and was in ground crew and technical supply for the P-38.

STANLEY W. WOJCIK, was born in Glassport, PA on Sept. 13, 1920 and lived in East McKeesport, PA. He enlisted in the Army Air Corp on June 4, 1942 and trained as a single-engine pilot in the Southeast Training Command Class of 43E, operational training in P-40s at Sarasota, FL. He was a replacement P-40 pilot arriving in Africa. The assigned Air Force could not be located so he was absorbed into the P-38 1st Fighter Gp., 94th Fighter Sdn., 12th Air Force.

He had just enough P-38 transition to participate in the invasion of Italy at Salerno on Sept. 9, 1943. On a patrol strafing mission of some 400 enemy vehicles, he was shot down and bailed out at 2000 feet. As a POW of 20 months, he was in a German hospital and in OF LAG VA (Wiensberg) and Stalag Luft I (Barth) camps.

Graduated from the University of Pittsburgh with a Bachelor of Science Degree in Mechanical Engineering in 1947, he has been employed by Goodyear Aerospace Corporation, which is now Loral Corporation, Akron, OH.

ROBERT W. WOOD, born in Wrentham, MA, May 20, 1922, and moved to California in 1927 where I attended elementary and high schools in Yuba City.

I was always interested in flying but never received any aviation training outside the military. Entered Army Air Corps in Sacramento, CA, in 1942 and received pre-flight training at Santa Ana, CA. Completed Primary Flight Training at Ryan Field, Tucson, AZ, Basic Flight Training at Pecos, TX, and graduated from Advanced Flying School at Williams Field, AZ with the Class of 43-D.

After P-38 transition training at Muroc and Orange County Airport, I was sent overseas in June, 1943, landing at Amberly Field outside of Brisbane, Australia. In July, 1943, I was assigned to the Ninth Fighter Sdn., 49th Fighter Gp., Fifth Air Force, which at that time was flying out of Dobodura Air Strip in Buna, New Guinea. I flew 188 missions of all kinds including bomber cover, convoy cover, strafing, dive bombing, ground support, etc. I was credited with one Oscar aircraft. These missions were flown in both P-38 and P-47 aircraft since the Ninth Sdn., during one period, flew P-37 due to the inability to obtain replacement P-38 aircraft. During my service with the Ninth, the unit moved up through New Guinea and the Philippine Islands. I returned to the United States from Lingayen Gulf in March 1945.

During my tour with the Ninth I was privileged to fly with Dick Bong and many others who I shall never forget. My longest missions was from the Halmahera Islands to Borneo which required seven hours, 45 minutes flying time, a long time in a fighter. The most "hairy" missions were those to Rabaul, New Britain in the latter part of 1943.

I was awarded the Distinguished Flying Cross and the Air Medal with four clusters.

In 1946, I was separated from the Service at McCord Field, WA, with the rank of Major.

Since September 1946, I have worked for the Pacific Gas and Electric Company, retiring as San Joaquin Division Superintendent of Electric Operations in 1982.

Married to Rose Marie, have one son and two grandchildren residing in Sherman, TX.

My current address is: Robert W. Wood, 5743 N. Briarwood Lane, Fresno, CA 93711, telephone (209) 431-1092.

WILTON EUGENE WYCHE, was born in Gustine, Comanche Co., TX, Sept. 16, 1919.

He worked for Comanche County REA as a survey engineer for two years before joining military service. On Sept. 17, 1940, Wilton enlisted in the U.S. Army Air Corps and was assigned to Brooks Field, TX for basic training and further assignment. In July 1941 he was assigned to McCarren Field, NV (now Nellis AF Base) and from here received an appointment as Aviation Cadet in class 42-H. After graduation from Williams Field, AZ in August, 1942, Wilton was assigned to 38th Fighter Sdn., 55th Fighter Gp., at Paine Field, WA.

While stationed at Paine Field, Wilton's name became "Gene" which is still used to this date.

In August 1943 Gene accompanied the 55th Fighter Gp. to the ETO-stationed at Nuthampstead, and later at Wormingford, England. While here Gene flew 101 sorties over occupied Europe. The most remembered of these missions was the one of Mar. 3, 1944, when Gene accompanied Col. Jenkins and the rest of the 55th Fighter Gp. over Berlin.

After completing combat tour Gene received a variety of duty stations and position assignments, retiring from service as Lieutenant Colonel in October 1963.

Next for Gene was seven years flying for Air America in Laos, and seven years flying for the Department of Interior at Boise, ID.

Menieres Syndrome terminated Gene's aviation career in August 1980.

In November 1942 Gene and Miss Clara Powers were married and they are the parents of one son. They now reside in Tucson, AZ.

ADELBERT YAGER, joined the Army Apr. 1, 1942 and went to Ft. Niagara. As a group, we were entrained to Harding Field in Baton Rouge. About three weeks later we were sent to Mojave Desert about the first of May, 1942. From there we went to Long Beach, CA.

We were shipped overseas on the Queen Mary, Sept. 29, 1942. The Queen Mary sunk a British escort ship with over 300 lives lost at sea.

We then left Ireland on January of 1943 for North Africa in a convoy. Our ship, the Franconia, sprung a prop shaft and lagged behind the convoy to go through the Straits of Gibraltar alone at just sunrise. It was a beautiful sight with every hue of the rainbow in the sky and the blue, blue Mediterranean sea.

As a carpenter of the squadron, I was sent on advanced echelon at times to get set up for the rest of the squadron. From Leccee, Italy, I went on ahead. We had had a tent for our quarters and workshop, while the other men slept in the barracks. The morning the rest of the squadron were on the road ready to leave Leccee, a lone German plane straffed where my tent had been. Axis Sally said she knew where we were going and would be up to see us. Thank goodness, I never did see her. I was returned to the States in July of 1945, but was still declared essential. The war with Japan ended while I was home on furlough. Did you ever see a grown man cry?

VERNE M. YAHNE, was born in Mitchell, SD. Joined the 147th Field Artillery which activated in November 1940. Completed CPTP at Dakota Wesleyan University. In September 1941 went to Texas for flying training. Graduated Mar. 7, 1942 from Kelly Field. Was a charter member of the 82nd Fighter Gp., 95th Sdn. flying the famous P-38.

Went to Ireland, then North Africa all the way across, with missions over Sicily, Sar-

dinia—escorted bomber on first missions over Rome and Naples-Invasion of Sicily and Pantelleria and many other firsts. Officially one ME-109.

Returned to U.S. September 1943. Became P-38 RTU instructor. Flew missions in Korea in F-51 and B-26s. Became Wing Commander of the 123rd TRW. Retired October 1972.

ROBERT G. ZIMMERMAN, Capt. USAF (ret.) 0-746244, was born Nov. 3, 1921 in Youngstown, OH. He enlisted in the U.S. Air Corps in February 1942; sworn in Apr. 6, 1942.

Active duty Santa Ana; pre-flight August 1942. Soloed in Stearman PT-17 September 1942, Cal-Aero-Tech, Ontario, CA, Primary. Basic at Lancaster, CA Nov. 12, 1942, Vultee BT-13. Advanced at Williams Field, Chandler, AZ, Curtiss AT-9. Checked out in RP-322. Graduated Class 43-E May 1943. Combat training at Muroc and Van Nuys, CA.

Joined 14th Fighter, 37th Sdn. as replacement pilot St. Marie Du Zit near Tunis, North Africa. First combat mission eight hours, 15 minutes over water (I joined Army rather than Navy to avoid combat over water) to escort bombers to Bolzano, Italy. Ball bearing factory, I believe. Moved to Foggia, Italy Christmas 1943. Flew various fighter sweeps and escort missions over Alps to Steyer, Reagansburg, Vienna, Southern France, Wenernustadt, Ploesti. Briefed for Bresleau. (Would have lost 30% of 15th AF because of lack of gas to return). Believe I shot down two ME-109s, one FW-190 and some probables. No gun camera. We were usually in combat every mission outnumbered at least two to one, many times four to one. On 27th mission to escort bombers to Ploesti Apr. 15, 1944, 16 of us jumped about 60- FW-190s and got nine without a scratch. C.O. after 20 minutes, called to break off combat and return 500 miles to base as we were low on gas. I was leading element in Green Flight—Tail End Charley in Blue Flight (Joe Garrity) called to say he was hit and bailing out. My wing man called—hit. I turned to help. He went straight down. I ran for clouds, was hit, on fire; bailed out 26,000 feet, speed of 720 mph. Landed in Bulgaria, captured and spent six months in POW Camp near Varna, on Black Sea. We were turned over to American Consul, Istanbul, Turkey and returned to Italy September 1944. Returned to States November 1944.

Married in April 1945; one son and a daughter. Attended Case-Western Reserve, Cleveland, OH and Youngstown University at Youngstown, OH. Studied electrical engineering. Retired from Ohio Edison Company, Akron, OH Feb. 1, 1983 as Senior Vice-President in charge of Division Operations and Customer Service. Reside in Atlantis, FL.

NICHOLAS ZINNI, was born in Batavia, NY, Aug. 22, 1922 graduated from Batavia High School, 1940.

Enlisted as Aviation Cadet May 1942, graduated May, 1943, from Single Engine Flight School, Marianne, FL.

Flew the P-39 Aircobra, 220 hours in California; then 20 hours in P-38.

In December, 1943, joined the 80th Fighter (Head hunters) Sdn., Fifth Air Force, Pacific Theatre of War.

Flew 170 Combat Missions in 18 months-from New Guinea to Philippine Island; downed two Japanese Fighters confirmed; numerous strafing and dive bombing missions; participated in night aerial attack against Japanese Naval Force near Mindoro, Dec. 26, 1944. Received Distinguished Flying Cross, Air Medal, eight Oak Leaf Clusters, rank of Captain. He confirmed 26th or 27th plane downed by Dick Bong—to break E. Richenbacker's record WW I.

In October 1944, instructed in P-51s and P-47s, in New Guinea.

Due to service-incurred malignancy, lost right arm and shoulder in 1947. Started civilian Flying in 1956; still owns and flies retractable-gear plane.

Graduated Buffalo Law School, 1951; practices Law in Batavia, NY. Married, three children; (son also Attorney and Private Pilot).

Arthur W. Heiden

John Silberman

WALTER E. ZURNEY, was born Dec. 27, 1918 in Davenport, IA. He graduated from high school in Kittanning, PA and enlisted in Army Air Corps June 17, 1939; assigned Albrook Field, Canal Zone, Panama, during which time he received 2nd A.M. rating and promoted to Sergeant. Entered Aviation Student Pilot Training in Class 42-H. Graduated from Ellington Field, TX Sept. 13, 1942, promoted to S/Sergeant Pilot and assigned to Hondo Navigation School, Hondo, TX. Promoted to Flight Officer December 1942. Promoted to 2nd Lieutenant November 1943 and completed B-24

transition at Smyrna, TN. Completed B-24 overseas training at Casper, WY. Picked up aircraft at Topeka, KS and delivered same to 15th AAF.

Crew was assigned to 456th Bomb Gp., May 1944 and flew first bomb mission against Ploesti oil refineries. I was area assigned to 82nd Fighter Gp. and flew first mission as a fighter pilot on a high altitude bomb mission in a modified P-38 which carried a Norden Bomb Sight and a bombardier in nose, in place of guns and was designated "Droop Snoot". A very high degree of bombing accuracy was achieved when fighter aircraft, each of which carried two 500 gal. drop tanks and four 500 lbs. bombs, would fly a close formation and would release bombs on count-down command from "Droop Snoot".

Completed 50th fighter pilot mission Apr. 12, 1945. Was separated from service November 1945, entered Purdue University. Graduated June 1949. Employed as an engineer by PPG Ind. and retired in June 1982 with 31 years' service. Retired from the Air Force as a Major in December 1978. Married 45 years to Mary; two children, William and Marsha. Reside in Ford City, PA.

Francis R. Goldsberry

Gerald A. Greene

G.F. Leinweber

Stanley A. Long

George I. Tripp, Jr.

Harold D. Rigney

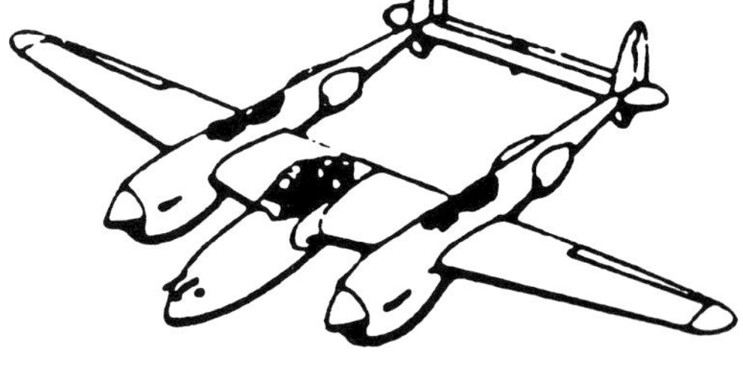

Kneeling, L to R: Kord, Highsmith, Hersey, Vardel, Hill. Standing, L to R: Lambie, Kocour, Hanna, Herrall and Jenkins. North Africa, 1943. (Courtesy of Gene Hersey)

P-38 NATIONAL ASSOCIATION ROSTER

ABBOTT, DONALD W.
ACKLEY, JOHN W.
ACKLING, CURTIS W.
ADAMS, CHAS. E., JR.
ADELSTEIN, SID
ADIE, LEWISE
AKESON, LLOYD R.
ALBER, GEO. D.
ALBER, STEVEN C.
ALEXANDER, LUCIEN B.
ALGER, MARTIN P.
ALLRED, WAYNE
ALTIS, RUSSELL V.
ALWORTH, FRANCIS E.
AMAYA, FRANK
AMOS, ART
AMUNDSEN, EARL
ANDERON, WAYNE D.
ANDERSON C. VAN
ANDERSON, JAMES W.
ANDERSON, MILTON R.
ANDES, LLOYD M.
ANDREWS, HARRY L.
ANTON, SOREN E.
ARBANAS, WALTER J.
ARMSTRONG, W.F.
ASBURY, ERNEST C.
ASCHENBRENER, ROBT. W.
ATTUTIS, JUSTIN S.
AUBELE, FRED
AUSTIN, ROBT. E.
BAILEY, DEBORAH
BAILLARGEON, ERNEST L.
BAIRD, VERNON
BAKER, BUFORD O.
BAKER JAS. H.
BALDWIN, ED
BALLARD, DEWEY E.
BALLARD, JOHN W.
BANNON, JOHN H.
BANTA, HAROLD MAX
BAQUET, EDWIN C.
BARBER, REX T.
BARLOW, HARLEY T.
BARNES, HERSCHEL M., DR.
BARR, JOS.
BARRETT, FRANK
BAYLESS, DUDLEY F.
BEACH, BEN, JR.
BEALL, JOS. H., MD
BECHTEL, PAUL S.
BECKTEL, BOB
BEEBE, PHOEBE M.
BEIMDIEK, ARTHUR O.
BELL, L'ROY O.
BELL, RAY F.
BELLOWS, D. C.
BELT, TAYLOR
BENDA, DANIEL B.
BENEDICT, TRUMAN
BENNETT, ROBT. M.
BENSON, JACK J.
BERGANDER, ROBT. B.
BERKHEIMER RAYMOND D.
BERNIER, RUDOLPH J.
BIEDUL, CHESTER A.
BIRTCIEL, FRANK E.
BISHOP, STAN
BLACK, ROBT. C.
BLAIR, HUBERT M.
BLAIR, WALTER J.
BLAKE, STEPHEN M.
BLANCHARD, MERLE E.
BLUM, ALLEN H.
BOARD BRADLEY J.
BODIE, WARREN M.
BOEHM, LESLIE R.
BOETTNER, CHARLES O.
BOHN, E.A.
BOONE, WILLIS D.
BORGE, JAMES
BOSTROM, MILTON R.
BOUFFARD, ALBERT J.
BOWER, LEROY C.
BOWERS, WM. H.
BOWLING, TEMPLE
BOYD, LEN
BOYD, WARREN D.
BRADDOCK, EDWARD I.
BRADFORD, GLEN E.
BRADLEY, LEONIDIS
BRADLEY, RICHARD W.
BRAMBLET, RUSS
BRAMMER, KENNETH L.
BRECKLER, JOS. J.
BRENNAN, DONALD J.
BRISTOW, C.T.
BROOKS, FRANK N.
BROOME, WM. G.
BROSKI, RICHARD E.
BROWN, GERALD
BRUSKRUD, WARREN
BRYANT, ELMER E.
BRYANT, OLIVER
BUCHANAN, JOS. B.
BUDIG, T.J.
BUNDY, ART
BUONAGUIDI, AIDO A.
BURKHARDT, PAUL L.
BURNS RICHARD D.
BUSCAINO, STEVE C.
BUSKIRK, CLAIR E.
BUSSE, KEN
BUTLER, BARRY
BYASSEE, ROBT. A.
CALLAHAN, JOHN E.
CAMERON, WM. A.
CAMPBELL, BRUCE L.
CAMPBELL, ROBT. F.
CAMPBELL, WARREN G.
CANADA, E.D.
CAPUTO, JOHN A.
CARHART, THOS. M.
CARLSON, L. S.
CARNACHAN, BRUCE
CARONTO, ELMER J.
CARROLL, JOHN J.
CARTER, OLIN B.
CARTER, ROBT. F.
CASTLE, HENRY G.
CAUGHLIN, W. H.
CAUTHEN, GEO. M.
CAVANAUGH, GERARD F.
CEFARATT, GIL
CELLINI, OLIVER S.
CERUTTI, DINO
CERVENY, DON
CHAFFEY, KAY GOTT
CHANDLER, GEORGE
CHAPMAN, FRANCIS H.
CHARLES, WM. F.
CHARLTON, WILFRED H.
CHASE, JIM R.
CHELCHOWSKI, THADDEUS F.
CHIARA, JOS.
CHILDS, LEO F.
CHILES, MARK
CHITTY, CHARLES D.
CHRISTIE, BARBARA
CICERO, MATHEW P.
CICERO, PAUL R.
CLARK, KEN B.

CLARKE, FRED W., JR.
CLAYBOURN, GUY R. JR.
CLEARY, RAYMOND M.
CLELAND, THOS. J.
CLIFTON, GARLAND F.
CLOYD, EARVIE T.
COBURN, A. L.
COFFEY, GEORGE L.
COFFMAN, MILLARD F.
COHN, JOHN A.
COLE, ARTHUR V., JR.
COLLINS, F. PATRICK
COLLINS, JAS. W.
COLLINS, THOS. G.
CONNOR, ROY J.
COOK, ROGER H.
COPENHAVER, KENT
CORBIN, RUSSELL B.
COSHOW JACK E.
COSTELLO, RAYMOND E.
COSTENBADER, RALPH K.
COSTLEY, JAS. H.
COUGHLIN, WM. H.
COUSINS, WELDON A.
COWIE, ROBT. P.
CRAIG, LAWRENCE H.
CRAIG, PAUL S.
CRAMER, DARRELL S.
CRANE, C. B.
CRAPNELL, MARK, D.
CROCKER, MRS. CHAS. J.
CROTHERS, HOWIE B.
CUMMINGS, ROBT. D.
CUMMINGS, URBAN K.
CURTIN, JOHN J.
CURTIS, JACK
CURTIS, JOHN A.
CWIEK, MITCH
DAHLQUIST, ERIC
DANAHER, JAS. E.
DANIELS, LLOYD T.
DARNELL, C. H., JR.
DARNELL, CECIL
DAVIS, JOHN R.
DAVIS, ROBERT E.
DAYE, JOHN F., JR.
DAYTON, WM. J., II
DE FRANCE, J. E.
DE GREY, RICHARD P.
DE LAMERE, WALTER E.
DE LONG, J. C.
DEAVERS, CLYDE J.
DECKER, CLAIR D.
DEECKEN, ROBT. H.
DEIGNAN, J. WERNER
DEMERS, JULIAN G.
DEMKOVICH, JOS. P.
DENMAN, ALDEN L., DDS
DENT, ELLIOTT E., JR.
DETTNER, WILBUR D.
DEUTSCH, MONROE K.
DI BELLO, STEPHEN L.
DICKENSON, ARNOLD H.
DICKMAN, MELVIN L.
DICKSON, MAX
DIEFENDORF, ALLEN J.
DILLER, FRANKLIN W.
DILLON, IRIS L.
DINGER, ROLAND E.
DOAN, JOS. H.
DOBBS, BYRON A.
DOBROWOLSKI, JOS. A.
DOBSON, JOHN R.
DODDS, WAYNE
DOERING, EDWARD W.
DONOHOO, THOS. W.
DOOLEY, JAS. E.
DOUGHERTY, CHAS. R.
DRECKMAN, H. GENE
DRISCOLL, DANIEL P.
DROSTE, LOUISE E., JR.
DUFFIE, CLAIRE
DUHON, TRUBY E.
DUNGAN, CHAS.
DUNN, DONALD A.
DUNSAY, M. DENISE
DUQUETTE, PAUL L.
DURKIN, JOHN J.
DUSARD, LEO F.
DUTKO, FRANK J.
EASTWOOD, DOUGLAS H.
EBERHARDT, D. E.
EBERHARDT, H. ALFRED
EDEN, RALPH M.
EDWARD, BRUCE
EDWARDS, J.J., JR.
EHLINGER, JOHN L.
ELBERTY, CHAS. W.
ELLIOTT, CHALMERS F.
ELLSWORTH, HOWARD B.
ELWOOD, JAS. J.
ENGLEHART, RALPH
ENING, FRED K.
ENRIQUEZ, F. ROD
ENTZ, DONALD C.
EPPERSON, GEO. M.
ERDMANN, ORVILLE L.
ERICSSON, JACK A.
ERNEST, WALT
ESHELMAN, GEO. T.
ESSER, EDWARD O.
ESTES, CARL A.
ESTES, HIAWATHA
ETHELL, E. C.
ETHELL, JEFFREY L.
EUBANK, R. H.
EVERETT, PATRICK C.
FARNHAM, CHAS. R.
FAZIO, FRANK M.
FESSLER, ROBT. C.
FETNER WM. W.
FIELD, WM.
FINCH, FRANK
FINE, ELIHU
FISHBURN, JAS. B.
FLECKENSTEIN, GEO. J.
FLOOD, STEPHEN H., JR.
FLORENTINE, WM. C.
FLOWER, ROBT. P.
FLYNN, WALTER F.
FODOR, JOS. N.
FOLLIS, TOM
FONG, ROY K.
FORSTER, JOS. M.
FORSYTH, EUGENE S., JR.
FOURNET, PAUL A.
FOWKES, WM.
FOWLER, THOS. R.
FRAKES, ROBT. L.
FREITCHEN, ALBERT L.
FRIEND, RAY
FRITZ, ZELLIN K.
FRUGE, FREEMAN J.
FUNK, LLOYD S.
FUQUA, GORDON W.
GALLAGHER, JAMES P.
GALLUP, FRANCIS C.
GANISTER, EDWARD J.
GARBETT, JOHN S.
GARDNER, JAMES F.
GARDNER, LEFTY
GARRATT, CHAS. S. III
GASKILL, WM. M.
GATES, CHAS. A., JR.
GATHERS, B.B., JR., MD.
GAUGHAN, JOHN E. JR.
GAWLER, JOHN, P.
GEDDES, JOHN K.
GETKER, M.J.
GIBNEY, DONALD C.
GILL, NORVAL D.
GILLILAND, ROBT. J.
GISOLO, HAROLD
GLASS, BURL E.
GLEASON, THOS. A.
GLENDENING, ROBT. E.
GLENN, JOHN GAIL
GNEWIKOW, MILD E.
GOAD, GEO.
GOELZ, ROBT. L.
GOFF, JAMES H.
GOFF, RICHARD L.
GOLDBERG, SEYMOUR M.
GOLDSBERRY, F. R.
GOLDSBOROUGH, E. SCOTT
GOLIN, TED
GONZALEZ, HECTOR R.
GOOD, ROBT. E.
GORDON TERRY A.
GORDON, THOS. F.
GORDY, JAMES F.
GOSS, HOWARD L.
GOSS, WM. N.
GOSSARD, PHILIP D.
GOUDEY, VICTOR
GOVIER, EVERETT
GRACE, JOHN T.
GRAHAM, HAROLD B.
GRAHAM, JIM
GRAHAM, PHILIP M.
GRAINGER, ROBT. L.
GRANT, JOHN B.
GRANTHAM, A. KEVIN
GRAVATT, PERCY O.
GRAVES, FRANK N.
GRAY, GEORGE
GRAY, RUFFIN W.
GRAY, THOS. W.
GREEN, JOHN FOREST
GREENBERG, HERBERT A.
GREENE, GERALD A. MRS.
GREENHALGH, WM. JR.
GREENWOOD, TRAVIS M.
GREGORY, WM. J.
GRICE, CHAS. A.
GRIESHABER, EVERETT E.
GRIEVES, JOHN P.
GRIFFIS, DONALD L.
GRIFFIS, JAS. W.
GRIFFITH, WM.
GRISHAM, ROLAND K.
GRISWOLD, STEELE T.
GUENLEY, WM. F.
GUERRY, ALEX
HACKMAN, CLAYTON P.
HAESSIG, CLARENCE GLEN
HAGENBACK, JIM
HAGERTY, HAROLD F.
HALE, ROBERT A.
HALL, ALFRED A.
HALL, GEO. W.
HALL, STEVEN K.
HALLBERG, CHESTER A.
HALLOCK, JUDSON
HAM, WM. A.
HAMBURGER, ROBT. N.
HAMBURN, JOHN E.
HAMILTON, ED
HAMMERICH, ROBT.
HAMMOND, FRANK W.
HAMMOND, WILBUR D.
HANEY, GEO. R.
HANSEN, JULIUS G.
HANSON, MARILYN F.
HANTON, JOHN T.
HARDEN, S. WAYNE
HARDY, ALONZO H.
HARMON, TOM
HARRIS, FRANCIS H.
HARRIS, JAS. W.
HARRIS, O. J.
HARROW, JACK R.
HARTT, RICHARD
HARVEY, WOODROW W.
HATCH, H. B.
HATTENDORF, WILL S.
HAUS, PAUL E., JR.
HAWES, ROBT. S.
HAWKS, RAYMOND S.
HAYES, JACK W.
HAYNER, FREDERICK C.
HAZLETT, JIM
HECK, WALTER J.
HEDENBERG, JOHN W.
HEDGES, EDDIE C.
HEDLUND, LT/GEN EARL C.
HEIDEN, ARTHUR W.
HENDERSON, DEXTER
HENDRICKS EDGAR A.
HENDRIX, GEO. D.
HERRALL, CHUCK
HERRICK, KENNETH W.
HERSEY, EUGENE
HESS, WM. P.
HEYN, DON G.
HICKOK, GENE G.
HIGGINS, PRESTON K.
HILDEBRANDT, R. W.
HILDERBRAND, MELVIN
HINCK, ROBT. B.
HINES, M. S.
HIRST, CHARLE L.
HOBAUGH, SCOTT K.
HOCHSTETLER, GEO. E.
HOEFER, ROBINSON K.
HOELZEL, MILES
HOFFMAN, CLARENCE W.
HOFFMANN, LARRY
HOLCOMB, GEO.
HOLCOMBE, RICHARD B.
HOLDREN, KENNETH H.
HOLLINGSWORTH, JAS. S.
HOLLISTER, B. E.
HONEYCUTT, HARVEY
HONSA, HENRY J.
HOOD, JOHN J.
HOOKER, VERNE E.
HOOVER, JOHN R.
HOSTETTER, MELVIN L.
HOUCK, NIEL A.
HOVERSTEN, HERMAN E.
HOWARD, RICHARD A.
HOWES, EDWAD B.
HOXEY, FREDERICK
HUDGENS, JOHN E. MRS.
HUFF, SHERIL D.
HULKA, DAVID H.
HULLSIEK, ROBT. E.
HUMPHREY, WALTER J.
HUNT, CLARENCE A.
HYDE, RUSSELL C.
HYMEL, JULES J.
ILFREY, JACK
ISAACSON, C. M.
JACKSON, ARCHIE G.
JACKSON, CHAS. D.
JACKSON, JOHN S. JR.
JACKSON, ROBT. J.
JAMES, JOHN G.
JEFFREY AUTHUR F.
JENKINS, JACK S.
JEPSON, BILL
JEPSON, JUNIOR R.
JEWELL, JAS. L.
JOHANNES, DONALD J.
JOHANSEN, FREDERICK W.

JOHNSON, CLAYTON A.
JOHNSON, KELLY
JOHNSON, LE ROY J.
JOHNSON, ROSS PATRICK
JOHNSTON, DONALD S.
JOLLY, JAS. H.
JONES, ALVIN H.
JONES, C. L.
JONES, LINCOLN D.
JONES, ROBT. B.
JONES, ROBT. E., JR.
JONES, ROBT. H.
JONES, STAN
JONES, WARREN F.
JONOSKI, TED
JORGENSEN, KENNETH B.
JOY, CARROLL
KAATZ, EMMETT V.
KAEMPFER, ROBT.
KAMIN, CARL S.
KAMMERER, DORSEY B.
KASPAZYK, STAN J., JR.
KEATING, ERNEST W.
KEEN, CLAY O.
KEEN, PATRICIA F.
KEENER, NELSON L.
KEITH, TROY
KELLY, DONALD L.
KELTON, STANLEY
KELTY, JACK A.
KEMPER, B. HOWARD
KENNEDY, FRED R.
KERMES, WM. J.
KERR, KANIEL J., JR.
KIELBASA, EMIL A.
KIENHOLZ, DEL L.
KIICK, RICHARD D.
KIME, DUANE
KING, LON A.
KING, ROY L.
KING, WARREN S.
KIRCHHOFER, KEN M.
KLEIN, ELLIOT B.
KLEIST, WM. A.
KLINKNER, JEROME L.
KNAPP, LESLIE E.
KRAMER, CECIL C.
KRISLE, LEMUEL
KUHN, DAVID A.
KUHN, JOSEPH A.
KUHN, KEVIN J.
KUNDRATH, LOUIS W.
KINKLE JAS. K.
LA CLARE, EDWARD F.
LAGIER, FRANCIS W.
LAMBERT, CHAS. R.
LAMPHERE, ROBT. M.
LARSEN, JOHN M. JR.
LARSON, LOUIS E.
LAURIE, JACK
LAVEN, GEO. JR.
LEARNED, ALBERT S.
LEDFORD, OTTO
LEDSON, B. NOBLE
LEE, DONALD H.
LEE, RICHARD J.
LEIDY, LESTER R., JR
LEINICKE, WAYNE F.
LEMONS, CHAS. M.
LETTE,LEROY L.
LEVIER, TONY
LEVINE, LLOYD L.
LEWELLING, ALFRED B.
LEWIS, GLENN M.
LEWIS, JOHN D.
LEWIS, JOHN K.
LEYMEISTER, GEO. F.
LINDLEY, EVERETT S.
LINDSEY, JAMES T.
LINDSTROM, JOHN W.
LINEHAN, RICHARD W.
LINEKER, ELBERT W.
LINGENFELTER, RICHARD L.
LITTLEFIELD ,ROBT. M.
LIVESEY, DONALD W.
LLOYD, KENNETH B.
LONDON, BARBARA E.
LONG, CHAS. N.
LONG, DAVID A.
LONG, VIRGINIA L.
LORUSSO, PAUL
LOVE, JACK E.
LOWE, BURTON M.
LOWNES, CARL R.
LUBER, VERNON N.
LUCAS, JOSEPH E.
LUCAS, RIO G.
LUOTKE, ROBT. E.
LUND, JOHN R.
LUNSTRUM, RICHARD M.
LUSHER, JAMES F.
LUTHY, ARTHUR E.
LUTTRELL, DONALD A.
MAACH, JAYSON C.
MACALLISTER, P. E.
MACARTHUR, DOUG J. B.
MACDOUGALL, COLIN D.
MACGOWAN, PETER
MACK, BLAINE L.
MACKENZIE, IAN B.
MACPHERSON, ERICKA B.
MADISON, HAROLD N.
MAHONEY, THOS. C.
MAIMONE, PETER J.
MALONE, DONALD F.
MALONEY, WILBUR B., JR.
MALVEN, C. J.
MANION, CHAS. W.
MARSISON, ROBERT L.
MARLING, MILBY
MARSHALL, REV. ANN
MARSHALL, RICHARD E.
MARTIN, ALEX
MARTIN, DONALD J.
MARTIN, EDWARD L.
MARTIN, ROBT. J., PH.D.
MARTIN, WM. E.
MARTINEAU, ROBT. S.
MARTINEZ, MARTIN
MARVIN, GEORGE C.
MASTERSON, JOHN S.
MATHIS, JAS. D.
MATTERN, JIMMIE
MATZEN, FRED W.
MAUSS, CHAS. A.
MAX, HOWARD
MAXWELL, GLENN
MAXWELL, ROBT. B.
MAY, MICHAEL
MAYBERRY, M. G.
MCAFEE, OLIVER S.
MCAFEE, THOMAS
MACAFFREY, JOS.
MCCALL, GEO. W.
MCCARTHY, ROYAL E.
MCCLANAHAN, FRED C.
MCCLOUD, DON
MCCONNELL, FRED B.
MCCULLOUGH, JAS. E.
MCDERMOTT, ROBT. F., B/G
MCDONNEL, JAS. J.
MCFARLAND, DICK
MCGUIRE, THOS. G.
MCKAY, GEORGE E.
MCKENZIE, H. G.
MCKETON, RUSSELL J.
MCKILLIP, ROBT. H.
MCLARY, MAURICE R.
MCMORROW, NORMAN S.
MCNICHOLAS, JOHN M.
MEDLEY, OLIN L.
MEEKER, WM. M.
MESSINSON, PERCY C.
MEISS, HENRY
MEISENBACH, ALBERT E., MD.
MEKEEL, DOUGLAS
MERCIER, DONALD L.
MERCIER, PAUL J.
MERRITT, EDWARD E.
MERRYMAN, ALBERT H.
MESSER, MAURICE B.
MICKEY, MELVIN S.
MIKSCH, RUSSELL S.
MILLER, EVERETT
MILLER, J. BEN
MILLER, LOREN W.
MILLER, ROBT. A.
MILLIREN, ROBT.
MILNE, BARRIE J.
MILNE, BRUCE W.
MILNE, DUNCAN M.
MILNE, GREGOR W.
MIMS, SHARON SHAY
MINGES, WM. T.
MIRON, ALBERT A.
MIRON, FRANCIS E.
MISCHKE, RICHARD M.
MISHLER, W. B.
MITCHELL, ROBT. H.
MONTEITH, PHILLIP B.
MONTGOMERY, FRED L.
MONTGOMERY, RAY G.
MOON, ALLEN J.
MOORE, CLARENCE G.
MOORE, CLIFFORD J., JR.
MOORE, FREDERICK
MOORMAN, GEO. E.
MOREING, JAMES
MORGAN, E. CLAUDE
MORGAN, WM. M.
MORRIS, DANNY
MOSELLE, RAY A.
MOSIER, ARLETT G.
MOYER, ROBERT P.
MRZEK, EMIL A.
MUELLER, C. J.
MUELLER, RUDY W.
MULLAN, JACK W.
MULLINS, JOHN D.
MULVANY, GEROLD MELVIN
MULVEY, JOHN RALPH JR.
MUNGER, PAUL O.
MURPHY, ARCHIE C.
MURPHY, JOHN F.
MURRAY, BLAINE
MURRAY, CLAUDE
MURRAY, ROBT. A.
MURTHA, JOHN R.
MYERS, D. W.
NAVARRE, FRED
NEAL, THOS. D.
NEALE, MARY LOU
NEFFINGER, G. G.
NELSON, A. R.
NELSON, C. RICHARD
NELSON, THOS. F.
NEMETH, ALFRED
NEUFELD, HAROLD L.
NEVILLE, TED
NEWMAN, R. D.
NICHOLS, FRANKLIN A.
NICHOLSON, CARLTON P.
NISH, ALBERT R., JR.
NOON, WM. J.
NORDGREN, ORVILLE J.
NORMAN, WM. E.
NORRIS, ROY E.
NOTZ, MATT
NUEHRING, LOUIS C.
O'DRISCOLL, THOS. B.
O'GRADY, PATRICK E.
O'TOOLE, LAWRENCE A.
OAKLEY, HARRY R.
OBERSTEBRINK, WALTER
OCHALEK, WM. J.
OLDIS, DON A.
OLSON, GEORGE
OLSON, MAURICE S.
OLSON, THOMAS R.
ONESTY, JOE
ORDWAY, STANLEY I.
ORREN, LOUIS M.
OSTEN, ELLIOTT J.
OSTRONIK, K. C.
OSTRONIK, RICHARD D.
OWENS, M. P.
OXFORD, THOS. M.
PACE, JOHN D.
PAFFORD, EUGENE E.
PAGAN, LEON H.
PALMER, JAS. M.
PARRY, ROBT. E.
PARSONS, RAYMOND N.
PATTERSON, CHET
PATTON, WALTER H., M.D.
PAUL, HARRY L.
PAUL, JOHN W., JR.
PAUL, ROBT. L.
PAVLOVSKY, JIM
PAYNE, W. LEE
PAYNE, WALTER G.
PEAK, J. P.
PEEPLES, LAMAR
PELLINGER, THOS.
PEMBERTON, JACK
PENA, ALFONSO R.
PERKS, ARTHUR CAMPBELL
PERLONGO, PAUL
PERRY, I. NEWTON
PERSON, JUNIUS W.
PETTUS, JACK D.
PHILLIPS, FRANK V.
PHILLIPS, FREDERICK S.
PHILLIPS, JOHN R.
PHILLIPS, WILBUR P.
PHIPPS, ROBT. JR.
PILKINTON, ROY L.
PINSON, CHAS.
POLLARD, SIDNEY D.
POLLOCK, JOHN
PORTA, JOHN B.
PORTER, LEROY
POWELL, GERALD L.
PRAHLER, ROBT. H.
PRATT, LEONARD W.
PREECE, WM.
PRELL, SEYMOUR V.
PREVATT, MICHELE
PRICE, ROBT. W.
PUGH, THOMAS
PURDY, GUY M.
PURDY, JOHN E.
PURLIA, SAM
PYLES, MAX D.
QUATTLANDER, FRANK P.
QUESSETH, CECIL H.
QUINN, FRED
RADER, CRANSTON B.
RAMSAY, JIM
RATAJSKI, CHAS. J.
READE, RICHARD
REED, GEO. F.
REED, RAYMNOND E.
REICHARD, JAS. N.
REID, JOHN C.
REID, WARREN E.

REINECK, CECIL V.
REISING, CARL E.
RESWEBER, FRANK
REYNOLDS, EUGENE F.
RHODEBECK, CHAS. E.
RICE, MELVIN
RICHARDS, H. R., JR.
RICHARDS, WM. L.
RICHARDSON, STANLEY P.
RIEAD, T. G.
RIEMENSNIDER, ROBT. H.
RIGBY, JUDSON L.
RIGNEY, H. D.
RILEY, ROBT. W.
RIVEST, GEO. L.
ROBBINS, JAY T., L/S
ROBBINS, RICHARD T.
ROBERTS, LEWIS V.
ROBERTS, RAYNOR H.
ROBERTSON, LEROY D.
ROBERTSON, PHILIP
ROBISHEAUX, EARL E.
ROCKER, BOB
ROOLF, JOHN H.
ROESTEL, RAY E.
ROGERS, FRANCES C.
ROOS, FREDERICK A.
ROSEN, JULIUS J.
ROTHGEB, WAYNE P.
ROUNDS, B. L.
ROWE, DAVID
RUED, HARRISON W.
RUFF, NORBERT C.
RUNYON, T. H.
RUPLEY, MARTHA D.
RUPP, FLOYD B.
RUSSELL, ROY C., JR.
RUSSO, LOUIS A.
RUTHERFORD, GERALD D.
SAFARIK, WM. F.
SALMON, W. K.
SAMAN, GEO.
SANFORD, DONALD D.
SANVIK ANITA H.
SAUER, ROBT. R.
SAVAGE, WALTER G.
SCHAFFHOUSER, JOS. F.
SCHMIDT, GUSTAV S.
SCHMIDT, WM. F.
SCHNEIDER, ALBERT B.
SCHNEIDER, E. H.
SCHNEIDER, WES
SCHOTT, JAS. A.
SCHOTTELKORB, ROBT. W.
SCHREFFLER, C. E.
SCHULTZ, HERBERT R.
SCHWARTZ, DAVID G.
SCILLEY, KENNETH G.
SCOTT, LEEROY H.
SCOTT, RICHARD M.
SEEBERG, GEORGE
SELFRIDGE, WALTER E.
SELLE, FRED A.
SENTER, LEWIS B.
SETH, JOHN DAVID
SHADE, WALTER D.
SHALLENBERGER, CHAS. F.
SHANNON, JOS. L.
SHAPIRO, LEONARD
SHARP, JOHN H.
SHARP, SIDNEY A.
SHEARIN, FRANK C., JR.
SHELTON, JACK M.
SHEPARD, BERT K.
SHERMAN, ELMER L.
SHOLTEN, WM.
SHOOP, RICHARD R.
SHY, JACK V.
SICKLESTEEL, MILTON E.
SIDWELL, GENE W.
SILBERMAN, JOHN
SILK, JOHN W.
SILVERS, ROY EL.
SIMERLY, WALLACE J.
SIMMONS, CHAS. L.
SIMMS, ESTON L.
SIMPSON, DICK
SIMMS, BOB
SIMS, EDWIN D.
SINGLETON, RONALD D.
SIRMON, REVIS
SKILLING, DAVE
SKINNER, DARRELL
SLAGLE, WARREN L.
SLATER, WM. F.
SLIFER, MAX
SLOAN, JACK M.
SLOUKA, GEO. E.
SMART, EDWIN J.
SMERCHEK, GEO. H.
SMITH, ARTHUR F.
SMITH, GARY P.
SMITH, GREGG A.
SMITH, GUY E.
SMITH, JOS. A.
SMITH, RICHARD EUGENE
SMITH, RONALD T.
SMITH, SHAFEEK
SMITTLE, WALTER, JR.
SMYTHE, RAYMOND E.
SNEDDON, WAYNE O.
SNOW, ERNEST A.
SNYDER, EUGENE P.
SOBAK, STEVE
SOMMERKAMP, FRANK M.
SOMMERS, DAVIS M.
SONDERMAN, CLARENCE H.
SORENSEN, WALLACE A.
SPARKS, R. T., JR.
SPARREVOHN, FRED D.R.
SPENADER, ROBT. G.
SPIKER, DEWEY
SPINDLER, DONALD D.
SPIVEY, PAUL
SPURGEON PHILIP R.
SPYCKABOER, JACOB W.
STABILE, FRANK M.
STANHOPE, JAS. H.
STANIFER, JOHN
STARKE, WM. H.
STAY, DON H.
STEGE, JOHN A.
STELL, JOSEPH
STEPHENS, JACOB S.
STEPHENS, LEON B.
STEVENS, J. LUCILE
STEWART, DOUGLAS D.
STEWART, JAS. A.
STICHA, JAS. A.
STINE, DONALD P.
STITT, JAS. M. JR.
STONE, DAYTON D.
STONE, FRANCIS A.
STONE, STEPHEN A., JR.
STOUTENBOROUGH, DON
STRADER, CHAS. R.
STRADER, ROBT. F.
STRONG, RICHARD W.
STROZIER, BUDDY A.
STUART, HOWARD C.
STYER, IRVINE M.
SUBLER, JAS. C.
SUGGS, L. S.
SUMPTER, WM. L.
SUBOTT, JOHN L.
SUTTON, MARK
SWART, WM. A.
SWARTZ, WM. A.
SWINFORD, ROSS
SWITZER, MALCOM E., MD
SYKES, GEO. K.
SYKES, RICHARD T.
TATE, JOHN C.
TATHAM, LOUIS I.
TAYLOR, DICK
TAYLOR, JOEL
TAYLOR, JOHN C.
TAYLOR, ROBT. F.
TENNANT, DAVE
TENNANT, HOMER J.
THISTED, DALE O.
THOMAS, DAVID W.
THOMPSON, DR. HUGH L.
THOMPSON, LUTE C.
THORSEN, GUNVALD B.
TILLEY, JOHN ALDEN
TIPTON, CARL W.
TOLIVER, RAYMOND F.
TOLL, HENRY C.
TOMASI, THOS. M.
TOMBERS, ROBT. M.
TOMKINS, FRANK D.
TOMLINSON, RICHARD A.
TOUMANOFF, GEO.
TRAYLOR, ROBT. F.
TREMPLER, ARNOLD J.
TRIBBY, DAVID
TRICK, MARION C.
TRIPP, GEO. I., JR.
TURNER, WM. W.
TYDEMAN, ROBT. E.
URBANK, W. J.
VAN BOVEN, PAUL W.
VAN SLYCK, ASHLEY
VANDENBERG, CHARLES
VASSAR, ED
VEDRANI, DAVID J.
VERODA, VICTOR J.
VOGELSONG, HARRY H., JR.
VRILAKAS, ROBT. A.
WAGNER, JACK
WAKEMAN, JOHN S.
WALKER, GEO. WAYNE
WALKER, JACK S.
WALKER, WM. A.
WALLACE, WM. E., MD.
WALLISCH, WM. B.
WALTER, GEO.
WANDREY, RALPH H.
WARD, BRUCE C.
WARD, CLINTON E.
WARD, RUSS
WASIL, EDWIN E.
WATSON, GUY
WATSON, HAROLD E.
WATTS, THOS. L. JR.
WEART, GEORGE S.
WEARY, ROBT. K.
WEBER, LEROY, JR.
WEBSTER, LEWIS M.
WEBSTER, LINDEL S.
WEEKS, JOHN S.
WEGHER, A. J.
WEIRICH, F. C.
WELCH, DARRELL S.
WELCH, GEO. R.
WELKE, ROBT. F.
WELTMAN, JOHN W.
WEMYSS, GEO. S.
WERNER, RICHARD K.
WEST, EDGAR R.
WEST, RICHARD L.
WHEELER, HAROLD W.
WHIPPLE, WARREN E.
WHITE, JAS. F.
WHITE, PERRY
WHITLEY, JOHN E.
WIESINGER, ROBT. L.
WILBER, WENDELL W.
WILDER, JOHN WATSON
WILKES, JOHN R.
WILLETT, RALPH P.
WILLHITE, ALVIN R.
WILLIAMS, ALFRED (REV)
WILLIAMS, CYRIL E.
WILLIAMS, DON E.
WILLIAMS, EUGENE S.
WILLIAMS, FRANK M.
WILLIAMS, WOODROW W.
WILLINGHAM, WM.
WILLIS, DONALD D.
WILLSIE, R. E.
WILSON, OSCAR RALPH
WINDOFFER, ROBT. A.
WINGROVE, MARVIN V.
WIRE, CALVIN C.
WISE, VICTOR H., JR
WITTERS, A. B.
WOLF, EDGAR JR.
WOLFE, FRED J.
WOLK, I. F.
WOOD, ROBERT W.
WOOLSON,PETER C.
WRIGHT, ROBT. C.
WRIGHTSON, VERNON E.
WYATT, JAS. L.
WYCHE, WILTON E.
YACKEY, DWIGHT A.
YAHNE, VERNE M.
YEABBLEY, KEN
YOCHUM, KENNETH H.
YODER, JOHN F.
YORK, GARLAND A.
YORK, RUSSELL
YOUNG, A. STUARD, JR.
YOUNG, JAS. S.
YOUNG, ROBT. H.
YOUTZ, HEWITT S.
ZAVERL, JACK R.
ZINNI, NICHOLAS
ZURNEY, WALTER E.

ADDITIONAL NOTES:

Page 101-Roy King, 18 FG
Page 130-Harry Spring, Lockheed Tech Rep.
Page 137-Jack Wilkes, 37 FS 14 FG

(Thad Jones, KIA, and B.A. Dobbs proudly display a Japanese flag. (Courtesy of Byron A. Dobbs) Opposite page top: P-38s in formation. (Courtesy of Gil Cefaratt) Below: Armament is loaded into a P-38, Mines Field, CA, 1942. (Courtesy of Robert Kietley, photograph from Los Angeles Bureau, Wide World Photos) Center: Lt. Len Boyd, 459th FS, stands in front of his P-38L awaiting ordinance repair at Dhudkundi, AFB, India, May 1945. (Courtesy of Len Boyd) Opposite page bottom: Formation flown for the funeral of Milo Burcham, a P-38 test pilot. (Courtesy of Stanley Long)